PAUL A. BARTLETT
DEPARTMENT OF CHEMISTRY
UNIVERSITY OF CALIFORNIA
BERKELEY, CALIFORNIA 94720

Use of X-Ray Crystallography in the Design of Antiviral Agents

Use of X-Ray Crystallography in the Design of Antiviral Agents

Edited by

W. Graeme Laver

Influenza Research Unit
John Curtin School of Medical Research
The Australian National University
Canberra, Australia

Gillian M. Air

Department of Microbiology
University of Alabama at Birmingham
Birmingham, Alabama

Academic Press, Inc.
Harcourt Brace Jovanovich, Publishers
San Diego   New York   Berkeley   Boston
London   Sydney   Tokyo   Toronto

Chemistry Library

Front cover photograph by Julie Macklin and Stuart Butterworth, Australian National University: Neuraminidase crystals from an influenza virus isolated from a noddy tern on Australia's Great Barrier Reef. Courtesy of W. Graeme Laver.

This book is printed on acid-free paper. (∞)

Academic Press, Inc.
San Diego, California 92101

United Kingdom Edition published by
Academic Press Limited
24–28 Oval Road, London NW1 7DX

Library of Congress Cataloging-in-Publication Data

Use of X-ray crystallography in the design of antiviral agents /
    edited by W. G. Laver, Gillian Air.
        p. cm.
      ISBN 0-12-438745-4 (alk. paper)
      1. Antiviral agents. 2. Viral antibodies. 3. Immune complexes.
    4. X-ray crystallography. 5. Drugs--Design.   I. Laver, William
    Graeme, Date. II. Air, Gillian.
    RS431.A66U74   1990
    616.9'25061--dc20   89-37194
                                                                CIP

Printed in the United States of America
90 91 92 93   9 8 7 6 5 4 3 2 1

Contents

17. Analysis of Antibody–Protein Interactions Utilizing Site-Directed Mutagenesis and a New Evolutionary Variant of Lysozyme

Thomas B. Lavoie, Lauren N. W. Kam-Morgan,
Corey P. Mallet, James W. Schilling, Ellen M. Prager,
Allan C. Wilson, and Sandra J. Smith-Gill

18. Approaches toward the Design of Proteins of Enhanced Thermostability

J. A. Bell, S. Dao-pin, R. Faber, R. Jacobson,
M. Karpusas, M. Matsumura, H. Nicholson,
P. E. Pjura, D. E. Tronrud, L. H. Weaver, K. P. Wilson,
J. A. Wozniak, X-J. Zhang, T. Alber, and
B. W. Matthews

Contributors

Numbers in parentheses indicate the pages on which the authors' contributions begin.

Ravi Acharya (161), Laboratory of Molecular Biophysics, Oxford OX1 3QU, United Kingdom

Gillian M. Air (13, 49), Department of Microbiology, University of Alabama at Birmingham, Birmingham, Alabama 35294

T. Alber[1] (233), Institute of Molecular Biology, University of Oregon, Eugene, Oregon 97403

P. M. Alzari (203), Département d'Immunologie, Institut Pasteur, 75724 Paris Cedex 15, France

A. G. Amit (203), Département d'Immunologie, Institut Pasteur, 75724 Paris Cedex 15, France

Edward Arnold (283), Center for Advanced Biotechnology and Medicine (CABM) and Department of Chemistry, Rutgers University, Piscataway, New Jersey 08855-0759

Gail Ferstandig Arnold (283), Center for Advanced Biotechnology and Medicine (CABM) and Department of Chemistry, Rutgers University, Piscataway, New Jersey 08855-0759

Francis K. Athappilly (35), Department of Biochemistry and Molecular Biophysics, Columbia University, New York, New York 10032

Paul A. Bartlett (247), Department of Chemistry, University of California, Berkeley, California 94720

J. A. Bell (233), Institute of Molecular Biology, University of Oregon, Eugene, Oregon 97403

G. Bentley (203), Département d'Immunologie, Institut Pasteur, 75724 Paris Cedex 15, France

Pamela J. Bjorkman[2] (19), Department of Microbiology and Immunology, Stanford University, Stanford, California 94305

---

[1] Present address: Department of Biochemistry, School of Medicine, University of Utah, Salt Lake City, Utah 84132

[2] Present address: Division of Biology, California Institute of Technology, Pasadena, California 91125

G. Boulot (203), Département d'Immunologie, Institut Pasteur, 75724 Paris Cedex 15, France

Fred Brown (161), Wellcome Biotech, Beckenham, Kent BR3 3BS, United Kingdom

Lorena E. Brown (75), Department of Microbiology, University of Melbourne, Parkville 3052, Australia

Charles E. Bugg (261), Department of Biochemistry, University of Alabama at Birmingham, Birmingham, Alabama 35294

Roger M. Burnett (35), The Wistar Institute, 3601 Spruce Street, Philadelphia, Pennsylvania 19104

Zhaoping Cai (35), The Wistar Institute, 3601 Spruce Street, Philadelphia, Pennsylvania 19104

W. Michael Carson (261), Center for Macromolecular Crystallography, University of Alabama at Birmingham, Birmingham, Alabama 35294

V. Chitarra (203), Département d'Immunologie, Institut Pasteur, 75724 Paris Cedex 15, France

S. Dao-pin (233), Institute of Molecular Biology, University of Oregon, Eugene, Oregon 97403

G. Darby (309), Department of Molecular Sciences, The Wellcome Research Laboratories, Beckenham, Kent, BR3 3BS, United Kingdom

Paul L. Darke (321), Department of Molecular Biology, Merck Sharp and Dohme Research Laboratories, West Point, Pennsylvania 19486

David R. Davies (199), Laboratory of Molecular Biology, National Institute of Diabetes and Digestive and Kidney Diseases, National Institutes of Health, Bethesda, Maryland 20892

Mark M. Davis (19), Department of Microbiology and Immunology and the Howard Hughes Medical Institute, Stanford University, Stanford, California 94305

Sonia M. Dayan (75), C.S.I.R.O. Division of Biotechnology, Parkville 3052, Australia

Guy D. Diana (187), Department of Virology and Oncopharmacology, Sterling Research Group, Rensselaer, New York 12144

David H. Drewry (247), Department of Chemistry, University of California, Berkeley, California 94720

Frank J. Dutko (187), Department of Virology and Oncopharmacology, Sterling Research Group, Rensselaer, New York 12144

Allen B. Edmundson (95), Department of Biology, University of Utah, Salt Lake City, Utah 84112

David A. Einfield (335), Department of Microbiology, University of Alabama at Birmingham, Birmingham, Alabama 35294

Kathryn R. Ely[3] (95), Department of Biology, University of Utah, Salt Lake City, Utah 84112

R. Faber (233), Institute of Molecular Biology, University of Oregon, Eugene, Oregon 97403

Andrea L. Ferris (297), Bionetics Research Inc.-Basic Research Program, National Cancer Institute, Frederick Cancer Research Facility, Frederick, Maryland 21701-1013

David J. Filman (139), Department of Molecular Biology, Research Institute of Scripps Clinic, La Jolla, California 92037

T. Fischmann (203), Département d'Immunologie, Institut Pasteur, 75724 Paris Cedex 15, France

Paula M. D. Fitzgerald (321), Department of Biophysical Chemistry, Merck Sharp and Dohme Research Laboratories, Rahway, New Jersey 07065

Graham Fox (161), Wellcome Biotech, Beckenham, Kent BR3 3BS, United Kingdom

M. Patricia Fox (187), Department of Virology and Oncopharmacology, Sterling Research Group, Rensselaer, New York 12144

Elizabeth Fry (161), Laboratory of Molecular Biophysics, Oxford OX1 3QU, United Kingdom

Paul S. Furcinitti[4] (35), Biology Department, Brookhaven National Laboratory, Upton, New York 11973

E. F. Garman (309), Laboratory of Molecular Biophysics, University of Oxford, Oxford OX1 3QU, United Kingdom

---

[3] Present address: La Jolla Cancer Research Foundation, La Jolla, California 92037

[4] Present address: Biophysics Research Division, University of Michigan, Ann Arbor, Michigan 48109

H. Mario Geysen (95), Coselco Mimotopes Party Limited, Parkville, Victoria 3052, Australia

R. Griest (309), Laboratory of Molecular Biophysics, University of Oxford, Oxford OX1 3QU, United Kingdom

Debra L. Harris (95), Department of Biology, University of Utah, Salt Lake City, Utah 84112

A. J. Hay (1), Division of Virology, National Institute for Medical Research, London NW7 1AA, United Kingdom

Jill C. Heimbach (321), Department of Molecular Biology, Merck Sharp and Dohme Research Laboratories, West Point, Pennsylvania 19486

Beverly A. Heinz (173), Institute for Molecular Virology, University of Wisconsin, Madison, Wisconsin 53706

James N. Herron (95), Department of Biology, University of Utah, Salt Lake City, Utah 84112

Amnon Hizi (297), Sackler School of Medicine, Tel Aviv University, Ramat Aviv, Israel

James M. Hogle (139), Department of Molecular Biology, Research Institute of Scripps Clinic, La Jolla, California 92037

Stephen H. Hughes (297), Bionetics Research Inc.-Basic Research Program, National Cancer Institute, Frederick Cancer Research Facility, Frederick, Maryland 21701-1013

Eric Hunter (335), Department of Microbiology, University of Alabama at Birmingham, Birmingham, Alabama 35294

David C. Jackson (75), Department of Microbiology, University of Melbourne, Parkville 3052, Australia

David Jacobson (139), Department of Molecular Biology, Research Institute of Scripps Clinic, La Jolla, California 92037

R. Jacobson (233), Institute of Molecular Biology, University of Oregon, Eugene, Oregon 97403

E. Y. Jones (309, 345), Laboratory of Molecular Biophysics, Oxford OX1 3QU, United Kingdom

Lauren N. W. Kam-Morgan (213), Department of Biochemistry, University of California, Berkeley, California 94720

M. Karpusas (233), Institute of Molecular Biology, University of Oregon, Eugene, Oregon 97403

Alex P. Korn (35), The Wistar Institute, 3601 Spruce Street, Philadelphia, Pennsylvania 19104

Lawrence A. Lamden (247), Department of Chemistry, University of California, Berkeley, California 94720

B. A. Larder (309), Department of Molecular Sciences, The Wellcome Research Laboratories, Beckenham, Kent BR3 3BS, United Kingdom

W. Graeme Laver (13, 49), Influenza Research Unit, John Curtin School of Medical Research, The Australian National University, Canberra City, ACT 2601, Australia

Thomas B. Lavoie (213), Laboratory of Genetics, National Cancer Institute, National Institutes of Health, Bethesda, Maryland 20892, and Department of Zoology, University of Maryland, College Park, Maryland 20742

Chih-Tai Leu (321), Department of Molecular Biology, Merck Sharp and Dohme Research Laboratories, West Point, Pennsylvania 19486

Derek Logan (161), Laboratory of Molecular Biophysics, Oxford OX1 3QU, United Kingdom

D. M. Lowe (309), Department of Molecular Sciences, The Wellcome Research Laboratories, Beckenham, Kent BR3 3BS, United Kingdom

Ming Luo (49), Department of Microbiology, University of Alabama at Birmingham, Birmingham, Alabama 35294

Corey P. Mallet (213), Laboratory of Genetics, National Cancer Institute, National Institutes of Health, Bethesda, Maryland 20892

R. A. Mariuzza (203), Département d'Immunologie, Institut Pasteur, 75724 Paris Cedex 15, France

M. Matsumura[5] (233), Institute of Molecular Biology, University of Oregon, Eugene, Oregon 97403

B. W. Matthews (233), Institute of Molecular Biology, University of Oregon, Eugene, Oregon 97403

[5] Present address: Department of Immunology, Research Institute of Scripps Clinic, La Jolla, California 92037

Brian M. McKeever (321), Department of Biophysical Chemistry, Merck Sharp and Dohme Research Laboratories, Rahway, New Jersey 07065

Mark A. McKinlay (187), Department of Virology and Oncopharmacology, Sterling Research Group, Rensselaer, New York 12144

Dennis W. Metzger (13), Department of Immunology, St. Jude Children's Research Hospital, Memphis, Tennessee 38101

Ramachandran Murali (35), The Wistar Institute, 3601 Spruce Street, Philadelphia, Pennsylvania 19104

Manuel A. Navia (321), Department of Biophysical Chemistry, Merck Sharp and Dohme Research Laboratories, Rahway, New Jersey 07065

H. Nicholson (233), Institute of Molecular Biology, University of Oregon, Eugene, Oregon 97403

Eduardo A. Padlan (199), Laboratory of Molecular Biology, National Institute of Diabetes and Digestive and Kidney Diseases, National Institutes of Health, Bethesda, Maryland 20892

Daniel C. Pevear (187), Department of Virology and Oncopharmacology, Sterling Research Group, Rensselaer, New York 12144

D. C. Phillips (309), Laboratory of Molecular Biophysics, University of Oxford, Oxford OX1 3QU, United Kingdom

P. E. Pjura (233), Institute of Molecular Biology, University of Oregon, Eugene, Oregon 97403

R. J. Poljak (203), Département d'Immunologie, Institut Pasteur, 75724 Paris Cedex 15, France

A. Portner (49), Department of Virology and Molecular Biology, St. Jude Children's Research Hospital, Memphis, Tennessee 38101

K. L. Powell (309), Department of Molecular Sciences, The Wellcome Research Laboratories, Beckenham, Kent BR3 3BS, United Kingdom

Ellen M. Prager (213), Department of Biochemistry, University of California, Berkeley, California 94720

D. J. M. Purifoy (309), Department of Molecular Sciences, The Wellcome Research Laboratories, Beckenham, Kent BR3 3BS, United Kingdom

Siegfried H. Reich (247), Department of Chemistry, University of California, Berkeley, California 94720

James M. Rini (87), Department of Molecular Biology, Research Institute of Scripps Clinic, La Jolla, California 92037

M.-M. Riottot (203), Département d'Immunologie, Institut Pasteur, 75724 Paris Cedex 15, France

Michael G. Rossmann (115), Department of Biological Sciences, Purdue University, West Lafayette, Indiana 47907

R. Scott Rowland (261), Department of Biochemistry, University of Alabama at Birmingham, Birmingham, Alabama 35294

Dave Rowlands (161), Wellcome Biotech, Beckenham, Kent BR3 3BS, United Kingdom

Roland R. Rueckert (173), Institute for Molecular Virology, University of Wisconsin, Madison, Wisconsin 53706

Nicole S. Sampson (247), Department of Chemistry, University of California, Berkeley, California 94720

F. A. Saul (203), Département d'Immunologie, Institut Pasteur, 75724 Paris Cedex 15, France

James W. Schilling (213), California Biotechnology, Inc., Mountain View, California 94043

Ursula Schulze-Gahmen (87), Department of Molecular Biology, Research Institute of Scripps Clinic, La Jolla, California 92037

Deborah A. Shepard (173), Institute for Molecular Virology, University of Wisconsin, Madison, Wisconsin 53706

Steven Sheriff (199), Laboratory of Molecular Biology, National Institute of Diabetes and Digestive and Kidney Diseases, National Institutes of Health, Bethesda, Maryland 20892

Irving S. Sigal (321), Department of Molecular Biology, Merck Sharp and Dohme Research Laboratories, West Point, Pennsylvania 19486

J. J. Skehel (1), Division of Virology, National Institute for Medical Research, London NW7 1AA, United Kingdom

Sandra J. Smith-Gill (213), Laboratory of Genetics, National Cancer Institute, National Institutes of Health, Bethesda, Maryland 20893

H. Souchon (203), Département d'Immunologie, Institut Pasteur, 75724 Paris Cedex 15, France

James P. Springer (321), Department of Biophysical Chemistry, Merck Sharp and Dohme Research Laboratories, Rahway, New Jersey 07065

D. K. Stammers (309), Department of Molecular Sciences, The Wellcome Research Laboratories, Beckenham, Kent BR3 3BS, United Kingdom

Robyn L. Stanfield (87), Department of Molecular Biology, Research Institute of Scripps Clinic, La Jolla, California 92037

David I. Stuart (161, 309, 345), Laboratory of Molecular Biophysics, Oxford OX1 3QU, United Kingdom

Enrico A. Stura (87), Department of Molecular Biology, Research Institute of Scripps Clinic, La Jolla, California 92037

R. J. Sugrue (1), Division of Virology, National Institute for Medical Research, London NW7 1AA, United Kingdom

Rashid Syed (139), Department of Molecular Biology, Research Institute of Scripps Clinic, La Jolla, California 92037

G. L. Taylor (309), Laboratory of Molecular Biophysics, University of Oxford, Oxford OX1 3QU, United Kingdom

D. Tello (203), Département d'Immunologie, Institut Pasteur, 75724 Paris Cedex 15, France

S. D. Thompson (49), Department of Virology and Molecular Biology, St. Jude Children's Research Hospital, Memphis, Tennessee 38101

M. Tisdale (309), Department of Molecular Sciences, The Wellcome Research Laboratories, Beckenham, Kent BR3 3BS, United Kingdom

Gordon Tribbick (95), Coselco Mimotopes Party Limited, Parkville, Victoria 3052, Australia

D. E. Tronrud (233), Institute of Molecular Biology, University of Oregon, Eugene, Oregon 97403

Jan van Oostrum[6] (35), Department of Biochemistry and Molecular Biophysics, Columbia University, New York, New York 10032

Edward W. Voss, Jr. (95), Department of Microbiology, University of Illinois, Urbana, Illinois 61801

---

[6] Present address: Department of Biotechnology, CIBA-GEIGY AG, CH-4002 Basel, Switzerland

Rebecca C. Wade (61), Department of Chemistry, University of Houston, Houston, Texas 77204-5641

N. P. C. Walker (345), Hauptlaboratorium, BASF Aktiengesellschaft, 6700 Ludwigshafen, Federal Republic of Germany

L. H. Weaver (233), Institute of Molecular Biology, University of Oregon, Eugene, Oregon 97403

Robert G. Webster (13, 49), Department of Virology and Molecular Biology, St. Jude Children's Research Hospital, Memphis, Tennessee 38101

W. I. Weis[7] (1), Department of Biochemistry and Molecular Biophysics, Harvard University, Cambridge, Massachusetts 02138

S. A. Wharton (1), Division of Virology, National Institute for Medical Research, London NW7 1AA, United Kingdom

D. C. Wiley (1), Department of Biochemistry and Molecular Biophysics, Harvard University, Cambridge, Massachusetts 02138

Allan C. Wilson (213), Department of Biochemistry, University of California, Berkeley, California 94720

Ian A. Wilson (87), Department of Molecular Biology, Research Institute of Scripps Clinic, La Jolla, California 92037

K. P. Wilson (233), Institute of Molecular Biology, University of Oregon, Eugene, Oregon 97403

J. A. Wozniak (233), Institute of Molecular Biology, University of Oregon, Eugene, Oregon 97403

Todd O. Yeates (139), Department of Molecular Biology, Research Institute of Scripps Clinic, La Jolla, California 92037

X-J. Zhang (233), Institute of Molecular Biology, University of Oregon, Eugene, Oregon 97403

Vilma M. Zubak (75), C.S.I.R.O. Division of Biotechnology, Parkville 3052, Australia

---

[7] Present address: Department of Biochemistry and Molecular Biophysics, Columbia University, New York, New York 10032

Foreword

Most of the progress in the past twenty years in the development of antiviral drugs has come from the application of empirical methods of drug discovery. This is a slow and tedious process that depends more on good fortune and persistence than efficiency and purpose. It is also extremely expensive since successful drug screening programs must rely on the ability to screen thousands of compounds each year using simple, but often insensitive, laboratory screening tests. This screening process has resulted in all of the antiviral drugs in use in the United States today. However, only seven antiviral drugs[1] have been approved to date by the Food and Drug Administration for their effectiveness in the therapy or prevention of viral infections as a result of this deliberate process. The process of screening large numbers of compounds is not the only factor. One of the other reasons for this disappointing record is that antiviral drug development is complex. Not only must the antiviral drug effectively interdict viral infection, it must do so without being unduly toxic for the host. Fortunately, this dismal picture may be about to change. The first International Workshop on the Use of X-Ray Crystallography in the Design of Antiviral Agents, held in Kona, Hawaii, provided convincing evidence that we may soon have many more antiviral drugs in the clinic. The antiviral drugs of the future will be found by the logical application of information on the structure of viral proteins and viral replication to the problem of antiviral drug research and development.

Much of the recent progress in this area is due to at least three factors. The sciences of virology and immunology have matured to the point that considerable information is now available about the molecular processes that are required for viral replication, viral neutralization, pathogenesis, and host response to viral infection. Another important catalyst has been the application of X-ray crystallography to a wider spectrum of biologic, including virologic, problems. It is clear that structural information is needed to understand biologic phenomena at the molecular level. This has profound implications for antiviral drug design, as was elegantly demonstrated during this scientific conference by the research done on the structure of rhinoviruses and the use of this information to understand how antiviral drugs can directly interfere with the replication of these viruses. Finally, the emerging epidemic due to human immunodeficiency virus

---

[1] The seven drugs are Ara-A, Iododeoxyuridine, Ara-AMP, Acyclovir, Ribavirin, Amantadine, and Azidothymidine.

(HIV) has stimulated interest in this area on a scale not witnessed before. This has resulted in increasing funding at all levels in the search for antiviral drugs that can inhibit the replication of HIV.

This international workshop clearly demonstrated not only the vitality and excitement created by the application of X-ray crystallography in the development of antiviral compounds but also the phenomenal opportunities in targeted antiviral drug design that are before us which made this meeting especially appropriate and timely. Antiviral drugs that are effective in the treatment and prevention of viral infections are badly needed, particularly for such severe infections as AIDS, rabies, and cytomegalovirus. Modern virology, immunology, and molecular biology have been applied to the problems of antiviral drug development with limited success. Viruses have always been tempting targets for crystallographers, and it was gratifying to see that this enthusiasm has not diminished but actually increased. I am confident that this picture is about to change and that we are about to enter an exciting new era in antiviral drug discovery, research, and development.

John R. La Montagne
Microbiology and Infectious Diseases Program
National Institute of Allergy and Infectious Diseases

Preface

This book describes material presented at an International Workshop held in Kona, Hawaii, on February 6–8, 1989, which was convened to discuss the use of X-ray crystallography in the design of antiviral agents. These agents include antiviral drugs, antibodies, and T lymphocytes.

With a few minor exceptions, no effective cure exists for any virus disease. Some viral diseases can be prevented by vaccination, but once infection is initiated, no effective cure exists for any virus infection (although some surface herpes infections can be controlled). For a therapeutic substance to be an effective antiviral agent it must inhibit viral replication without damaging the host.

While effective vaccines have been developed for diseases such as smallpox, polio, measles, and yellow fever, vaccines made against many other viruses are only marginally useful (influenza), totally ineffective (common cold), or dangerous (respiratory syncytial virus). As more three-dimensional structures of viruses and viral proteins are being elucidated, we are starting to understand why some viruses may be resistant to efficient control by vaccines. Much remains to be learned about inhibition of viral replication by antibodies, but the recent determination of structures of antibodies complexed with protein antigens (including influenza neuraminidase) is giving insight into mechanisms of viral neutralization by antibodies. However, it has become apparent that many antibodies do not neutralize virus infectivity, even though they bind tightly.

For those viruses that exist in large numbers of serotypes (rhinoviruses, adenoviruses, HIV) or which change antigenic character in response to antibody pressure (influenza), it may never be possible to develop a truly effective vaccine, and drugs which inhibit viral functions may be necessary to control such infections.

Two approaches to the development of antiviral agents exist. The first, which has been used for 50 years with limited success, is the empirical approach. Thousands of substances have been screened for antiviral activity, but very few of these have proved useful.

The other approach, which is receiving increasing attention, is to determine the three-dimensional structures of viruses and virus proteins with biological activity. Once the active site(s) are defined, computer molecular modeling is used to design inhibitors specific for the virus concerned.

The three-dimensional structures of a number of viruses and functionally important virus proteins are currently known and more discoveries are anticipated in the near future. Knowledge of these structures will allow the rational design of virus replication inhibitors which are able to control virus infections in man and in economically important domestic animals.

The three-dimensional structures of a number of immune complexes are also known, and again, more discoveries are anticipated in the near future. These immune complex structures (involving complexes of antigen with antibodies or of peptide antigens complexed with an MHC molecule) will allow us to see how the immune system deals with pathogens and will enable the rational design of effective synthetic vaccines for those viruses for which the empirical approach has failed.

The meeting was convened to discuss these possibilities.

Steve Wharton discussed the influenza virus hemagglutinin (HA) as a target for antiviral drugs, particularly in terms of its membrane fusion activity. He also described current knowledge of the mechanism of action of amantadine. Membrane fusion is totally dependent on a low pH-induced conformational change of the HA which involves an opening of the trimer and extrusion of the previously buried hydrophobic fusion sequence. The requirements for this conformational change have been studied by mapping mutations which alter the pH at which the change occurs. Many of these mutants were selected by growth in high concentrations of amantadine. The structure of one mutant has been determined at 3-Å resolution and shows the side chain difference but no distant effects of the mutation. The ability of this mutant to fuse at higher pH than the wild type appears to result from the loss of one or two hydrogen bonds.

Amantadine appears to inhibit influenza virus replication by raising the pH in intracellular compartments and thus preventing the conformational change in the HA which leads to fusion. At high concentrations, the basic character of amantadine may itself raise the pH, but at low concentrations it appears to act on the M2 protein, which is proposed to function in controlling proton movement and hence internal pH.

Rob Webster described the use of monoclonal antibodies to influenza virus neuraminidase (NA) to select escape mutants. These mutants had single amino acid sequence changes which were located on the three-dimensional structure of the NA. Many of the changes occurred on a surface loop of the NA within the epitope recognized by NC41 antibody (see below). He also described experiments in which a seven-residue peptide corresponding to the central loop of the NC41 epitope was synthesized. This peptide did not compete in neuraminidase inhibition tests with NC41 antibody and did not bind to the antibody; antiserum raised against the peptide had no N1 activity.

Pamela Bjorkman reviewed the crystal structure of a human class I MHC molecule, with an emphasis on what the structure reveals about the ways histocompatibility molecules bind antigenic peptides and interact with T cell receptors. She also presented a hypothetical model of MHC-restricted T cell receptor binding to the peptide–MHC complex. The three-dimensional structure of a T cell receptor has not yet been determined, but because of the similarity between T cell receptors and antibodies, the known structure of an Fab fragment was used to serve as a first-order T cell receptor model structure in order to make an educated guess about the mechanism of T cell recognition of the peptide–MHC complex.

It is hoped that an increased understanding of the physical nature of antigenic peptide interactions with MHC molecules will allow the design of peptides to stimulate the immune response against a viral infection or the design of high-affinity ligands to block the self-reactive recognition of MHC molecules involved in autoimmune diseases. Before this can be accomplished, the forces comprising the peptide–MHC complex and the way in which T-cell receptors bind to it will have to be understood.

Graeme Laver described crystals of monoclonal antibody Fab fragments complexed with influenza virus neuraminidase of the N9 subtype. A number of crystalline complexes, two of which diffract X rays to beyond 3 Å, were obtained. In each case, four Fab fragments are bound to the NA tetramer. The two antibodies NC41 and NC10 recognize overlapping epitopes on the NA.

The structure of the NC41 Fab–N9 neuraminidase complex shows an epitope composed of four surface loops of the neuraminidase. Sequence changes in these loops diminish or abolish antibody binding.

Fab fragments from four different monoclonal antibodies have also been complexed with influenza B virus neuraminidase (B/Lee/40) and the complexes have been crystallized. One of the complex crystals (B/Lee/40 NA–B1 Fab) forms large crystals which diffract X rays to 3.0-Å resolution. These crystals are well ordered, in contrast to the crystals of uncomplexed NA from two strains of influenza B, which were both disordered. A progress report on the structure determination of the NA–Fab complex was given by Ming Luo.

The hemagglutinin neuraminidase (HN) protein of Sendai virus (a paramyxovirus) has been isolated from virus particles in a biologically active soluble form after removal by proteolytic digestion of the hydrophobic amino-terminal anchor sequence. The soluble HN exists as both dimers and tetramers, and crystallization trials with each of these forms have so far yielded amorphous material.

Dimers complexed with Fab fragments of a monoclonal antibody formed long needle crystals. These so far are not suitable for X-ray diffraction analysis but the results suggest that, as in the case of some influenza NAs, paramyxovirus HN molecules which do not crystallize may, when complexed with Fab fragments, yield crystals suitable for X-ray diffraction analysis.

Rebecca Wade described the design of ligands to block and so prevent the attachment of influenza viruses to cells. The influenza hemagglutinin coat glycoprotein is responsible for the attachment of the virus to the host-cell receptors by binding to a terminal sialic acid, and also for mediating a membrane fusion event leading to the release of the viral nucleocapsid into the cytoplasm. The hemagglutinin may therefore be a suitable target for the design of anti-influenza agents which would inhibit its action by binding at a functionally important region of the molecule.

From the already known three-dimensional studies of the hemagglutinin, the host-cell receptor binding site was studied using the GRID method. This is a procedure for determining energetically favorable binding regions on molecules of known structure. By mapping the interaction with hemagglutinin of a variety of small chemical groups which could be incorporated into antiviral compounds, regions of the host-cell receptor binding site where particular groups might bind strongly and specifically to hemagglutinin were identified. The exploitation of these binding regions and GRID energy maps in designing anti-influenza agents was discussed.

David Jackson showed a poster describing crystals of Fab fragments of a monoclonal antibody to a peptide from the influenza virus hemagglutinin (residues 305–328 of HA_1) and also crystals grown from a mixture of the Fab and peptide which might be involved in complex formation.

Allen Edmundson described the structures and binding properties of three proteins: (1) a human Bence–Jones dimer (Mcg); (2) the Fab from a murine monoclonal autoantibody (BV04-01) which binds single-stranded DNA; and (3) the Fab from a murine monoclonal antibody (4-4-20) with high affinity for fluorescein in aqueous solutions.

Michael Rossmann discussed the three-dimensional structures of a number of small animal RNA viruses which are now known to near-atomic resolution. All these viruses have similar structures and have in all probability, diverged from a common ancestral virus able to infect a variety of organisms. Knowledge of their structures has elucidated the site of attachment to host-cell receptors and the mode of protection of this site against host immune pressure. An internal hydrophobic pocket in rhinoviruses is the site for binding of antiviral drugs which inhibit uncoating and can inhibit attachment of some viruses. The pocket is probably a functional necessity and thus is a suitable target for well-designed antiviral agents for many viruses.

Dave Stuart described the three-dimensional structure of foot-and mouth-disease virus at 2.9-Å resolution. The structure was determined at close to atomic resolution by X-ray diffraction without experimental phase information. The virus shows similarities with other picornaviruses but also several unique features. The canyon or pit found in other picornaviruses is absent; this has important implications for cell attachment. The most immunogenic portion of the capsid, which acts as a potent peptide vaccine, forms a disordered protrusion on the virus surface.

Roland Rueckert showed how WIN compounds block attachment of human rhinovirus 14 (HRV14) to HeLa cells. This block appears to result from deformation of a specific region of the canyon floor that occurs when the drug binds in an underlying pocket of VP1. Escape mutant analysis was used to identify amino acids in the drug-binding pocket that affect the activity of WIN 52084. Two classes of drug resistance were recognized: high resistance (HR) and low resistance (LR). Nucleotide sequence analysis of over 80 independent mutants showed that the two classes comprised mutations at different amino acids. HR mutations were confined to two positions, V188 and C199 in VP1, and the substitutions were invariably bulkier side chains pointing into the drug-binding pocket. The LR mutations, however, occurred in a wider variety of positions. More significantly, all fell in the drug-deformable regions of the canyon, suggesting they might be attachment-altered. These data, and crystallographic analyses of two HR mutants, suggested that HRV14 might acquire resistance by either reducing drug binding or by compensating for the inhibitory effect of bound drug.

Further evidence showed that both mechanisms of resistance exist: Most HR mutations inhibited binding of the compound, whereas most LR mutations relieved the attachment-blocking effect of bound drug. The ability of some mutations to attach to cells despite the presence of bound drug was verified using single-cycle growth assays.

Frank Dutko described a variety of oxazolinylphenyl isoxazoles, an interesting group of antiviral compounds that inhibit the replication of picornaviruses. These compounds bind to a specific site in VP1 of the viral capsid and cause conformational changes which result in inhibition of adsorption of HRV14. Both an analysis of compound binding using X-ray crystallography and the generation of structure–activity relationships using *in vitro* antiviral assays have contributed to the design of agents with improved antiviral activity. While this information may not be itself helpful in directing the synthesis of broad-spectrum antipicornavirus agents, it is the essential first step in understanding the "ground rules" for designing hydrophobic capsid-binding compounds.

David Davies described how antibodies interact with protein antigens using the three-dimensional structures of two complexes of lysozyme with anti-lysozyme–Fabs (HyHEL-10 and HyHEL-5). He pointed out that the nature of the forces involved in antibody–antigen interaction are basically no different from those observed in the general area of protein–protein interactions. In both complexes there is remarkable complementarity in shape between the interacting surfaces, so that water is almost entirely excluded from the interface. In both complexes the epitope is discontinuous, comprising several segments of polypeptide chain. Only small conformational changes occur in the lysozyme on binding Fab. In the HyHEL-5 complex, charge neutralization appears to play an important (but not exclusive) role, while in HyHEL-10 there is only one weak salt bridge formed, and this is at the edge of the epitope. The epitopes recognized by HyHEL-5 and HyHEL-10 are virtually nonoverlapping, and they are also distinct from the epitope recognized by antibody D1.3 (discussed by Roy Mariuzza). The three epitopes cover about 40% of the lysozyme surface; this supports the idea that any part of a protein surface accessible to antibody is potentially antigenic.

Sandi Smith-Gill described site-directed mutagenesis experiments to analyze further the interactions in binding antibodies HyHEL-5, HyHEL-8, or HyHEL-10 to lysozyme. Mutations were introduced into the epitopes on lysozyme or into the complementarity determining regions (CDRs) of the antibodies. Docking experiments using results of epitope mapping and molecular dynamics did not accurately predict the crystallographically defined epitope. Chain recombination experiments indicated that the H3 CDR was a determinant of specificity. The fine specificity differences between HyHEL-8 and -10 could be traced to H chain position 101.

Roy Mariuzza summarized the three-dimensional structure of the lysozyme–Fab D1.3 complex in the context of the other known protein antigen–antibody complex structures and discussed the structures of epitopes and the role of conformational flexibility in antigen–antibody recognition. He also described recent results on an idiotype–anti-idiotype complex along with efforts to generate a water-soluble form of the T cell antigen receptor for eventual use in crystallographic studies.

The factors involved in making proteins more stable were discussed by Brian Matthews. The lysozyme from bacteriophage T4 is being used as a model system to test ways in which the stabilities of proteins might be increased. Possible approaches that are being explored include the use of hydrophobic interactions, improvements in hydrogen bonding, stabilization of α-helix dipoles, introduction of substitutions that decrease the entropy of unfolding, the removal of strain, and the introduction of disulfide bridges. It appears that there are many ways in which the thermostability of proteins might be increased. The results also provide quantitative information on the contributions that different types of interaction make to the stability of proteins and as such are relevant to the importance of such interactions in enzyme–inhibitor and drug–receptor complexes.

Paul Bartlett described two approaches to the rational design of enzyme inhibitors. In order to design mechanism-derived inhibitors, knowledge of the enzyme substrate and mechanism is required. However, mechanism-derived approaches are for the most part inapplicable to the design of ligands for noncatalytic protein binding sites such as receptors or allosteric sites.

Given the accelerating pace at which structures of enzymes and other biologically interesting proteins are being determined in the crystalline state by X-ray diffraction and in solution by NMR, it is clear that another approach to inhibitor design is emerging. From a knowledge of protein structure alone, inhibitors will be designed without reference to the mechanism of an enzyme or its normal substrate. Such compounds have been called ''structure-derived inhibitors'' in order to distinguish them from the more traditional kinds. The type of compounds that can arise from such a design approach would be mimics of natural, perhaps peptidic, ligands for the target binding site, or de novo inventions devised to complement the binding cavity sterically and electronically. In this presentation, Paul Bartlett used the serine proteases as examples of enzymes for which inhibitors of both types have been devised.

Charlie Bugg described ways in which stages of molecular design can be improved by utilizing the vast amount of structural information in crystallographic databases. He provided examples to show how the databases, for example, can be used to predict preferred interactions with phenylalanine residues of proteins. The results from both the Cambridge Structural Database and the Brookhaven Protein Database clearly showed that there is a preference for carbonyl oxygen atoms to interact with Phe residues. The carbonyl oxygen atoms tend to be in the plane of the aromatic ring and the observed distributions are statistically different from the distribution expected by chance. It is expected that crystallographic databases will permit the preferred interaction patterns to be established for a variety of other chemical groups and that this information will play an increasingly important role in molecular design processes.

Eddie Arnold suggested that AIDS will continue to be a major health problem in the foreseeable future. However, detailed information about the three-dimensional structures of human immunodeficiency virus (HIV) components should greatly enhance the current drug design efforts targeted against HIV. In the absence of such information, it may be useful to consider the details of functionally and evolutionarily related cellular and viral products. The current base of information about other possibly related structures gives both a framework for thinking about retroviral protein structures in detail and a possible start for experimental structure determinations by methods such as molecular replacement.

Steve Hughes described the expression of HIV reverse transcriptase in _Escherichia coli_. Since the reverse transcriptase is synthesized as part of a large polyprotein, modifications must be made for efficient expression in heterologous systems. A termination codon can be introduced at precisely the proteolytic recognition site at the C terminus, but at the N terminus it is necessary to add a methionine codon. This is conveniently done by embedding the ATG within the recognition site for the enzyme Nco1, thus enabling expression of the unfused protein. Large amounts can be made in and purified from _E. coli_. The expressed protein has been used to generate monoclonal antibodies and to study the mechanism of proteolysis of the 66-kDa form to generate the 55-kDa form.

Dave Stammers discussed the problem of trying to obtain usable crystals of HIV reverse transcriptase (RT). A wide-ranging survey of crystallization conditions for recombinant HIV RT was undertaken. Approximately 3000 crystallization conditions were tried. A number of apparently different crystal forms of HIV RT were grown that are of distinct morphology and have different subunit compositions. It was observed that 66-kDa RT, during the course of crystallization, undergoes cleavage at the C terminus to give a 51-kDa subunit that forms, in combination with a 66-kDa subunit, a stable heterodimer. Crystals of this form of RT are orthorhombic and also were grown as long rods. Also observed were prismlike crystals that contain a 3 : 1 ratio of 66-kDa to 51-kDa subunits. This probably represents cocrystallization of equimolar amounts of 66-kDa–51-kDa heterodimer and 66-kDa–66-kDa homodimer. Under different conditions needlelike crystals containing only 66-kDa subunits have been obtained. All these crystals show disorder when examined in the X-ray beam. The maximum resolution of the observed diffraction pattern is 6 Å.

Cocrystallization of RT with DNA oligomers produces well-formed platelike crystals that show disorder. A number of mutant forms of RT produced by site-directed mutagenesis have been crystallized. These involve changes in single amino acids or truncations of the C-terminal region.

Paul Darke and Jim Springer described the three-dimensional structure of the HIV protease, which is solved at 3-Å resolution, and also studies on the enzyme activity and its inhibition. The HIV protease has been chemically synthesized and also expressed as a 99-amino acid protein in _E. coli_.

The purified protein is capable of hydrolyzing its natural substrate, gag p55, and peptides representing all the known processing sites in the HIV polyproteins gag and pol. The peptide hydrolysis activity exhibits a maximum at pH 5.5 and is inhibited by pepstatin with a K_i of 1.4 μM. Minimal peptide substrate analysis indicates that seven amino acids are required for efficient hydrolysis.

The protease has been crystallized and the structure solved to 3-Å resolution, which has shown that the protein is a dimer. While large regions of each monomer correspond roughly to the N- and C-terminal domains of pepsinlike aspartyl proteases, significant differences do exist. A high degree of structural homology is found in the immediate active site region, which is assembled around the dimer interface and includes one characteristic Asp-Thr-Gly sequence from each protease monomer. The structure places constraints on the mechanism of autoproteolytic excision of the protease from the gag–pol polyprotein.

Eric Hunter has studied the glycoproteins of some retroviruses and shown by cross-linking and sedimentation analysis that both the Rous sarcoma virus and Mason–Pfizer monkey virus glycoproteins are oligomerized, most likely to trimers. The M-PMV oligomer is markedly pH-sensitive.

Dave Stuart described work done by Yvonne Jones in Oxford on the crystal structure of tumor necrosis factor (TNF). TNF, a polypeptide mediator of the cellular immune system, has the structural motif of a viral jelly roll. In TNF three such subunits tightly associate to form the biologically active trimer in a form of packing echoed in adenovirus. There is obviously much of interest to be learned about receptor binding both of viruses and of host mediators such as TNF. It is certainly a strange twist in the tale to note that the jelly roll motif in rhinovirus binds to intercellular adhesion factor-1 (ICAM-1), expression of which is controlled by TNF, a jelly roll mediator of the host immune system. Further studies of these systems may highlight common structural–functional principles of use in the pharmacological applications of TNF, itself possibly a natural antiviral agent, and in the search for antivirals directed against the viral jelly roll structures.

In addition to the work summarized above and described in detail in this book, two topics of particular interest were introduced at the meeting. Steve Harrison described progress on the structure determinations of two small DNA viruses, SV40 and polyoma. These viruses are fundamentally different from the well-characterized small RNA viruses in that while the RNA virus capsid is a vehicle to deliver mRNA to the ribosomes the DNA virus capsid must deliver a minichromosome to the nucleus. At the current resolution (3.8 Å for SV40), there is no similarity to the ubiquitous structure element of small RNA viruses. The other topic was the description by Steve Marlin of how a cell adhesion molecule, ICAM-1, has been shown to be the receptor for the major group of rhinoviruses. Since ICAM-1 is inducible in the cells so far examined, interesting questions are raised about initial events in rhinovirus infection.

We gratefully acknowledge support for the meeting provided by National Institute of Allergy and Infectious Diseases, National Institutes of Health; Department of Industry, Technology & Commerce, Australia; The International Union of Biochemistry; International Union of Crystallography; Sterling-Winthrop Research Institute; The Wellcome Foundation, U.K.; E. I. DuPont DeNemours & Company; Merck & Company; Schering Corporation; Burroughs Wellcome Company, U.S.A.; Eli Lilly and Company; Smith Kline Beckman Corporation; Wyeth-Ayerst Research; and Australian Overseas Telecommunications Commission.

Gillian M. Air
W. Graeme Laver

1 Membrane Fusion by Influenza Viruses and the Mechanism of Action of Amantadine

S. A. Wharton, A. J. Hay, R. J. Sugrue, and J. J. Skehel
Division of Virology, National Institute for Medical Research, London NW7
1AA, United Kingdom

W. I. Weis* and D. C. Wiley
Department of Biochemistry and Molecular Biophysics, Harvard University,
Cambridge, Massachusetts 02138

I. Studies on Membrane Fusion

A. The Structure of the Hemagglutinin Molecule

Influenza viruses are segmented, negative strand, RNA-enveloped viruses. The hemagglutinin (HA) and neuraminidase (NA) molecules are the two types of glycoproteins in the viral membrane. Apart from being the primary antigenic determinant for the virus, the HA is responsible for receptor binding and possesses the membrane fusion activity which is required for viral infection (reviewed in Wharton, 1987; Wiley and Skehel, 1987; White *et al.*, 1983). The entire ectodomain of HA can be released from the viral membrane by the action of the protease bromelain, and the three-dimensional structure of the soluble product, bromelain-released hemagglutinin (BHA), has been determined to 3 Å resolution by X-ray diffraction (Wilson *et al.*, 1981). BHA is a trimer of identical subunits of 70-kDa molecular mass; a schematic diagram of one subunit is shown in Fig. 1. Each subunit consists of two polypeptide chains, HA_1 and HA_2, which are disulfide linked and are formed by posttranslational cleavage of a single polypeptide precursor HA_0. This cleavage is essential for viral infectivity (Lazarowitz and Choppin, 1975; Klenk *et al.*, 1975). Most of HA_1 forms the globular membrane-distal region of the molecule and contains the major antigenic sites which surround the receptor binding site. The majority (six of seven) of the glycosylation sites are on HA_1; HA_2 and the remainder of HA_1 form the fibrous stem of the molecule, the major feature of which is a 10.5-nm-long coiled coil of three α helices. The highly conserved hydrophobic amino terminus of HA_2 is involved in membrane fusion activity (see Section I,B). The molecule is attached to the viral membrane via the carboxy terminus of HA_2.

*Present address: Department of Biochemistry and Molecular Biophysics, Columbia University, New York, New York 10032

Use of X-Ray Crystallography in the Design of Antiviral Agents

1

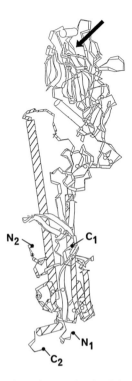

Figure 1. Schematic representation of one subunit of the BHA trimer. The carboxy and amino termini of HA_1 and HA_2 (striped) are shown (C_1N_1 and C_2N_2). The position of the receptor binding site is shown by an arrow. The trimeric interface is to the left of the diagram and the viral membrane to the bottom. The diagram was generated using the computer program designed by Lesk and Hardman (1982).

B. The Acid-Induced Conformational Change of Hemagglutinin

Influenza virus utilizes the receptor-mediated endocytotic pathway when infecting a cell. The virus binds to sialic acid-containing receptors on the cell plasma membrane and is taken into endosomes. These are acidified by the action of cellular proton pumps. At a pH specific to each influenza strain (usually between 5.5 and 6.0), the HA undergoes a conformational change which is required for the fusion of the viral membrane with the endosomal membrane, resulting in the release of the transcription complex into the cell. The nature of the conformational change has been extensively studied since inhibitors of this essential step in the infection process would be valuable antiviral agents.

The low pH-induced conformational change in BHA results in an increase in hydrophobicity such that the BHA molecules aggregate into rosettes. This increase in hydrophobicity has, to date, precluded the formation of suitable crystals for X-ray diffraction studies. The site of aggregation is the amino terminus of HA_2; removal of this region by the protease thermolysin resolubilizes the molecules (Daniels *et al.*, 1983a; Ruigrok *et al.*, 1988). In addition to the extrusion of the hydrophobic amino terminus of HA_2 from its buried location in the trimeric interface, the conformational change involves decreased contacts between subunits at the membrane-distal end of the molecule. This has been shown by a number of studies: Trypsin (Skehel *et al.*, 1982) and reducing agents (Graves *et al.*, 1983) both result in the release of HA_1 from low-pH BHA as monomers; epitopes at the trimeric interface are destroyed when the HA undergoes conformational change (Daniels *et al.*, 1983b; Webster *et al.*, 1983); and regions in the trimeric interface become accessible to antipeptide antibody binding (White and Wilson, 1987). The stalk of HA_2 remains trimeric (Ruigrok *et al.*, 1988) although some rearrangement in the tertiary structure may occur (Wharton *et al.*, 1988a). Negligible alterations in secondary structure occur during conformational change as judged by circular dichroism (Skehel *et al.*, 1982). The conformational change is irreversible and membranes must be present when the pH is lowered in order for fusion to occur, since HA previously incubated at low pH is no longer fusogenic.

Additional identification of the regions involved in triggering the conformational change was achieved by mapping amino acid substitutions in mutant HAs of viruses selected by growth in the presence of high doses (500 μM) of amantadine, which acts as a lysosomotropic agent (Daniels *et al.*, 1985). These HAs undergo the conformational change and fuse membranes at elevated pH, probably because the substitutions destablize the regions of the molecules involved in the change, making the transition from the native form to the low pH form more favorable. The substitutions fall into two groups: substitutions at regions of intersubunit contacts resulting in the loss or weakening of mainly salt bridges that stabilize the native trimer; and substitutions that destablize the structure of the amino terminus of HA_2 either by loss or weakening of hydrogen bonds to the amino terminal region or substitutions in the amino terminal residues that produce unfavorable packing interactions. These classes of substitutions are consistent with the description of the conformational change presented above.

It is thought that the extruded hydrophobic amino terminus of HA_2 interacts with either the endosomal or the viral membranes, resulting in their destablization and subsequently promoting membrane fusion. Support for this has come from the observation that synthetic peptides corresponding to the sequence of the amino terminus of HA_2 interact with and cause fusion of artificial membrane vesicles (Lear and De Grado, 1987; Wharton *et al.*, 1988b).

C. The Structure of a Mutant Hemagglutinin Which Fuses Membranes 0.5 pH Units above the Wild Type

The three-dimensional structure of a HA molecule with a substitution of D→G at position 112 of HA_2 (D112→G) (Daniels *et al.*, 1985; Wharton *et al.*, 1986) was determined to 3 Å resolution by X-ray crystallographic analysis as detailed elsewhere (Weis *et al.*, 1989). A schematic diagram of the amino terminus of HA_2 and of the region of the amino acid substitution is shown in Fig. 2. The difference in electron density between D112→G and wild-type HA showed that the structural effects of the D→G substitution are completely localized with no detectable change distant from the residue. The results suggest that the position occupied by the side chain of Asp 112 in the wild-type molecule is occupied by a water molecule in the D112→G mutant. In the wild-type HA the carboxyl group of Asp 112 forms hydrogen bonds with the amide nitrogens of HA_2 residues 3, 4, 5, and possibly residue 6. Even if the water molecule in D112→G HA is in a position to form hydrogen bonds with the amide nitrogens of residues 3 and 4 of HA_2, which is not clear from the structural determination, there are at least one or two fewer hydrogen bonds in the D112→G mutant. In addition bonds formed with the water molecule as acceptor will be weaker than the hydrogen bonds formed with the negatively charged aspartic acid carboxylate oxygens. Therefore less energy would be required to extrude the amino terminus of HA_2, resulting in the conformational change occurring at a higher pH than that required for the change in the wild-type HA. The structural analysis of the HA_2 amino-terminal region of the molecule also revealed the presence of electron density tentatively identified as three water molecules in both wild-type and D112→G HAs, and in the latter case, the middle water of the three is in a position to form a hydrogen bond with the water molecule that substitutes for the Asp 112 side chain. The water molecule to the left in the diagram is in a position to form hydrogen bonds with the guanidinim N_ϵ of HA_1, arginine 321, and the amino terminus of HA_2. It is possible that stabilizing and destabilizing forces in this region of the HA are balanced so that the molecule is not prematurely triggered and the pH required for the conformational change is not too low to be physiological. It is also clear from heat denaturation studies that native HA is a metastable molecule since in its low-pH form HA denatures at a higher temperature than the native molecule (Ruigrok *et al.*, 1986).

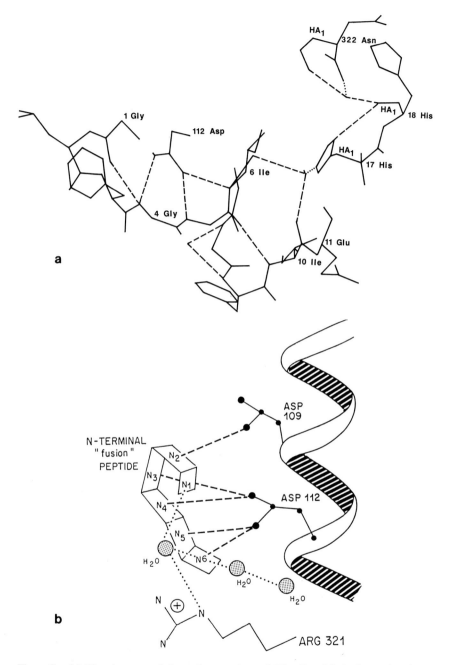

Figure 2. (a) The structure of the amino terminus of HA_2. Possible hydrogen bonds are shown as dashed lines. (b) A closer view of the vicinity of the aspartate residue 112 of HA_2. Again dashed lines illustrate hydrogen bonds and those formed with bound water molecules are shown as dotted lines.

II. The Mechanism of Action of Amantadine

A. The Genetic Basis of Drug Resistance

Amantadine (1-amino adamantane hydrochloride) and the related compound rimantadine (1-methyl-1-adamantane methylamine hydrochloride) are currently the only effective antiviral drugs against influenza A virus, both in prophylactic and therapeutic treatments (Dolin *et al.*, 1982). The replication of many influenza A strains is inhibited by these agents. However, resistant viruses readily arise *in vitro* in cell culture (Hay *et al.*, 1985) and have arisen *in vivo* in chickens, mice, and humans (Belshe *et al.*, 1988; Webster *et al.*, 1985). The doses required to inhibit viral replication are much lower than those required for generating the high-pH HA mutants discussed in the previous section, and the nature of the mutations in low-dose amantadine mutants is different from that of high-dose amantadine mutants. Identification of the mutations in low-dose amantadine-resistant strains have shown that the HA (gene 4) and matrix protein (M_1–M_2) (gene 7) genes are important in conferring resistance (Hay *et al.*, 1985; Scholtissek and Faulkner, 1979). In human viruses the M_1–M_2 gene is the sole determinant of sensitivity whereas in certain H5 or H7 avian or equine viruses, although the M_1–M_2 gene is the prime determinant, experiments with reassortant viruses have shown that properties in the HA gene have some importance in conferring sensitivity to amantadine and rimantadine. Nucleotide sequence analyses have shown that in every case a mutation in the M_2 protein, the spliced product of the M_1–M_2 gene, is responsible for resistance (Hay *et al.*, 1985). M_2 is a 97-residue unglycosylated membrane protein (Lamb *et al.*, 1985) originally thought to be nonstructural but now thought to be present in the virus in low numbers (Zebedee and Lamb, 1988). The substitutions which confer resistance are in the unique part of the M_2 protein and not in the sequence which this protein shares with the matrix M_1 protein; they occur only in the hydrophobic, membrane-spanning region of the molecule, specifically at residues 27, 30, 31, and 34 (Fig. 3). The position and nature of the substitutions is similar for variants isolated *in vitro* or *in vivo*. If the membrane-spanning region adopts an α-helical structure then all the substitutions would fall on one face of the helix near the luminal side of the membrane (Hay, 1989).

B. The Site of Action of Amantadine

Studies to elucidate the site of action of amantadine have been done on cells grown in culture and the block in virus replication was observed at either of two stages. In permissive infections of H5 and H7 subtypes in

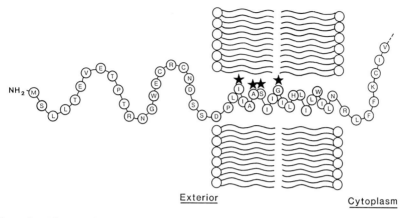

Figure 3. Diagramatic representation of the amino-terminal part of the M_2 protein. The positions of substitutions in amantadine-resistant viruses are illustrated by stars. The proton flux caused by the M_2 protein would be from left to right, from the lumen to the cytoplasm.

which HA_0 is cleaved to HA_1 and HA_2 intracellularly, the block is at a "late," post-protein synthetic step. In other sensitive strains the block is at a pre-primary transcription, "early" step, probably involving virus uncoating. Most studies have been done on the late effect. Experiments with conformation-specific antibodies, reducing agents, or proteases have shown that administration of amantadine results in the appearance of low pH-like HA on the cell surface; the release of virus particles is inhibited (Sugrue *et al.*, 1989). The amount of HA, its posttranslational modification, and its transport are not affected by the action of the drug. Pulse–chase experiments indicate that the amantadine block occurs 20 min after HA synthesis coincident with its intracellular cleavage to HA_1 and HA_2, probably in the trans-Golgi or post-Golgi vesicles. The appearance of low pH-like HA in amantadine-treated cells suggests that amantadine affects the pH of an intracellular compartment, and since the M_2 protein is the sole determinant of amantadine sensitivity or resistance, M_2 itself appears to be responsible for influencing the pH of the compartment. As yet it is not clear how M_2 functions. It is possible that it interacts with cellular proton pumps to alter their activity or that M_2 itself acts as a proton channel. Although it cannot be ruled out that M_2 interacts with HA to prevent its conformational change, an interaction which may be blocked by amantadine, the next section discusses the various lines of evidence which support the suggestion that M_2 influences the pH of an intracellular compartment.

C. Evidence that M_2 Influences the pH of Intracellular Compartments

Figure 4 is a schematic representation of the Golgi (or possibly post-Golgi vesicles) and endosomes showing the putative action of M_2. Considering the effect of amantadine upon the Golgi, which are acidified by cellular proton pumps as are endosomes, in this model the presence of M_2 results in protons being pumped out of the vesicle, thus elevating the luminal pH. This would prevent the low pH-induced conformational change of the cleaved HA. In amantadine-treated cells this pumping would be inhibited, the pH of the Golgi would be lower, and the HA would be triggered giving rise to low pH HA as observed experimentally.

There are at present four other experimental observations that can be explained in terms of the model:

1. The proton ionophore monensin antagonizes the action of amantadine. Since monensin equilibrates all pH gradients within the cell, even with the action of M_2 blocked the Golgi would never be acidic and therefore low pH-like HA would not be formed.

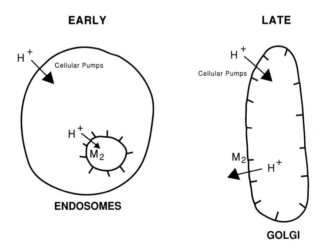

Figure 4.   A model of how M_2 could influence intracellular compartments both in the early virus uncoating stage and in the late stage during transport through the Golgi or post-Golgi vesicles. The ectodomains of HA molecules are illustrated by dashes.

2. The dose dependence of amantadine inhibition is such that virus replication is inhibited by low doses $(0.5–5\mu M)$ but is not inhibited at very high doses $(50–500\mu M)$ (Hay *et al.*, 1985). This may be because at higher doses amantadine itself raises the pH of the Golgi by virtue of being a weak base and therefore the HA would not be exposed to low pH.

3. The Rostock virus [A/chicken/Germany/34(H7N1)] HA, which undergoes the conformational change at pH 6.0, is affected by amantadine in MDCK cells and chick embryo fibroblasts whereas the Weybridge virus [A/chicken/Germany/27(H7N7)] HA, which undergoes the conformational change at pH 5.6, is only affected by amantadine in chick cells. This may be because fibroblasts have a lower pH in their intracellular compartments than do epithelial cells (Anderson and Orci, 1988) and even in the presence of amantadine, the pH of the Golgi in chick cells is not sufficiently low to trigger the conformational change in Weybridge HA but is low enough to trigger the Rostock HA. In the fibroblasts the pH is sufficiently low to trigger both molecules.

4. Considering the effect of amantadine upon virus uncoating, amantadine has been observed to slow the rate of virus–liposome fusion (a model system for virus endosome fusion) by those viruses whose replication is inhibited by amantadine, whereas the rate of fusion of amantadine-resistant strains is unaffected (Wharton *et al.*, 1989). The rate-limiting step of fusion is at present unknown but since monensin speeds up virus–liposome fusion in a dose-dependent manner (Wharton *et al.*, 1989), acidification of the virus interior could be rate-limiting.

A low internal pH of the virus could destablize the viral membrane permitting the onset of fusion. It is known, for example, that matrix protein becomes detergent-extractable at low pH, so possibly this interacts with the viral membrane at low pH or becomes dissociated from the ribonuclear particles. The topography would be such that the M_2 protein could be responsible for allowing protons to enter the virus (see Fig. 4), thus facilitating fusion. Amantadine, by blocking this effect, may decrease the rate of acidification of the viral interior, which may be reflected in the slower rate of fusion observed. Since this effect is observed with purified virus, M_2 itself rather than a cellular component would be responsible for the proton flux.

Further observations consistent with this sort of model have been made in other fields of research. Amantadine blocks the ionic channel of the nicotinic acetylcholine receptor (Albuquerque *et al.*, 1978) and studies with noncompetitive blocking agents of the receptor (Changeux et al., 1987) and with synthetic peptides (Lear et al., 1988) have indicated that serine residues in the transmembrane region are important in ion conductance. In the amino acid substitutions in amantadine-resistant viruses changes involving serine or threonine are common.

Finally, influenza virus B infections are not susceptible to amantadine treatment. It could be that the NB protein, a spliced product of the neuraminidase gene, carries out the same functions as the M_2 protein in influenza A but is insensitive to the action of amantadine.

III. Conclusions

Inhibition of the low pH-induced conformational change of HA and inhibition of the function of the M_2 protein provide the two potential sites for the action of antiviral agents. Since the conformational change, particularly the membrane fusion aspect, involves highly conserved sequences, it is unlikely that spontaneous virus mutations would generate HA molecules resistant to the action of antivirals directed against this process. The problem of vaccines is that antigenic drift limits the duration of efficacy. As amantadine-resistant viruses do arise in vitro, it is unclear whether extensive use of this drug would lead to the circulation of resistant viruses in the human population. Such viruses have been observed in chicken farms where antivirals are administered (Webster et al., 1985). However, understanding the molecular basis of the effect of amantadine and the function of M_2 may make development of novel compounds which would be more effective possible, and resistant viruses would not arise so readily.

References

Albuquerque, E. X., Eldefrawi, A. T., Eldefrawi, M. E., Mansour, N. A., and Tsai, N.-C. (1978). Amantadine: Neuromuscular blockade by suppression of ionic conductance of the acetylcholine receptor. Science **199**, 788–790.

Anderson, R. G. W., and Orci, L. (1988). A view of acid intracellular compartments. J. Cell Biol. **106**, 539–543.

Belshe, R. B., Hall-Smith, M., Hall, C. B., Betts, R., and Hay, A. J. (1988). Genetic basis of resistance to rimantadine emerging during treatment of influenza virus infection. J. Virol. **5**, 1508–1512.

Changeux, J.-P., Giraudet, J., and Dennis, M. (1987). The nicotinic acetylcholine receptor: Molecular architecture of a ligand regulated ion channel. Trends Pharm. Sci. **8**, 459–465.

Daniels, R. S., Douglas, A. R., Skehel, J. J. Waterfield, M. D., Wilson, I. A., and Wiley, D. C. (1983a). Studies of the influenza virus haemagglutinin in the pH5 conformation. In "The Origin of Pandemic Influenza Viruses" (W. G. Laver, ed.), pp. 1–7. Elsevier, New York.

Daniels, R. S., Douglas, A. R., Skehel, J. J., and Wiley, D. E. (1983b). Analyses of the antigenicity of influenza haemagglutinin at the pH optimum for virus-mediated membrane fusion. *J. Gen. Virol.* **64,** 1657–1661.

Daniels, R. S., Downie, J. C., Knossow, M., Skehel, J. J., Wang, M.-L., and Wiley, D. C. (1985). Fusion mutants of influenza virus haemagglutinin glycoprotein. *Cell* **40,** 431–439.

Dolin, R., Reichman, R. C., Madore, H. P., Maynard, R., Linton, P. M., and Webber-Jones, J. (1982). A controlled trial of amantadine and rimantadine in the prophylaxis of influenza A infection. *N. Engl. J. Med.* **307,** 580–584.

Graves, P. N., Schulman, J. F., Young, J. F., and Palese, P. (1983). Preparation of influenza virus subviral particles lacking the HA$_1$ subunit of HA: Unmasking of cross-reactive HA$_2$ determinants. *Virology* **126,** 106–116.

Hay, A. J. (1989). The mechanism of action of amantadine and rimantadine against influenza viruses. *In* "Concepts of Viral Pathogenesis III" (A. L. Notkins and M. B. A. Oldstone, eds) pp 361–367. Springer Verlag, New York.

Hay, A. J., Wolstenholme, A. J., Skehel, J. J., and Smith, M. D. (1985). The molecular basis of the specific anti-influenza action of amantadine. *EMBO J.* **4,** 3021–3024.

Klenk, H.-D., Rott, R., Orlich, M., and Blodorn, J. (1975). Activation of influenza A viruses by trypsin treatment. *Virology* **68,** 426–439.

Lamb, R. A., Zebedee, S. L., and Choppin, P. W. (1985). Influenza virus M$_2$ protein is an integral membrane protein expressed on the infected cell surface. *Cell* **40,** 627–633.

Lazarowitz, S. G., and Choppin, P. W. (1975). Enhancement of infectivity of influenza A and B viruses by proteolytic cleavage of the haemagglutinin polypeptide. *Virology* **68,** 440–454.

Lear, J. D., and de Grado, W. F. (1987). Membrane binding and conformational properties of a peptide representing the amino terminus of influenza virus HA$_2$. *J. Biol. Chem.* **262,** 6500–6505.

Lear, J. D., Wasserman, Z. R., and de Grado, W. F. (1988). Synthetic amphiphilic peptide models for protein ion channels. *Science* **240,** 1177–1181.

Lesk, A. M., and Hardman, K. D. (1982). Computes generated schematic diagrams of protein structures. *Science* **216,** 539–540.

Ruigrok, R. W. H., Martin, S. R., Wharton, S. A., Skehel, J. J., Bayley, P. M., and Wiley, D. C. (1986). Conformational changes in the haemagglutinin of influenza virus which accompany heat-induced fusion of virus with liposomes. *Virology* **155,** 484–497.

Ruigrok, R. W. H., Aitken, A., Calder, L. J., Martin, S. R., Skehel, J. J., Wharton, S. A., Weis, W., and Wiley, D. C. (1988). Studies on the structure of the influenza virus haemagglutinin at the pH of membrane fusion. *J. Gen. Virol.* **69,** 2785–2795.

Scholtissek, C., and Faulkner, G. P. (1979). Amantadine-resistant and sensitive influenza A strains and recombinants. *J. Gen. Virol.* **44,** 807–815.

Skehel, J. J., Bayley, P. M., Brown, E. M., Martin, S. R., Waterfield, M. D., White, J. M., Wilson, I. A., and Wiley, D. C. (1982). Changes in the conformation of influenza virus haemagglutinin at the pH optimum of virus-mediated membrane fusion. *Proc. Natl. Acad. Sci. USA* **79,** 968–972.

Sugrue, R., Bahadur, G., Zambon, M. C., and Hay, A. J. (submitted for publication).

Webster, R. G., Brown, L. E., and Jackson, D. C. (1983). Changes in the antigenicity of the haemagglutinin molecule of H3 influenza virus at acidic pH. *Virology* **126,** 587–599.

Webster, R. G., Kawaoka, Y., Bean, W. J., Beard, C. M., and Brugh, M. (1985). Chemotherapy and vaccination: A possible strategy for the control of highly virulent influenza virus. *J. Virol.* **55,** 173–176.

Weis, W., Cusack, S. C., Brown, J. H., Daniels, R. S., Skehel, J. J., and Wiley, D. C. (1989). The structure of a membrane fusion mutant of influenza virus haemagglutinin (submitted for publication).

Wharton, S. A. (1987). The role of influenza virus haemmaglutinin in membrane fusion. *Microbiol. Sci.* **4**, 119–124.

Wharton, S. A., Skehel, J. J., and Wiley, D. C. (1986). Studies of influenza haemagglutinin-mediated membrane fusion. *Virology* **149**, 27–35.

Wharton, S. A., Ruigrok, R. W. H., Martin, S. R., Skehel, J. J., Bayley, P. M., Weis, W., and Wiley, D. C. (1988a). Conformational aspects of the acid-induced fusion mechanism of influenza virus haemagglutinin. *J. Biol. Chem.* **263**, 4474–4480.

Wharton, S. A., Martin, S. R., Ruigrok, R. W. H., Skehel, J. J., and Wiley, D. C. (1988b). Membrane fusion by peptide analogues of influenza virus haemagglutinin. *J. Gen. Virol.* **69**, 1847–1857.

Wharton, S. A., Martin, S. R., and Skehel, J. J. (manuscript in preparation).

White, J. M., and Wilson, I. A. (1987). Anti-peptide antibodies detect steps in a protein conformational change: Low-pH activation of the influenza virus haemagglutinin. *J. Cell Biol.* **105**, 2887–2896.

White, J., Kielian, M., and Helenius, A. (1983). Membrane fusion proteins of enveloped animal viruses. *Rev. Biophys.* **16**, 151–195.

Wiley, D. C., and Skehel, J. J. (1987). The structure and function of the haemagglutinin membrane glycoprotein of influenza virus. *Annu. Rev. Biochem.* **56**, 365–394.

Wilson, I. A., Skehel, J. J., and Wiley, D. C. (1981). Structure of the haemagglutinin membrane glycoprotein of influenza virus at 3Å resolution. *Nature (London)* **289**, 366–373.

Zebedee, S. L., and Lamb, R. A. (1988). Influenza A virus M_2 protein: Monoclonal antibody restriction of virus growth and detection of M_2 in virions. *J. Virol.* **62**, 2762–2772.

2 Epitope Mapping and Idiotypy of the Antibody Response to Influenza Neuraminidase

Dennis W. Metzger
Department of Immunology, St. Jude Children's Research Hospital,
Memphis, Tennessee 38101

Gillian M. Air
Department of Microbiology, University of Alabama at Birmingham,
Birmingham, Alabama 35294

W. Graeme Laver
Influenza Research Unit, John Curtin School of Medical Research, The
Australian National University, Canberra City, ACT 2601, Australia

Robert G. Webster
Department of Virology and Molecular Biology, St. Jude Children's Research
Hospital, Memphis, Tennessee 38101

I. Introduction

Influenza A virus is a negative-strand RNA virus that causes primarily an infection of the upper respiratory tract in humans, horses, and swine, and a respiratory and intestinal tract infection in many avian species. The ability of influenza to cause catastrophic pandemics is well known, the most dramatic example being the 1918–1919 Spanish influenza that killed over 20 million people worldwide. Recent outbreaks with high rates of mortality in chickens and seals have served as warnings that an extremely virulent human virus may reappear in the future.

Vaccination attempts against influenza have concentrated on the two surface glycoproteins: the 77-kDa hemagglutinin (HA) that has 13 antigenic subtypes (H1–H13) and the 56-kDa neuraminidase (NA) with 9 subtypes (N1–N9). Antibodies to HA are directly neutralizing while antibodies to NA limit viral spread by preventing release of virions from infected cells. The appearance of new strains that are able to escape established immunity is due primarily to changes in these proteins. These can include minor changes, usually somatic mutations, within subtypes (antigenic drift), as well as genetic reassortment between subtypes (antigenic shift).

Structural knowledge of antigenic sites on HA and NA will allow us to determine how to mimic these sites and design new strategies for vaccine development. Both proteins, as well as antibody–NA complexes, have been crystallized and their three-dimensional structures established. In this report, we discuss recent experiments in the avian N9 NA system that exploit the structural information for testing the immunogenic efficacy of anti-idiotype monoclonal antibodies (MAbs).

II. Identification of a Dominant Antigenic Site on N9 Neuraminidase

In an attempt to map the N9 NA antigenically, we prepared 35 MAbs to the NWS-tern influenza virus (H1N9) and determined that all but one of these MAbs recognize a dominant N9 site located near the rim of the substrate binding cavity (Webster *et al.,* 1987). The specificity of the MAbs was assessed by the ability to react with solubilized crystals of NA and by inhibition of NA enzymatic activity. All of the MAbs were of the IgG1 or IgG2a isotype. The majority of the MAbs inhibited NA enzymatic activity when fetuin (molecular weight, 50,000) was used as the substrate, presumably by blocking access of the substrate to the active site. Some of the MAbs (11 out of 35) also inhibited NA activity when the small substrate N-acetylneuramin-lactose (molecular weight 600) was used as substrate. The MAbs were used to select N9 escape mutants with single point mutations that markedly reduced or abolished MAb binding. Correlation of the amino acid substitutions in the variant molecules with the three-dimensional structure of NA revealed their location to be in a series of five polypeptide loops on or near the rim of the concave enzymatic site (Fig. 1).

The antibody contact residues were further defined by X-ray diffraction analysis of N9 NA complexed with the NC41 MAb (Colman *et al.,* 1987a). The results confirmed the position of the dominant N9 epitope that had been assigned by analysis of escape mutants. In addition, the crystal structure of the NC41 complex showed that while the MAb does not appear to block access of the small neuraminyl-lactose substrate to the active site, its binding does alter the conformation of one of the surface loops of NA, distorting the position of an active site residue (Arg 371). There is also an unusual pairing pattern between the domains of the NC41 V_H and V_L regions, implying either that the NC41 Fab differs in its three-dimensional structure from other Fab fragments or perhaps that binding to antigen induces small changes in the quaternary structure of the MAb, through sliding of domains at the V_H–V_L interface. This study, representing the first X-ray diffraction analysis of antibody contact residues on a disease agent, suggested that the structure of an antigen, and possibly also the structure of an antibody, can change upon binding. Another MAb, termed NC10, reacts with the same antigenic site as NC41 but does not inhibit neuraminyl-lactose recognition. This finding indicates that MAb attachment does not always result in conformational alterations.

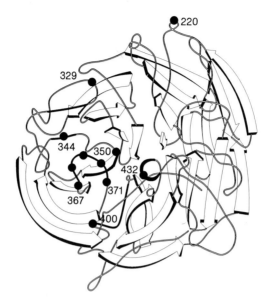

Figure 1. Schematic diagram of the N9 NA monomer viewed down the fourfold axis. The side chains of amino acids 368–370 point towards the viewer, while that of Arg 371 points away and into the catalytic site located above and to the right of this residue. Mutations at positions 367, 369, 370, 372, 400, and 432 abolish binding of NC41 MAb to NA, while a mutation at residue 220 has no effect. Solid chain segments represent the NC41 contact residues as determined by X-ray diffraction analysis (Colman *et al.*, 1987a). All except one of the 35 anti-NA MAbs recognize an epitope(s) overlapping with the NC41 epitope. (From Colman *et al.*, 1987b.)

Taken together, the results demonstrate the presence of a dominant, discontinuous antigenic site on the upper surface of N9 NA that involves approximately 16 amino acids and that can display a degree of structural flexibility upon interaction with antibody.

III. Anti-Idiotypes for Induction of Anti-N9 Neuraminidase Immune Responsiveness

Several investigators have shown that anti-idiotype antibodies can be used to elicit immune responses to pathogenic agents, including polio (UyteHaag and Osterhaus, 1985), hepatitis B (Kennedy *et al.*, 1986), rabies (Reagan, 1985), and reoviruses (Kauffman *et al.*, 1983). In many cases, these anti-idiotypes appear to bear internal images of the viral epitopes (UyteHaag and Osterhaus, 1985; Kennedy *et al.*, 1986; Kauffman *et al.*, 1983). However, conventional anti-idiotypes with paratopes that recognize combining site determinants (Ab2γ molecules) can also induce antiviral responses (Reagan, 1985), probably through direct activation of the relevant B cell clones. There are numerous advantages of an anti-idiotype approach to vaccine development:

1. An anti-idiotype vaccine overcomes the potential hazards of injecting attenuated organisms or their products, particularly when such materials are intrinsically cytotoxic, as in the case of influenza virus.
2. Discontinuous epitopes of proteins can in theory be effectively mimicked by an anti-idiotype.
3. Only a single epitope, or limited set of epitopes, is presented to the host, thus allowing induction of responses to determinants that are antigenically silent in the intact molecule and the avoidance of potential "suppressor" determinants.

The rational design of such vaccines, however, will depend upon a precise understanding of the structural basis for anti-idiotype recognition. We have therefore utilized the well-characterized N9 NA system to investigate structure–function relationships in anti-idiotype binding and induction of immune responsiveness.

Anti-idiotype antibodies were prepared in CAF_1 mice against NC41 and NC10 anti-N9, BALB/c MAbs that have been previously crystallized in complex with N9 NA and that display similar, but slightly different, epitope specificities. Serum samples from the injected mice were tested for the ability to inhibit binding of ^{125}I-labeled NC10 and ^{125}I-labeled NC41 to virus-coated microtiter plates. It was found that all of the injected mice produced anti-idiotype antibody, most of which was specific for the immunizing MAb. However, a fraction of the serum antibody cross-reacted with both NC10 and NC41, suggesting the presence of an NA-like internal image.

Spleen cells from these mice were fused to the Sp2/0 myeloma cell line to produce hybridomas, and three anti-idiotype MAbs were obtained. One MAb [3-2G12 (IgG1)] was derived from mice immunized with the NC10 MAb; two MAbs [11-1G10 (IgG2b) and 11-1C9 (IgG1)] were obtained from mice immunized with the NC41 MAb. Each of the anti-idiotype MAbs prevented binding of homologous anti-NA MAb to virus, demonstrating reactivity with a determinant within, or close to, the antigen-binding site. Furthermore, the anti-idiotypes prevented NC10- and NC41-mediated inhibition of NA enzymatic activity. However, each anti-idiotype MAb appeared to recognize a private idiotype unique to the immunogen, since they did not react with other anti-N9 antibodies. For example, binding of NC10 to plates coated with purified 3-2G12 anti-idiotype MAb was inhibited by NC10 but not by 13 other anti-NA MAbs with similar epitope specificities. Polyclonal mouse and chicken anti-NA antisera also failed to inhibit binding of NC10 to 3-2G12. Identical results were obtained with

Table I. Reactivity of Anti-Idiotype Mabs with Anti-N9 NA Antibodies

Anti-idiotype	Anti-N9 MAbs[1]						Polyclonal Anti-N9	
	NC10	NC41	NC11	NC20	NC34	NC47	Rabbit serum	BALB/c ascites
3-2G12	+	−	−	−	−	−	−	−
11-1C9	−	+	−	−	−	−	−	−
11-1G10	−	+	−	−	−	−	−	−

[1]All MAbs tested react with the same antigenic site on N9 NA.

NC41 binding to 11-1G10 and 11-1C9 anti-idiotype-coated plates (Table I). Thus, all three anti-idiotype MAbs appear to be Ab2γ molecules, that is, antibodies that recognize combining site idiotopes, rather than internal images.

Normal CAF$_1$ mice were injected intraperitoneally with 100 μg of each anti-idiotype mAb conjugated to KLH. After two injections, all of the animals possessed activity in their sera which inhibited the binding of NC10 or NC41 to the respective anti-idiotype. When tested by inhibition of NA activity, it was found that a proportion of these mice also expressed serum NA-specific antibody. In response to injection with 3-2G12 MAb (anti-NC10), one of four mice produced a high level of anti-NA antibody while two other mice produced marginal levels of antibody. Similar results were obtained from mice immunized with the NC41-specific anti-idiotype MAbs.

Our results show that anti-idiotype MAb immunization can induce anti-N9 NA antibody in a proportion of normal, syngeneic mice. This finding contrasts with a study by Mayer and colleagues (1987) who reported that treatment of syngeneic mice with an anti-idiotype MAb which reacts with anti-N1 and anti-N2 antibodies failed to induce anti-NA activity; however, anti-idiotype pretreatment did result in increased anti-NA titers after boosting with virus. Further experiments to determine the ability of our anti-idiotype MAbs to induce protective immunity against lethal influenza challenge will yield important insight into the potential value of antibody vaccines for the control of viral diseases. In addition, crystallization of anti-idiotype complexed with anti-NA MAb will allow us to compare this binding to that of the nominal antigen and will provide an understanding of the structural principles that govern immunogenicity.

Acknowledgments

This work was supported by NIH grants AI18880, AI08831, and CA21765, and by the American-Lebanese-Syrian Associated Charities. We thank Gail McClure and Robin Reed for excellent technical assistance.

References

Colman, P. M., Laver, W. G., Varghese, J. N., and Baker, A. T., Tulloch, P. A., Air, G. M., Webster, R. G. (1987a). *Nature (London)* **326,** 358–363.
Colman, P. M., Air, G. M., Webster, R. G., Varghese, J. N., Baker, A. T., Lentz, M. R., Tulloch, P. A., and Laver, W. G. (1987b). *Immunol. Today* **8,** 323–326.
Kauffman, R. S., Noseworthy, J. H., Nepom, J. T., Finberg, R., Fields, B. N., and Greene, M. I. (1983). *J. Immunol.* **131,** 2539–2541.
Kennedy, R. C., Eichberg, J. E., Landford, R. E., and Dreesman, G. R. (1986). *Science* **232,** 220–223.
Mayer, R., Ioannides, C., Moran, T., Johansson, B., and Bona, C. (1987). *Viral Immunol.* **1,** 121–134.
Reagan, K. J. (1985). *Curr. Top. Microbiol. Immunol.* **119,** 15–30.
UyteHaag, F. C. G. M., and Osterhaus, A. D. M. E. (1985). *J. Immunol.* **134,** 1225–1229.
Webster, R. G., Air, G. M., Metzger, D. W., Colman, P. M., Varghese, J. N., Baker, A. T., and Laver, W. G. (1987). *J. Virol.* **61,** 2910–2916.

3 Structure of a Human Histocompatibility Molecule: Implications for its Interactions with Peptides and T Cell Receptors

Pamela J. Bjorkman*
Department of Microbiology and Immunology, Stanford University, Stanford, California 94305

Mark M. Davis
Department of Microbiology and Immunology and the Howard Hughes Medical Institute, Stanford University, Stanford, California 94305

I. Introduction

The immune response against a viral infection is mediated by two different types of cells known as B and T lymphocytes. The receptor on the B cell is the well-characterized antibody molecule, which exists in a membrane-bound form and in a secreted form involved in the initiation of complement-mediated killing and the inactivation of viral particles by direct binding. The recognition molecule on T cells is the membrane-bound T cell antigen receptor, which has specificity for a combination of foreign antigen with a molecule of the major histocompatibility complex (MHC), as first demonstrated by Zinkernagel and Doherty (1974). MHC proteins exist in two closely related forms called class I and class II MHC molecules, both of which are cell surface glycoproteins that are highly polymorphic in the human population. In general, class II MHC molecules are involved in interactions with T helper cells, which cooperate with B cells to make antibody. Class I MHC molecules are recognized by T killer cells, or cytotoxic T lymphocytes (CTL), that lyse virally infected cells. In both cases, T cell recognition of antigen together with a "self" MHC molecule is termed MHC-restricted recognition.

It has now been established the MHC-restricted T cell receptors recognize peptide fragments of antigens (presumably derived from intracellular processing) bound to an MHC molecule at what appears to be a single site. Two major lines of evidence point to the involvement of peptides in T cell recognition: Peptide fragments of an antigen added to the outside of fixed class II MHC-bearing target cells can be recognized by T helper cells specific for the appropriate combination of antigen and MHC type (Shimonkevitz et al., 1983). Subsequently, short synthetic peptides (8–30 resi-

* Present address: Division of Biology, California Institute of Technology, Pasadena, California 91125

dues long) were shown to bind to purified class II proteins (Babbitt *et al.*, 1985). In some cases, the binding of a particular peptide by an MHC molecule correlated with the ability of an animal of that MHC type to mount an immune response against the antigen from which the peptide was derived (Buus *et al.*, 1987), suggesting a reason for the observation that there is a correlation between immunological response to an antigen and certain specificities of histocompatibility molecules (McDevitt *et al.*, 1972). It has not yet been possible to demonstrate peptide binding to purified class I molecules, although class I and class II MHC molecules are similar in domain organization, sequence, and presumably three-dimensional structure (Brown *et al.*, 1988). However, virus-specific T killer cells have been shown to lyse an uninfected target cell of appropriate class I specificity to which peptide fragments of a viral protein have been added (Townsend *et al.*, 1986). By analogy to the work described for class II MHC molecules, it is assumed that the class I molecules on the target cell bind peptides, and that it is a peptide–MHC complex that is recognized by T cell receptors on the cytotoxic killer cells. Some T cells recognize foreign or "nonself" MHC molecules in the apparent absence of antigen, although in these cases, it is possible that the peptide binding site is occupied by an endogenous peptide. The reactivity of T cells and antibodies against foreign MHC molecules leads to host rejection of transplanted tissue.

The discovery that MHC molecules bind antigenic peptides and present them to T cells has allowed the correlation of susceptibility to autoimmune disease with certain MHC alleles to be understood on a more molecular basis. Any given MHC molecule binds only a subset of peptides tested, and as discussed above, the ability to bind a peptide can determine the immune response to an antigen. Increased susceptibility to autoimmune diseases such as ankylosying spondylitis and insulin-dependent mellitus are found in individuals of certain MHC types, and such diseases arise when the body's immune system attacks its own proteins (reviewed in Todd *et al.*, 1988). It is thought that the MHC molecules correlated with autoimmune disease bind peptides from "self" proteins, leading to tissue damage by self-reactive T cell clones.

One hopes that an increased understanding of the physical nature of antigenic peptide interactions with MHC molecules will allow the design of peptides to stimulate the immune response against a viral infection, or the design of high-affinity ligands to block the self-reactive recognition of MHC molecules involved in autoimmune diseases. Before this can be accomplished, we will need to understand not only the forces comprising the peptide–MHC complex, but also consider how T cell receptors bind to it. In this article, we will briefly review the crystal structure of a human class I MHC molecule, with an emphasis on what the structure reveals about the way histocompatibility molecules bind antigenic peptides and interact with T cell receptors. We will also present a hypothetical model for MHC restricted T cell receptor binding to the peptide–MHC complex. The three-dimensional structure of a T cell receptor has not yet been determined, but because of the similarity between T cell receptors and antibodies, we can use the known structure of an Fab to serve as a first-order T cell receptor model structure in order to make an educated guess about how T cell recognition of the peptide–MHC complex occurs.

II. The Structure of HLA-A2

The three-dimensional structure of HLA-A2, a human class I histocompatibility molecule, has been determined (Bjorkman *et al.*, 1987a,b). Class I MHC molecules are composed of two polypeptide chains. The heavy chain spans the membrane bilayer and is composed of the polymorphic α_1 and α_2 domains followed by the immunoglobulin-like α_3 domain, a transmembrane region, and a short cytoplasmic tail. The light chain is β_2-microglobulin (β_2m), a molecule with sequence similarity to immunoglobulins and to α_3. Prior to crystallization, HLA-A2 was cleaved from the membrane with papain, yielding a soluble heterodimer composed of the α_1, α_2, and α_3 domains of the heavy chain bound to β_2m. As shown schematically in Fig. 1a, the molecule is composed of two structural motifs: The membrane-proximal α_3 and β_2m domains are folded like immunoglobulin constant regions but pair with a novel interaction not previously seen in antibody structures, and the membrane-distal α_1 and α_2 domains consist of a platform of 8 β-strands topped by two α-helices. For the orientation of the molecule shown, the cell membrane is assumed to be horizontal at the bottom of the figure.

A. The Immunoglobulin-Like Domains

As predicted from their sequences (Orr *et al.*, 1979), the α_3 and β_2m domains are both immunoglobulin-like structures composed of a four-stranded β-pleated sheet connected to a three-stranded β-sheet by a disulfide bond. Surprisingly, however, these domains do not pair with each other in the interaction found between pairs of constant domains in the known antibody structures (e.g., CH1 and CL in an Fab, or the CH3 dimer in Fc). Antibody constant domains are typically related by a nearly exact twofold symmetry axis, with their four-stranded β-sheets forming the

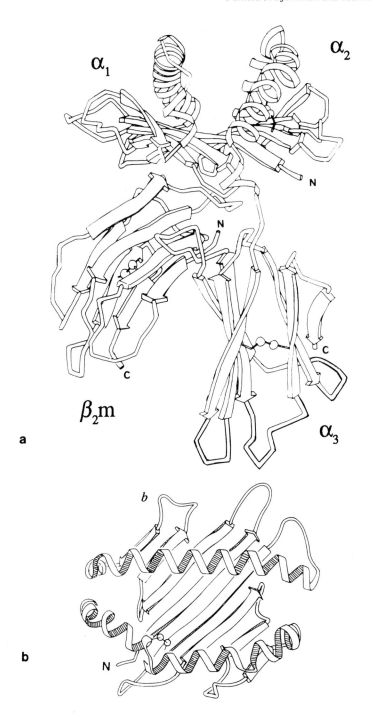

contact interface (Davies and Metzger, 1983). In HLA-A2, the two immunoglobulin-like domains contact each other predominantly with their four-
stranded β-sheets but are related by a 146° rotation followed by a 13 Å
translation. The effect of the translation is to move β_2m up (direction
defined by Fig. 1a) closer to the $\alpha_1-\alpha_2$ platform than is α_3.

B. The Peptide Binding Domains

The α_1 and α_2 domains fold together to form a single eight-stranded
β-pleated sheet spanned on its top side by two α-helices. Each individual
domain is composed of a four-stranded antiparallel β-sheet comprising its
N-terminal half, followed by a long helical region. When these two structurally similar domains are paired in the HLA molecule, they are related by
a nearly perfect twofold axis of symmetry, causing the two four-stranded
sheets to meet and form a single continuous sheet topped by the α-helices
from each domain. A large cleft separates the two α-helices. The sides of
the cleft are formed by residues from the α-helices that point in, and its
bottom is formed by residues on the β-sheet that point up. Figure 1b shows
a schematic view of the peptide binding domains as viewed from the top of
the structure.

Several lines of evidence suggest that the deep groove separating the
two helices on the top of the $\alpha1-\alpha2$ platform is the binding site for antigenic
peptides:

1. The groove is located on the top surface of the molecule (as oriented
   in Fig. 1a), and is thus in a likely position to interact with a receptor
   on the surface of the T cell.
2. The groove is ~25 Å long and ~10 Å wide and deep, dimensions
   consistent with the expectation that MHC molecules bind a
   processed (i.e., peptide) form of an antigen.
3. Many of the residues that form the sides and the bottom of the site
   are highly polymorphic or have been identified to be critical for T cell
   recognition of class I molecules, fitting with the expectation that
   MHC polymorphisms affect the peptide binding and T cell reactivity.
4. The crystal structure shows that the site appears to be occupied by a

Figure 1. Schematic representation of the structure of HLA-A2. The β-strands are depicted
as thick arrows in the amino–carboxy direction, α-helices are helical ribbons, and connecting
loops are thin lines. Disulfide bonds are shown as two connected spheres. (a) The entire
extracellular portion of HLA-A2. The four domains are labeled. (b) The top surface of
HLA-A2 showing the $\alpha1$ and $\alpha2$ domains with the peptide binding site located between the
two α-helices.

molecule or mixture of molecules that evidently copurified and co-crystallized with HLA-A2. Although other interpretations are possible, it seems most likely that these molecules represent a heterogeneous mixture of endogenous peptides, perhaps added during synthesis of HLA.

The structure of HLA-A2 therefore represents that of a molecule with an occupied peptide-binding site. The very fact that the binding site occupants remained during the entire purification of HLA (which takes ~2 weeks and requires extensive dialysis) suggests that removal is difficult, perhaps requiring a conformational change. We do not know if empty MHC molecules exist *in vivo,* but there is no structural reason why they could not. The residues of the peptide binding site are not particularly hydrophobic, thus the site could be filled with water in the absence of peptide with the result that an empty MHC structure might be conformationally quite similar to that of HLA-A2. Alternatively, one can imagine a closed structure with the top two helices closer together to occlude the site and maximize the favorable antiparallel alignment of their helical dipoles. It should therefore be kept in mind by those interested in designing ligands for the peptide-binding site that our knowledge of MHC structure presently only pertains to a structure with something already in the site; we do not know the structures of any empty MHC molecule or a specific peptide–MHC complex.

Currently, there is no physical evidence regarding the secondary structures of peptides bound to MHC molecules. Substitution of residues in a peptide, followed by testing of MHC binding and activation of T cell clones, has suggested an α-helical conformation for one peptide bound to a class II molecule (Allen *et al.,* 1987), and an extended β-strand conformation for another (Sette *et al.,* 1987). A study of the peptide binding site of HLA-A2 does not immediately suggest what conformation a specific peptide would adopt when bound to HLA. The site would accommodate an α-helical peptide of about 20 residues, or an extended polypeptide chain of about 8 residues. Longer peptides could bind to the site if one or both ends projected out. A peptide in the standard α-helical conformation is a very tight fit in the HLA-A2 binding site, and thus some minor movements of side chains or even opening up of the site by displacing the α-helices might be necessary before binding or releasing such a peptide. The movement need not result in a large structural change. It could be caused by a flexing or straightening of the β-sheet, resulting in a very slight moving of the α helices. However, the α_2 helix is tethered to the β-sheet by a disulfide bond and would not be free to move extensively. Another structural feature that may influence peptide entry or release from the site is a salt bridge located at one end of the site. The salt bridge is made up of Arg 55 and Glu 170, both of which are conserved (with one exception in an HLA-C specificity) in 22 human class I sequences.

Within any given protein antigen, there will be one or two peptide fragments capable of binding to an MHC molecule and eliciting a T cell response. It is not yet possible to predict accurately from a protein sequence which fragments will be antigenic, but two methods have been suggested. One method searches for a somewhat conserved motif within the primary sequence. The motif identified to be common among many T cell epitope peptides is a glycine or charged residue followed by two or three hydrophobic residues and then a charged residue (Rothbard and Taylor, 1988). When this motif has been used to search through protein sequences, some antigenic peptides have been correctly predicted. The other prediction method supposes that antigenic peptides form amphipathic α-helical structures, with the hydrophobic side of the helix in contact with the MHC molecule, and the hydrophilic side interacting with the T cell receptor (DeLisi and Berzofsky, 1985). However, one thing we know from examining the HLA-A2 structure is that because of the depth of the binding site, a peptide in any conformation will contact the MHC molecule on three of its sides, leaving the fourth side exposed to solvent or in a position to interact with a T cell receptor. It is also clear that peptides need not be amphipathic structures, with a nonpolar side interacting with the MHC, because the HLA-A2 binding site contains a number or polar and charged residues, as well as nonpolar amino acids. Examination of other class I sequences and a hypothetical model for the structure of class II MHC molecules suggests that other MHC binding sites are not particularly hydrophobic either.

C. Location of Polymorphic and Conserved Residues

Polymorphism at some class I residues is assumed to be responsible for MHC-restricted recognition by T cells, and for the variation in responsiveness to particular antigens. On the HLA-A2 structure, 15 of the 17 most highly polymorphic positions are located in the peptide-binding site. Some of the polymorphic residues are located on the central β-strands of the α_1–α_2 platform and would be buried if the binding site were occupied, suggesting that their polymorphism affects interactions with a bound peptide rather than with a T cell receptor. The other polymorphic residues are located on the sides of the helices pointing either in towards the site in positions to affect peptide binding, or projecting up into solvent, where they could interact directly with a T cell receptor.

Ten residues (five of which are tyrosines) pointing into the site are completely conserved in 22 human sequences. These residues may interact with the backbone of peptides binding to the site or recognize some other common feature of antigenic peptides. A pattern of every third or fourth residue is conserved on both helices of the peptide binding site. These residues form completely conserved faces on their respective helices, which point up into solution or away from the peptide binding site. It is not known if these residues are important for recognition by constant features on T cells, or if they are conserved to maintain the HLA structure.

III. T Cell Receptor Structure

T cell receptors (TCRs) are membrane-bound, disulfide-linked heterodimers that resemble the Fab fragments of immunoglobulins in sequence and domain organization (reviewed in Kronenberg *et al.*, 1986). Each polypeptide chain contains a domain with sequence similarity to antibody variable domains, followed by a domain similar to an antibody constant domain. C-terminal to the constantlike domain in each chain, there is a hinge region containing a cysteine residue involved in the formation of an interchain disulfide bond, and then a hydrophobic membrane-spanning sequence and short cytoplasmic tail. Figure 2 shows a schematic representation of the primary sequence organization for the four TCR polypeptide chains that make up the two different types of receptor heterodimers (α–β and γ–δ). Like immunoglobulins, both the α–β and γ–δ TCRs are assembled by the relatively random joining of different coding segments (V and J in the case of α and γ chains; V, D, and J in the case of β and δ chains) to constant region genes. Although variability in TCR V regions is less localized than in immunoglobulin V regions, hypervariability is found in the locations corresponding to the three classic immunoglobulin hypervariable regions or complementarity determining regions (CDRs), which are known in antibodies to form the principal points of contact with antigen. In both antibodies and TCRs, the first and second CDR are encoded within the V gene segment itself, and the third CDR is formed by the junction of the V gene segment with D and J gene segments (in the case of immunoglobulin heavy chains and TCR β and δ chains) or with J gene segments alone (immunoglobulin light chains and TCR α and γ).

A. Sequence Diversity in TCRs and Antibodies

Although TCRs and immunoglobulins share a similar organization of diversity and mechanisms for its generation, a closer look at TCR diversity shows a striking concentration of sequence polymorphism in the CDR3-

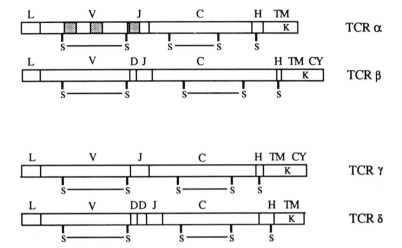

Figure 2. Primary sequence organization of T cell receptor polypeptides. The primary sequence organization for the four TCR chains involved in forming the two types of hetero-dimers (α: β and γ: δ) is shown. Leader (L), variable (V), diversity (D), and joining (J) segment contributions are indicated for the variable region domains. The constant portions of the T cell receptor protein is divided into the immunoglobulin-like constant region (C), a hinge (H), and the transmembrane (TM) and cytoplasmic (CY) domains. The approximate location of cysteine residues are indicated by an external S. There is a disulfide bond in the variable and constant domains, and cysteine residues within the hinge region form an interchain disulfide bond in both types of TCR heterodimer. The approximate location of CDR1-, CDR2-, and CDR3-equivalent regions are indicated in the primary structure of the TCR α-chain.

equivalent region as compared to this region in antibodies (reviewed in Davis and Bjorkman, 1988). By contrast, TCR diversity within the CDR1- and CDR2-equivalent regions is far less pronounced than in immunoglobulins. The primary reason for decreased TCR diversity in the first and second CDR regions is that there are far fewer TCR V gene segments than immunoglobulin V gene segments. Assuming that the V domains of TCRs pair to form a combining site (as is the case for antibody V regions), the combinatorial diversity resulting from a random pairing of TCR V domains shows even greater disparity from the amount of diversity resulting from antibody V region pairing. Estimates for the amounts of sequence diversity possible for immunoglobulins and TCRs are compared in Table I. As a result of the many mechanisms generating diversity within the CDR3-equivalent or junctional region in TCRs, the potential diversity is estimated to be between four and seven orders of magnitude higher in TCRs as compared to immunoglobulins. On the other hand, the amount of potential TCR nonjunctional diversity (within the CDR1- and CDR2-equivalent

Table I. Sequence Diversity in T Cell Receptor and Immunoglobulin Genes[a]

Region	IG		TCR I		TCRII	
	H	κ	α	β	γ	δ
Variable segments	250–1000	250	100	25	7	10
Diversity segments	10	0	0	2	0	2*
Ds read in all frames	rarely	—	—	Often	—	often
N-region addition	V–D,D–J	None	V–J	V–D,D–J	V–J	V–D1,D1–D2,D1–J
Joining segments	4	4	50	12	2	2
Variable region combinations	62,500–250,000		2500		70	
Junctional combinations	$\sim 10^{11}$		$\sim 10^{15}$		$\sim 10^{18}$	

[a]Calculated potential amino acid sequence diversity in T cell receptor and immunoglobulin genes without allowance for somatic mutation. The magnitude of combinatorial diversity resulting from the CDR1- and CDR2-equivalent regions encoded within the V gene segments is lower for TCRs than for immunoglobulins because there are far fewer TCR V gene segments than immunoglobulin V gene segments (listed as variable region combinations). However, the amount of calculated diversity encoded within the junctional or CDR3-equivalent region is far greater for TCRs than for immunoglobulins (listed as junctional combinations).

regions) is estimated to be between 10 and 1000 times less than that possible for antibodies.

B. A Model for TCR Interactions with Peptide–MHC Complexes

In order to rationalize the striking concentration of TCR diversity within the junctional region and relative lack of diversity elsewhere, it should be relevant to consider what TCRs are known to do, namely, to recognize a large number of small molecules (i.e., peptides) embedded in physically larger and much less diverse MHC molecules. The simplest interpretation of the skewing of diversity in the TCR case towards the CDR3-equivalent region is that the amino acids in this region are mainly interacting with peptide determinants and that residues within the much less diverse CDR1- and CDR2-equivalent regions are primarily involved with contacts to MHC determinants.

At this point, it would be useful to compare the three-dimensional structure of a TCR to HLA-A2, in order to see if there is any structural reason to hypothesize the alignment of CDR1 and 2 with the MHC and CDR3 with a bound peptide. Although the crystal structure of a TCR is unknown, we can use the large amount of available structural data on antibodies to predict how the three CDR regions will be arranged around a TCR combining site. Sequence data suggest that TCR variable regions are folded into β-sandwich structures resembling immunoglobulin V-regions. In antibodies, the variable regions from the heavy and light chains (V_H and

V_L) are paired such that CDR1, 2, and 3 from each domain are clustered at the ends of the Fab arms of the molecule, forming the antigen-binding site (reviewed in Davies and Metzger, 1983). Many of the residues in the antibody V-regions interface are conserved, as is the overall geometry of pairing, with the result that the CDRs are arranged similarly about the binding sites of the molecules whose structures are known (Chothia *et al.*, 1985). A study of immunoglobulin sequences and structures has identified conserved amino acids that are critical for maintaining the V_H–V_L contact surface, and most of these amino acids are found in homologous positions in TCR V-region sequences (Novotny *et al.*, 1986; Chothia *et al.*, 1988). It is therefore likely that TCR V-domains will fold into tertiary structures similar to antibody V-regions, and that the chain pairing and resulting combining sites of TCRs will be similar to those described for antibodies (Chothia *et al.*, 1988).

In antibodies, the first and second CDRs on V_L are separate from their counterparts on V_H, and the space between them is occupied by the CDR3 regions from each chain (Davies and Metzger, 1983). In Figure 3a, the relative locations of the six CDR regions in an immunoglobulin combining site are shown schematically, here representing the approximate structure of a TCR combining site. A schematic view of the MHC peptide binding site is shown in Fig. 3b and is drawn to scale with respect to the V-regions combining site (Fig. 3a). Both the V-regions combining site and the MHC–peptide complex have relatively flat surfaces. Structural studies of antibodies complexed with protein antigens have shown that side chains from all six CDRs contact the antigen and that the entire interface is a rather flat surface with protrusions and depressions in the antibody being complementary to the antigen surface (Amit *et al.*, 1986; Colman *et al.*, 1987; Sheriff *et al.*, 1987). Based on the similarities between TCRs and antibodies, we assume this will also be true for the TCR combining site. In the case of the MHC–peptide complex, the top surface recognized by a TCR also appears to be quite flat if the peptide binding site is occupied (as indeed is the case for HLA-A2, which cocrystallized with unknown molecule(s) in the site). When the (hypothetical) TCR combining site is compared to the top surface of the MHC–peptide complex that is its ligand, it is interesting to note that the α helices that make up the sides of the MHC peptide-binding site are separated by about the same distance as separates the CDR1 and 2 regions of one V-domain from another (18 Å). Thus the relatively flat surface of an MHC–peptide complex could interact with the combining site of an immunoglobulin-like TCR such that the limited diversity in the CDR1- and CDR2-equivalent regions on TCR V_α and V_β contacts the side chains of the MHC α-helices, leaving the centrally located and very diverse CDR3-equivalent regions to interact with the peptide. A

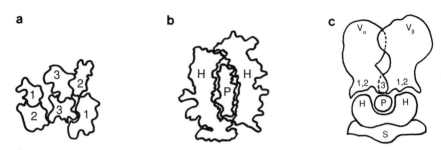

Figure 3. Schematic representations of (a) the immunoglobulin combining site, (b) the peptide binding site of an MHC molecule, and (c) the alignment of CDRs in a hypothetical TCR structure over a peptide–MHC complex. Color figures using the carbon-α coordinates of an Fab V-region and HLA-A2 are shown in Davis and Bjorkman (1988). (a) The arrangement of CDRs in an immunoglobulin antigen binding site as viewed from above (the direction of antigen). In this figure, the size and arrangement of CDRs is determined by tracing around a van der Waals surface of the residues within each CDR. The first and second CDRs from one variable domain (labeled with the numbers 1 and 2) are separate from their counterparts on the partner variable domain, with the space between them occupied by the third CDR from each domain (labeled with the number 3). Similarities between antibodies and TCRS suggest that TCRs may have a combining site that preserves the same general features. (b) Top surface of an MHC molecule with a (hypothetical) bound peptide. The outline of the van der Waals surface of the residues on the two α helices of HLA-A2 is shown (labeled with the letter H), with the van der Waals outline of a 12-mer peptide fitted into the peptide binding site (labeled with the letter P). This figure is to scale with respect to Fig. 3a. Notice that the distance between the MHC α-helices is approximately the same as the distance separating the CDR1 and CDR2 regions of one variable domain from these regions in the partner variable domain (Fig. 3a). The orientation of the molecule in this schematic representation of the peptide binding site is the same as for the line drawing of the α_1 and α_2 domains in Fig. 1a. (c) Model for TCR interaction with a peptide–MHC complex. The combining site CDRs are aligned over the peptide–MHC complex. The TCR V regions are labeled Vα and Vβ, and the approximate locations of the three CDR-equivalent regions are labeled with their respective numbers. The MHC–peptide complex is shown schematically with the space occupied by residues in the two α-helices labeled with the letter H, the peptide labeled with the letter P, and the residues in the β-sheet labeled with the letter S. (The orientation of the MHC α_1 and α_2 domains in this figure is similar to the view of this part of the molecule in Fig. 1a.) The molecules in this figure are rotated ~90° with respect to their orientations in parts (a) and (b). The relatively flat surface of the V-regions combining site is complementary to the relatively flat surface of the peptide–MHC complex allowing the limited diversity within the CDR1- and CDR2-equivalent regions to interact with MHC determinants and the much greater diversity within the CDR3-equivalent region to interact with peptide determinants.

schematic view of this interaction is shown in Fig. 3c. (Color pictures using α-carbon coordinates of HLA-A2 and Fab V-regions are shown in Davis and Bjorkman, 1988.)

This model for TCR and MHC–peptide interactions is consistent with some recent work correlating TCR α–β sequences with known antigen–

MHC specificities. For example, in some cytochrome c-specific T cell clones, changes in the junctional region alter the specificity for peptide without altering MHC specificity (Fink *et al.*, 1986; Winoto *et al.*, 1986). Also, some of the TCRs from these clones show a selection for certain amino acids within the CDR3-equivalent region, suggesting that junctional residues are important for peptide recognition (Hedrick *et al.*, 1988). This idea was tested by changing one of the conserved junctional residues by site-directed mutagenesis, with the result that the mutated TCR displayed a different fine specificity for antigen (Engel and Hedrick, 1988). However, even within the confines of our model for TCR interaction with MHC–peptide complexes, one would not always expect a direct correlation of TCR junctional residues with specificity for a particular peptide, for the simple reason that changes in MHC residues could alter the conformation or orientation of a bound peptide, thus requiring a compensatory change in the TCR residues contacting the peptide. We would also not expect that the V_α and V_β gene segments used by T cells restricted to the same MHC molecule to always be the same, because the surface of a peptide MHC complex is large enough to allow a TCR (assuming an immunoglobulin-like binding site) to bind in different registers along the MHC α helices. In addition, because antibody (and presumably TCR) V-regions pair with approximate dyad symmetry, the interaction of CDR1 and CDR2 residues with MHC determinants and CDR3 with peptide determinants can be accomplished in either of two orientations related to each other by 180°. (In other words, the interaction depicted in Fig. 3c would look the same if the V-regions dimer were rotated by 180° about its pseudodyad axis, which is vertical in this figure.) It is likely that an exact correlation between TCR hypervariable region sequences and MHC and peptide specificites will never be possible, due to the spatial proximity of CDR residues in the combining site. However, the structural complementarity between the MHC–peptide binding site and the (hypothetical) TCR combining site depicted in Fig. 3c may provide a molecular explanation for the original suggestion by Jerne (1971) that T cell antigen receptors and MHC molecules coevolved to have some affinity for each other. The value of this model for TCR/MHC–peptide interactions will only be proven by the structure determination of an MHC–TCR complex.

IV. Future Work and Implications for Drug Design

If one could accurately predict which fragments of a virus will bind to an MHC molecule and stimulate a T cell response, it might be possible to use synthetic peptides to stimulate the antiviral immune response. The current approach of analyzing sequences of known T cell epitope peptides for common features has been somewhat successful in the identification of new peptides capable of invoking T cell reactivity. However, the structural details of how any specific peptide interacts with an MHC molecule are unknown. Thus it will be necessary to determine the structure of a peptide–MHC complex before one can attempt to rationalize the physical reasons behind a common sequence motif in antigenic peptides. In the meantime, it may be possible to identify both common and differing properties of MHC peptide binding sites by analyzing their amino acid sequences and thereby work towards an understanding of the forces that allow binding of a peptide to one particular MHC molecule, but not to another.

A great deal of sophistication will be required before it is possible to design high-affinity ligands to block the peptide binding sites of MHC molecules implicated in autoimmune diseases. It is not known if all antigenic peptides adopt a similar secondary structural conformation when bound to an MHC molecule, and if so, what that conformation is. Therefore, it would be currently very difficult to design a peptide sequence for MHC binding. It is simpler to think about binding some sort of small organic molecule capable of existing in only one conformation. One could then possibly use the properties of the peptide binding site in the known structure of HLA-A2 to assist in predicting what sorts of molecules might bind. Even this approach is complicated by the fact that the peptide binding site in HLA-A2 is occupied, and this structure may be somewhat different than a structure capable of binding added material. Thus, the future years should see a large body of work in the fields of crystallography and immunology that will increase our understanding of the interactions between peptide–MHC complexes and T cells, and this work should assist in the design of antiviral and other therapeutic agents.

References

Allen, P. M., Matsueda, G. R., Evans, R. J., Dunbar, J. B., Marshall, G. R., and Unanue, E. R. (1987). Identification of the T-cell and Ia contact residues of a T cell antigenic epitope. *Nature* (*London*) **327**, 713–715.

Amit, A. G., Mariuzza, R. A., Phillips, S. E. V., and Poljak, R. J. (1986). Three-dimensional structure of an antigen–antibody complex at 2.8A resolution. *Science* **233**, 747–753.

Babbitt, B. P., Allen, P. M., Matsueda, G., Haber, E., and Unanue, E. R. (1985). Binding of immunogenetic peptides to immunoglobulin histocompatibility molecules. *Nature* (*London*) **324**, 317, 359–361.

Bjorkman, P. M., Saper, M. A., Samraoui, B., Bennett, W. S., Strominger, J. L., and Wiley, D. C. (1987a). Structure of the human class I histocompatibility antigen, HLA-A2. *Nature* (*London*) **329**, 506–512.

Bjorkman, P. J., Saper, M. A., Samraoui, B., Bennett, W. S., Strominger, J. L., and Wiley, D. C. (1987b). The foreign antigen binding site and T cell recognition regions of class I histocompatibility antigens. *Nature (London)* **329**, 512–518.

Brown, J. H., Jardetsky, T., Saper, M. A., Samraoui, B., Bjorkman, P. J., and Wiley, D. C. (1988). A hypothetical model of the foreign antigen binding site of class II histocompatibility molecules. *Nature (London)* **332**, 845–850.

Buus, S., Sette, A., Colon, S. M., Miles, C., and Grey, H. M. (1987). The relation between major histocompatibility complex (MHC) restriction and the capacity of Ia to bind immunogenic peptides. *Science* **235**, 1353–1358.

Chothia, C., Novotny, J., Bruccoleri, R., and Karplus, M. (1985). Domain association in immunoglobulin molecules. The packing of variable domains. *J. Mol. Biol.* **186**, 651–663.

Chothia, C., Boswell, D. R., and Lesk, A. M. (1988). The outline structure of the T-cell $\alpha\beta$ receptor. *EMBO J.* **7**, 3745–3755.

Colman, P. M., Laver, W. G., Varghese, J. N., Baker, A. T., Tulloch, P. A., Air, G. M., and Webster, R. G. (1987). Three-dimensional structure of a complex of antibody with influenza virus neuraminidase. *Nature (London)* **326**, 358–363.

Davies, D. R., and Metzger, H. (1983). Structural basis of antibody function. *Annu. Rev. Immunol.* **1**, 87–117.

Davis, M. M., and Bjorkman, P. J. (1988). T-cell antigen receptor genes and T-cell recognition. *Nature (London)* **334**, 395–402.

DeLisi, C., and Berzofsky, J. A. (1985). T-cell antigenic sites tend to be amphipathic structures. *Proc. Natl. Acad. Sci USA* **82**, 7048–7052.

Engel, I., and Hedrick, S. M. (1988). Site-directed mutations in the VDJ junctional region of a T cell receptor β chain cause changes in antigenic peptide recognition. *Cell* **54**, 473–484.

Fink, P. J., Matis, L. A., McElligott, D. L., Bookman, M., and Hedrick, S. M. (1986). Correlations between T-cell specificity and the structure of the antigen receptor. *Nature (London)* **321**, 219–226.

Hedrick, S. M., Engel, I., McElligott, D. L., Fink, P. J., Hsu, M.-L., Hansburg, D., and Matis, L. A. (1988). Selection of amino acid sequences in the beta chain of the T cell antigen receptor. *Science* **239**, 1541–1544.

Jerne, N. K. (1971). The somatic generation of immune recognition. *Eur. J. Immunol.* **1**, 1–9.

Kronenberg, M., Siu, G., Hood, L. E., and Shastri, N. (1986). The molecular genetics of the T-1 cell antigen receptor and T-cell antigen recognition. *Annu. Rev. Immunol.* **4**, 529–591.

McDevitt, H. O., Deak, B. D., Shreffler, D. G., Klein, J., Stimpfling, J. H., and Snell, G. D. (1972). Genetic control of the immune response. Mapping of the Ir-1 locus. *J. Exp. Med.* **128**, 1–11.

Novotny, J., Tonegawa, S., Saito, H., Kranz, D. M., and Eisen, H. N. (1986). Secondary, tertiary, and quaternary structure of the T-cell specific immunoglobulin-like polypeptide chains. *Proc. Natl. Acad. Sci. USA* **83**, 742–746.

Orr, H. T., Lancet, D., Robb, R. J., Lopez de Gastro, J. A., and Strominger, J. L. (1979). The heavy chain of human histocompatibility antigen HLA-B7 contains an immunoglobulin-like region. *Nature (London)* **282**, 266–270. Rothbard, J. B., and Taylor, W. R. (1988). A sequence pattern common to T cell epitopes. *EMBO J.* **7**, 93–100.

Rothbard, J. B., and Taylor, W. R. (1988). A sequence pattern common to T cell epitopes. *EMBO J.* **7**, 93–100.

Sette, A., Buus, S., Colon, S., Smith, J. A., Miles, C., and Grey, H. M. (1987). Structural characteristics of an antigen required for its interaction with Ia and recognition by T cells. *Nature (London)* **328**, 395–399.

Sheriff, S., Silverton, E. W., Padlan, E. A., Cohen, G. H., Smith-Gill, S. J., Finzel, B. C., and Davies, D. R. (1987). Three-dimensional structure of an antibody–antigen complex. *Proc.Natl. Acad. Sci. USA* **84,** 8075–8079.

Shimonkevitz, R., Kappler, J. W., Marrack, P., and Grey, H. M. (1983). Antigen recognition by H-2 restricted T cells. Cell-free antigen processing. *J. Exp. Med.* **158,** 303–316.

Todd, J. A., Acha-Orbea, H., Bell, J. I., Chao, N., Fronek, Z., Jacob, C. O., McDermott, M., Sinha, A. A., Timmerman, L., Steinman, L., and McDevitt, H. O. (1988). A molecular basis for MHC class II-associated autoimmunity. *Science* **240,** 1003–1009.

Townsend, A. R. M., Rothard, J., Gotch, F. M., Behador, G., Wraith, D., and McMichael, A. J. (1986). The epitopes of influenza nucleoprotein recognized by cytotoxic T lymphocytes can be defined with short synthetic peptides. *Cell* **44,** 959–968.

Winoto, A., Urban, J. L., Lan, N. C., Goverman, J., Hood, L., and Hansburg, D. (1986). Predominant use of a Vα gene segment in mouse T-cell receptors for cytochrome c. *Nature (London)* **324,** 679–682.

Zinkernagel, R. M., and Doherty, P. C. (1974). Restriction of *in vitro* T-cell mediated cytotoxicity in lymphocytic choriomeningitis within a syngeneic or semi-allogeneic system. *Nature (London)* **248,** 701–702.

4 Structure of the Adenovirus Virion

Roger M. Burnett
The Wistar Institute, 3601 Spruce Street, Philadelphia, Pennsylvania 19104

Francis K. Athappilly
Department of Biochemistry and Molecular Biophysics, Columbia University,
New York, New York 10032

Zhaoping Cai
The Wistar Institute, 3601 Spruce Street, Philadelphia, Pennsylvania 19104

Paul S. Furcinitti*
Biology Department, Brookhaven National Laboratory, Upton, New York
11973

Alex P. Korn and Ramachandran Murali
The Wistar Institute, 3601 Spruce Street, Philadelphia, Pennsylvania 19104

Jan van Oostrum†
Department of Biochemistry and Molecular Biophysics, Columbia University,
New York, New York 10032

I. Introduction

Adenovirus is a mammalian virus containing at least 10 different proteins in an architecture that is complex but creates the virion's characteristically simple icosahedral shape (Fig. 1). Adenovirus has been extensively investigated in the 35 years since its discovery by Rowe *et al.* (1953) and is now an important model system in several areas of molecular biology (see reviews in Philipson, 1983; Ginsberg, 1984). The major coat protein of adenovirus, hexon, forms the 20 interlocking facets of the outer capsid,

* Present address: Biophysics Research Division, University of Michigan, Ann Arbor, Michigan 48109
† Present address: Department of Biotechnology, CIBA-GEIGY AG, CH-4002 Basel, Switzerland

Use of X-Ray Crystallography in the Design of Antiviral Agents

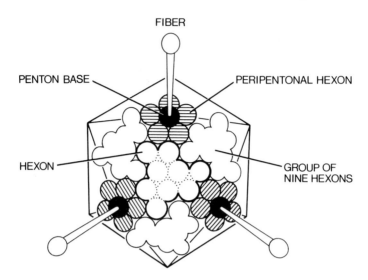

Figure 1. Diagram showing the organization of the major proteins of the icosahedral adeno-virus capsid. Dissociating the virion under mild conditions first releases the penton complex of fiber and penton base from the twelve vertices. Next, the five peripentonal hexons surrounding each penton base are released. Finally, the planar groups-of-nine hexons (GONs) at the center of each of the 20 facets of the icosahedron are released as a unit from the viral core. The GON is left-handed when viewed from outside the virion. (From Burnett, 1984.)

each of which contains 12 hexons. Penton, a complex between penton base and fiber, occupies the 12 vertices. The ad2 virion contains about 2700 polypeptides, derived from at least 10 different proteins, whose stoichiometric ratios are known (van Oostrum and Burnett, 1985). Progress in understanding the detailed structure of the whole virion by X-ray crystallographic methods is limited by the presence of fibers and the size and complexity of the virion, which has an insphere diameter of 900 Å and a particle mass of 150×10^6 Da (van Oostrum and Burnett, 1985). We describe below a different approach to the study of the structure of the virion that, using a combination of crystallography, electron microscopy, and biochemistry, is providing the desired atomic detail.

Our approach is made possible by the presence in infected cells of a 10–100-fold excess mass of soluble structural proteins over that of completed virions (White *et al.*, 1969). The availability of the major coat protein, hexon, in soluble form has permitted its purification and crystallographic structure determination (Burnett, 1984). The 6 Å resolution structure (Burnett *et al.*, 1985), later extended to 2.9 Å (Roberts *et al.*,

1986), has provided the molecular morphology and led to a model for the adenovirus capsid (Burnett, 1985). Conventional electron microscopy of groups-of-nine hexons (GONs) (Fig. 1) and other capsid fragments then has been used to determine the relative hexon positions in the capsid facet to an accuracy of 5 Å (van Oostrum *et al.*, 1987a,b). More precise electron microscopy, using the Brookhaven Scanning Transmission Electron Microscope (STEM), has refined the hexon positions to within 1 Å (Furcinitti *et al.*, 1986, 1987, 1989). Subsequently, difference images were calculated between real GONs and an artificial array of nine hexons, calculated from the X-ray structure. The difference images have revealed the binding sites of a small cementing protein, polypeptide IX, that stabilizes the hexon array in the capsid. The number and location of the experimentally derived binding sites agree with our stoichiometric determination for the amount of polypeptide IX within the virion (van Oostrum and Burnett, 1985) and earlier predictions (Burnett, 1984). The binding of polypeptide IX explains the dissociation pattern of the capsid into GONs, and GONs into individual hexons. The STEM study is of interest as it not only combines X-ray crystallography and electron microscopy but has succeeded in imaging a minor component with less than 6% of the total fragment mass. We plan to extend these studies to map other virion components.

II. Virion Proteins

Precise information on the representation of the various structural proteins in viruses is often lacking, but it is the prerequisite for developing an architectural model. We thus undertook an analysis of the molecular composition of the adenovirus virion (van Oostrum and Burnett, 1985), made possible by the availability of the complete ad2 DNA sequence (Aleström *et al.*, 1984; Roberts *et al.*, 1984). Table I shows the characteristics of the structural proteins whose stoichiometry was determined. Minor components (IVa$_1$ and IVa$_2$) and smaller polypeptides (X, XI, and XII) also have been observed (Everitt *et al.*, 1973), with the latter thought to be remnants of precursor polypeptides after maturation of the virion by proteolytic cleavage. Our study gave an accurate value for the molecular mass of the virion (148.7×10^6 Da, of which 22.2×10^6 Da is DNA) (van Oostrum and Burnett, 1985). It also resolved outstanding questions concerning the subunit composition of the penton complex by showing that its base is pentameric, and its fiber trimeric. It is noteworthy that this composition is in disagreement with a previously convincing theoretical model for the fiber as a dimer (Green *et al.*, 1983), which now must be amended. The study also cast some light on how the symmetry mismatch between

Table I. Adenovirus Structural Proteins[a]

Polypeptide	Location	Number of residues	Molecular mass (Da)	Number of copies	
				Experimental	Model
II	Hexon	967	109,077[b]	720	720
III	Penton base	571	63,296[b]	56 ± 1	60
IIIa	Vertex region	566	63,287	74 ± 1	
IV	Fiber	582	61,960[b]	35 ± 1	36
V	Core	368	41,631[b]	157 ± 1	
VI	Hexon associated?	217	23,449	342 ± 4	
VII	Core	174	19,412	833 ± 19	
VIII	Hexon associated?	134	14,539	211 ± 2	
IX	Groups-of-nine	139	14,339[b]	247 ± 2	240

[a]For full experimental details see van Oostrum and Burnett (1985).
[b]The molecular mass includes 42 Da for the N-terminal acetyl group.

the pentameric penton base and the threefold fiber is resolved. Two penton base polypeptides were observed, which we presume to arise from two closely spaced initiation codons. We hypothesize that the penton complex consists of a trimeric fiber, bound to three copies of the short base, and completed with two copies of the full-length base. We demonstrated the overall symmetry of the penton complex directly, in related work, by image analysis of electron micrographs showing a novel fragment, the "quarter-capsid," containing the penton surrounded by five complete capsid facets (van Oostrum *et al.*, 1987a,b).

III. Hexon Structure

The crystal structure of the trimeric hexon molecule has been solved at 2.9 Å resolution (Roberts *et al.*, 1986) (Fig. 2). Four rounds of model-building have been accomplished and over 90% of the 967-amino acid polypeptide has been defined. The missing regions include the N-terminus (1–43), and portions of the loops ℓ_1 (192–203 and 269–272) and ℓ_2 (444–453). The hexon subunit is composed of two very similar basal domains and so the pseudohexagonal symmetry revealed at low resolution (Burnett *et al.*, 1985) extends to the tertiary structure of the molecule. Moreover, the secondary structure of these basal domains, P1 and P2, is related to those of the RNA plant viruses (Olson *et al.*, 1983; Rossmann *et al.*, 1983) and the human RNA viruses, rhinovirus (Rossmann *et al.*, 1985) and poliovirus

(Hogle *et al.,* 1985). In adenovirus, the β-barrel axes are perpendicular rather than parallel to the virion surface since the β-barrels form the walls of the tubular hexon base. The β-barrels so far seem to be universal elements in the outer shells of spherical viruses, which presumably are descendants of an early primitive precursor. However, β-barrels have evolved to form shells of protein in different ways (Roberts and Burnett, 1987). The strength of a shell of β-barrel elements, as in the RNA viruses, possibly is not sufficient for a virion the size of adenovirus. A larger building block, the size of hexon, is the answer to the increased structural demands.

Hexon is an extremely stable molecule and is also unusual in requiring the aid of another protein, adenovirus 100K, to fold into its tertiary structure (Cepko and Sharp, 1982). We are seeking the basis for the stability by analyzing the intersubunit interactions in the trimer (Korn *et al.,* 1989). The topology of the molecule is clearly an unusual and striking contributor as the three subunits are "entangled" (Color Plate 1). The contribution of chemical factors also is being assessed. We first isolated the external surface of the subunit using Connolly's solvent accessibility algorithm (Connolly, 1983). Next, we searched for differences between the parts of that external surface that contact neighbors and those that remain contact-free. The surfaces then were classified in terms of hydrophobicity, charge concentration, and density of hydrogen bonds.

The importance of hydrophobicity in hexon intersubunit binding is illustrated in Color Plate 2. This is a cross section of the whole trimer, showing the hydrophobicity distribution of the interior as well as the interface regions. Hydrophobic amino acids are sequestered to the interior of the lower P1 and P2 domains. Hydrophilic amino acids predominate at the inside surface of the central cavity, the outer surface of the trimer, and some parts of the interior of the towers. A remarkable aspect of the hydrophobicity distribution is the *hydrophobic plug* situated between and below the towers. It is an interface between all three subunits and seals the central cavity in the hexon, and thus the virion interior, from the outside. It is probably the main source of hydrophobicity-driven subunit–subunit binding stability. We find also that the contact interfaces are bound to each other by matching hydrophobicities; that is the scattered hydrophobic and hydrophilic centers on one surface match up to the scattered but complementarily distributed hydrophobic and hydrophilic centers on the other surface. We have quantitated this aspect of protein–protein interfaces by a function we have called the *hydrophobicity complementarity function* (Korn and Burnett, 1989). The overall description of hexon can be rationalized in terms of the subunits' requirements for partial solubility, optimum binding, and accurate alignment. The hexon positions, determined from

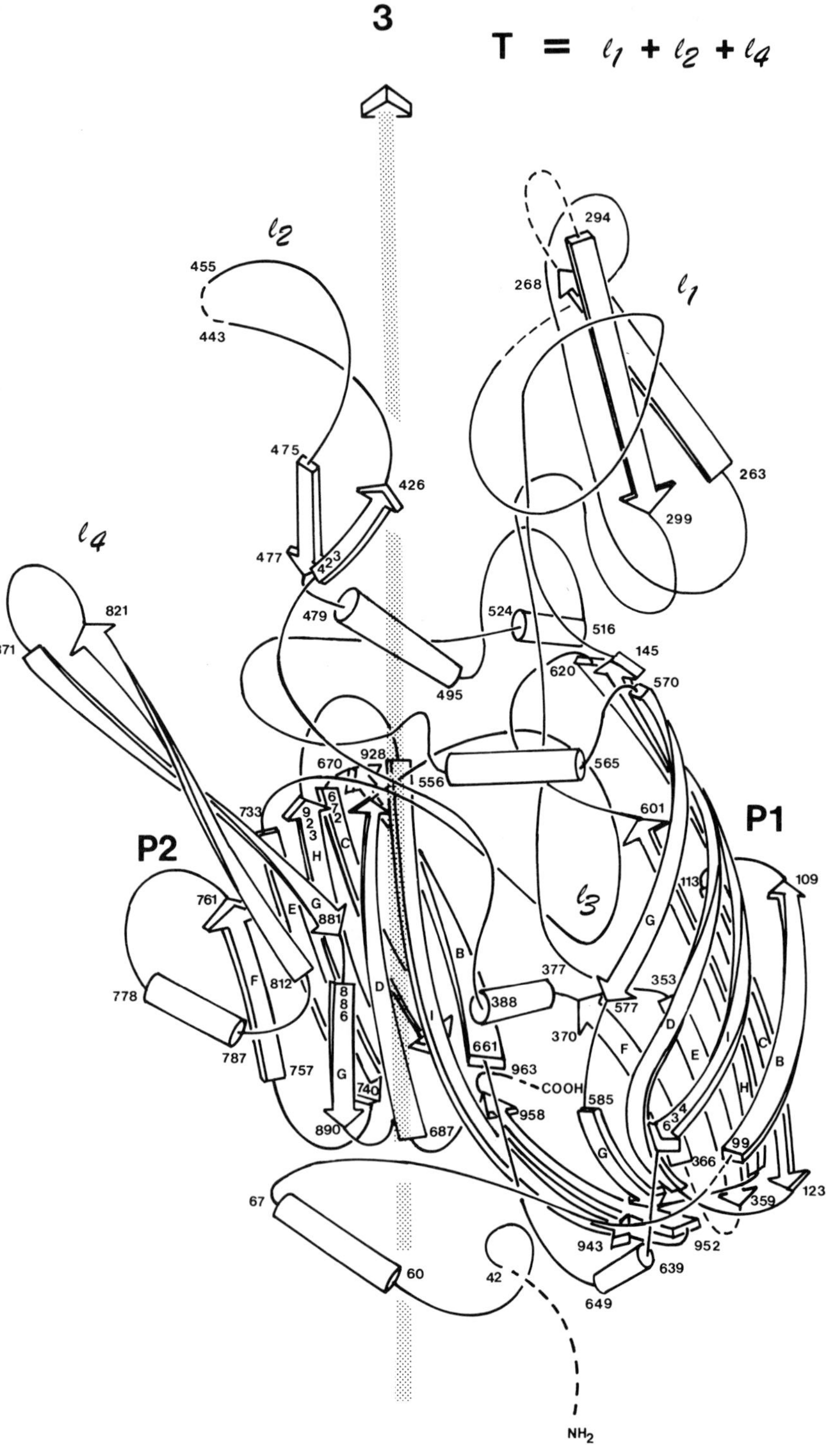
3
T = l1 + l2 + l4
l1
l2
l3
l4
P1
P2
294
268
263
299
455
443
475
426
477
423
821
871
479
524
516
145
495
620
570
928
670
556
565
601
733
923
672
H
C
113
109
761
E
G
881
B
G
377
353
778
F
812
388
577
886
370
D
F
E
I
C
B
H
787
757
661
963
COOH
585
G
740
958
890
687
634
99
366
123
G
359
67
943
952
639
60
42
649
NH2

electron microscopy, will be used to juxtapose molecular models of hexon so that these studies can be extended to interhexon interactions in the virion.

We have commenced work on hexons from other adenovirus species and are defining the regions of structural change in the molecule as new sequences become available. We have grown ad5 hexon crystals and shown them to be isomorphous to those from ad2 hexon with a diffraction pattern extending to 1.9 Å resolution (Roberts *et al.*, 1987). We now are pursuing this structure determination, in which we can use the ad2 phases directly. Ronald T. Hay's group (St. Andrews, Scotland) has been sequencing the hexon genes of ad40 (Toogood and Hay, 1988) and ad41 (Toogood *et al.*, 1989), which are associated with the enteric dysentery that causes high mortality in Third World infants. We have mapped the ad40 and ad41 hexon sequences onto the ad2 hexon structure and then have compared the various alterations with those altered in ad5 hexon (Roberts *et al.*, 1986; Toogood *et al.*, 1989). We plan crystallographic studies on all these species so that we can develop an understanding of the structural basis for their immunological differences.

IV. Arrangement of Hexons in the Capsid

Although the structure of hexon is known, the relative orientations of hexons in crystals are quite different from those in the adenovirus capsid. The hexon structure alone could not reveal how the capsid was constructed but could provide the key to understanding adenovirus architecture when combined with information from electron microscopy. The 6 Å resolution structure of hexon was used to interpret electron micrographs of adenovirus fragments and to develop a model for the capsid (Burnett, 1985). We then obtained direct confirmation for the capsid model by forming hexon arrays (van Oostrum *et al.*, 1986) and capsid fragments (van Oostrum *et al.*, 1987a,b) in the electron microscope and subjecting the resultant electron micrographs to image analysis. This work defined the positions of hexons within a capsid facet to approximately 5 Å (Fig. 3).

Figure 2. A sketch of the hexon subunit viewed from within the central cavity of the trimer. Most of the chain lies behind the vertical threefold molecular symmetry axis (marked 3). The secondary structure has been emphasized for clarity by drastically shortening the connecting loops. The N-terminus is partially disordered and lies underneath the molecule. Two domains, P1 and P2, form the base of the subunit. The three loops (ℓ_1, ℓ_2, and ℓ_4) from one subunit each contribute to one of the three towers. Thus the tower domain, seen only in the complete trimer, is formed from three loops (ℓ_1, ℓ_2', and ℓ_4''), each from a different subunit. (From Roberts *et al.*, 1986. Copyright American Association of Science.)

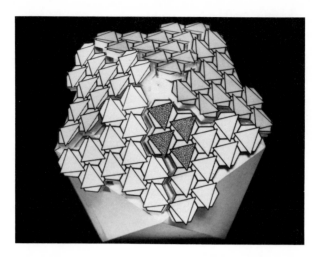

Figure 3. A model of the adenovirus capsid, showing the arrangement of hexons. Exact icosahedral symmetry relates the 60 asymmetric units of the icosahedral hexon shell, one of which is shaded. The four hexons in each asymmetric unit lie in locations with similar chemical characteristics as they are related by translational symmetry. (Reproduced from the *Biophysical Journal*, 1986, Vol. 49, pp. 22–24 by copyright permission of the Biophysical Society.)

Hexons have a pseudohexagonal base that allows the formation of a close-packed *p*3 net on each capsid facet with their threefold symmetry axes normal to the facet (Burnett, 1985; van Oostrum *et al.*, 1987b). The vertices of the triangular top of hexon are rotated by approximately 10° with respect to the pseudohexagonal base. When the hexons are closely packed to form the capsid facet, the outer surface thus contains four large and six small cavities at the local threefold axes. The asymmetric unit consists of four hexons, each of whose topography is different. The chemical interactions of each hexon thus are formally distinct. However, the translational symmetry, combined with the icosahedral point group symmetry, provides a very similar chemical environment for each hexon (Burnett, 1985). The bonding is identical for at least four of the six inter-hexon contact faces, but differs at the other faces, leading to overall conservation. The crystallike construction explains the flat facets and sharp edges that characterize adenovirus in electron micrographs of negatively stained specimens. The construction of adenovirus illustrates how a faceted impenetrable protein shell can be formed while highly conserved intermolecular bonding is maintained. This contrasts with the continuously curved shell that would result from a capsid constructed using quasi-equivalence (Caspar and Klug, 1962).

V. Position of Polypeptide IX

Our investigation of the virion has been advanced by using the STEM to determine the location of polypeptide IX (14,339 Da) (Aleström *et al.*, 1980), which is a minor component (6%) of the GONs. These fragments are released on dissociation of the virion under mild conditions (Smith *et al.*, 1965; Laver *et al.*, 1969; Prage *et al.*, 1970) (see Fig. 1). The GON is propeller-shaped and is left-handed when viewed from outside the virion (Pereira and Wrigley, 1974). Maizel *et al.* (1968) originally suggested that polypeptide IX stabilizes the capsid. This idea was confirmed by Colby and Shenk (1981), who showed that an ad5 deletion mutant in the gene for polypeptide IX produced virions that are less stable than the wild type and do not form GONs when dissociated. As polypeptide IX is not required for viral assembly, it is a capsid "cement."

The capsid model shows that there is the same number of large cavities in the facet, as in the GON. Burnett (1984) suggested that trimers of polypeptide IX in these positions would cement hexons within the GON but would not attach the peripentonal hexons to the GON (Fig. 4). The idea was supported by stoichiometric measurements showing 12 copies of polypeptide IX per GON and 240 per virion (van Oostrum and Burnett,

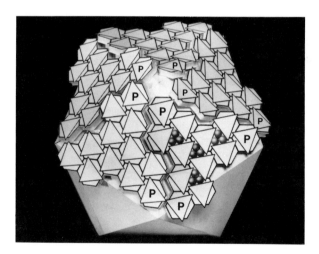

Figure 4. A model of the adenovirus capsid showing the binding sites of the capsid cement. Polypeptide IX binds in the central cavity, between the hexon towers, and in the three symmetry-related locations elsewhere in the facet. Polypeptide IX stabilizes the central nine hexons so that they dissociate as an independent entity, the GON, which then has its own subsequent dissociation pattern. The peripentonal hexons (P) are not cemented into the facet and are free to dissociate independently from the capsid.

1985) and explains the anomalous release of peripentonal hexons separate from the GONs (Fig. 1).

The STEM (Wall and Hainfeld, 1986) was used to image polypeptide IX in GONs completely embedded in negative stain (Furcinitti *et al.*, 1986, 1987, 1989). Image analysis was used to align the GON images and to identify well-preserved specimens with threefold power greater than 80%. Correspondence analysis (Frank and van Heel, 1982; Frank, 1984) then was used to identify similar images. The rotationally filtered average of 57 images, which has a resolution of 15 Å, is shown in Color Plate 3a, in which the contours depict the protein distribution throughout the GON. The excellent agreement between the three independent views of hexon, and the clear threefold molecular symmetry, indicate the excellence of the imaging. The density at the two different types of local threefold axes in the GON should be approximately identical as the density at each arises from three β-barrels of approximately equal height. The additional contour levels in the four large cavities clearly reveal the postulated binding sites of polypeptide IX.

The X-ray model of the hexon then was used to construct a two-dimensional image of its protein distribution (Color Plate 3b). This was used as a probe within the averaged STEM image to determine the precise relationship of the GON hexons. Rotational and translational alignment with the STEM GON image gave nine peaks with an average hexon separation of 90.7 $\pm$ 1.2 Å, in good agreement with the value of 89 $\pm$ 5 Å reported by van Oostrum *et al.* (1987b). Artificial GONs then were constructed by placing the X-ray hexon model at these positions (Color Plate 3c). The resultant image shows that the density at the two different local threefold axes is identical in the absence of polypeptide IX (cf. Color Plate 3a). Difference images then were calculated by subtracting the scaled X-ray GON (Color Plate 3c) from the STEM GON (Color Plate 3a). The difference image (Color Plate 3d) reveals additional density at the positions of polypeptide IX proposed earlier (Burnett, 1984). The internal consistency of the result is indicated by the excellent preservation of symmetry about each local threefold axis and by the absence of density in the unoccupied partial large cavity at the periphery of the GON (upper right in Color Plate 3a,d).

The difference image reveals that individual monomers of polypeptide IX extend along the hexon–hexon interfaces. The results are consistent with earlier observations on polypeptide IX and GONs. The difference image shows that polypeptide IX is an elongated monomer with an axial ratio of 2.4, in approximate agreement with the value of 3.5 found by Lemay and Boulanger (1980) and consistent with its role as a capsid cement. The monomers bond pairs of hexons and cluster as four trimers of polypeptide IX embedded in the large cavities in the upper surface of the GON to cement hexons into a highly stable assembly. The distribution directly confirms the arrangement suggested by Burnett (1984). The arrangement of polypeptide IX agrees with the stoichiometric determination (van Oostrum and Burnett, 1985), which found 12 copies within the GON and none elsewhere in the virion. The distribution of polypeptide IX also explains the nonrandom dissociation pattern of GONs (Pereira and Wrigley, 1974), the lack of reactivity of anti-IX antibodies to intact virions, and the low antibody response to GONs (Cepko *et al.*, 1981).

VI. Conclusions

We hope that our approach to understanding the architecture of adenovirus, which is approximately 20-fold larger than the RNA viruses whose detailed crystallographic structures are currently emerging, will be a model for other investigations of large assemblies. In particular, the utility of combining X-ray crystallography and electron microscopy to span the enormous scale from atomic resolution to the dimensions of small organelles has been demonstrated.

The difference imaging technique succeeded in locating polypeptide IX (14,339 Da) between hexon pairs (654,462 Da), a mass difference of only 2.2%. Using correspondence analysis to average the STEM images gave a resolution of 15 Å and allowed us to determine hexon positions in the adenovirus capsid to within 1 Å. With the coordinates from the refined X-ray structure, this study delineates the atomic positions of 62% of the protein in the ad2 virion (78.5 MDa). We propose to extend the investigation to locate polypeptide VI (23,449 Da), which remains attached to GONs under different experimental conditions (Everitt *et al.*, 1973).

A full understanding of complex macromolecular assemblies is required for the rational design of therapeutic agents. The architectural principles that have emerged from the adenovirus investigation (Burnett, 1985) explain the construction of viruses whose organization is too complex to be explained by earlier ideas (Caspar and Klug, 1962). Polypeptide IX is an example of a class of proteins that has received little attention but is important, as it resolves the conflicting demands for weak interactions during accurate assembly of multicomponent systems and the strength required for their ultimate stability. We believe that this class of proteins will be of increasing importance as targets for drugs that act by modifying their binding, or even replacing them, to enhance or diminish the overall stability of the macromolecular assembly.

Acknowledgments

Most of this work was performed while R. M. B., Z. C., A. P. K., and R. M. were at Columbia University. We thank our collaborators Manijeh Mohraz and P. Ross Smith for their aid with conventional electron microscopy, and Joseph S. Wall for his support and advice while P.S.F. was a member of the STEM team at Brookhaven National Laboratory. Timothy S. Baker kindly helped in producing Color Plate 3. We are grateful to our colleagues Donatella Pascolini and Phoebe L. Stewart for their critical reading of the manuscript. The Brookhaven STEM Biotechnology Resource is supported by National Institutes of Health Grant RR 0177. Research support was also provided at the Brookhaven National Laboratory by the Office of Health and Environmental Research of the U.S. Department of Energy. The work at Columbia University and The Wistar Institute is supported by grants to R.M.B. from the National Science Foundation (DMB 84-18111 and DMB 87-21431) and the National Institute of Allergy and Infectious Diseases (AI 17270).

References

Aleström, P., Akusjärvi, G., Perricaudet, M., Mathews, M. B., Klessig, D. F., and Pettersson, U. (1980). The gene for polypeptide IX of adenovirus type 2 and its unspliced messenger RNA. *Cell* **19,** 671–681.

Aleström, P., Akusjärvi, G., Lager, M., Yeh-kai, L., and Pettersson, U. (1984). Genes encoding the core proteins of adenovirus type 2. *J. Biol. Chem.* **259,** 13980–13985.

Burnett, R. M. (1984). Structural investigations on hexon, the major coat protein of adenovirus. *In* "Biological Macromolecules and Assemblies. Vol. 1: Virus Structures" (F. A. Jurnak and A. McPherson, eds.), pp. 337–385. Wiley, New York.

Burnett, R. M. (1985). The structure of the adenovirus capsid. II. The packing symmetry of hexon and its implications for viral architecture. *J. Mol. Biol.* **185,** 125–143.

Burnett, R. M., Grütter, M. G., and White, J. L. (1985). The structure of the adenovirus capsid. I. An envelope model of hexon at 6 Å resolution. *J. Mol. Biol.* **185,** 105–123.

Burnett, R. M., van Oostrum, J., and Roberts, M. M. (1986). Progress in understanding adenovirus architecture. *Biophys. J.* **49,** 22–24.

Caspar, D. L. D., and Klug, A. (1962). Physical principles in the construction of regular viruses. *Cold Spring Harbor Symp. Quant. Biol.* **27,** 1–24.

Cepko, C. L., and Sharp, P. A. (1982). Assembly of adenovirus major capsid protein is mediated by a nonvirion protein. *Cell* **1,** 407–415.

Cepko, C. L., Changelian, P. S., and Sharp, P. A. (1981). Immunoprecipitation with two-dimensional pools as a hybridoma screening technique: Production and characterization of monoclonal antibodies against adenovirus 2 proteins. *Virology* **110,** 385–401.

Colby, W. W., and Shenk, T. (1981). Adenovirus type 5 virions can be assembled *in vivo* in the absence of detectable polypeptide IX. *J. Virol.* **39,** 977–980.

Connolly, M. L. (1983). Solvent-accessible surfaces of proteins and nucleic acids. *Science* **221,** 709–713.

Eisenberg, D. (1984). Three-dimensional structure of membrane and surface proteins. *Annu. Rev. Biochem.* **53,** 595–623.

Everitt, E., Sundquist, B., Pettersson, U., and Philipson, L. (1973). Structural proteins of adenoviruses. X. Isolation and topography of low molecular weight antigens from the virion of adenovirus type 2. *Virology* **52,** 130–147.

Frank, J. (1984). The role of multivariate image analysis in solving the architecture of the *Limulus polyphemus* hemocyanin molecule. *Ultramicroscopy* **13,** 153–164.

Frank, J., and van Heel, M. (1982). Correspondence analysis of aligned images of biological particles. *J. Mol. Biol.* **161**, 134–137.

Furcinitti, P. S., van Oostrum, J., Wall, J. S., and Burnett, R. M. (1986). Correlation averaging of STEM images of negatively stained adenovirus groups-of-nine hexons. *Proc. Annu. Meet. Electron Microsc. Soc. Am., 44th* pp. 152–153.

Furcinitti, P. S., van Oostrum, J., Wall, J. S., and Burnett, R. M. (1987). Visualization of a small polypeptide within an adenovirus capsid fragment by subtracting the hexon X-ray structure from STEM images. *Proc. Annu. Meet. Electron Microsc. Soc. Am., 45th* pp. 738–739.

Furcinitti, P. S., van Oostrum, J., and Burnett, R. M. (1989). Adenovirus polypeptide IX revealed as capsid cement by difference images from electron microscopy and crystallography. *EMBO J.* **12**, in press.

Ginsberg, H. S., ed. (1984). "The Adenoviruses." Plenum, New York.

Green, N. M., Wrigley, N. G., Russell, W. C., Martin, S. R., and McLachlan, A. D. (1983). Evidence for a repeating cross-β sheet structure in the adenovirus fibre. *EMBO J.* **2**, 1357–1365.

Hogle, J. M., Chow, M., and Filman, D. J. (1985). Three-dimensional structure of poliovirus at 2.9 Å resolution. *Science* **229**, 1358–1365.

Korn, A. P., and Burnett, R. M. (1989). The sequestering of hydropathy in multi-subunit proteins and their interfaces. Submitted for publication.

Korn, A. P., Murali, R., Cai, Z., and Burnett, R. M. (1989). The molecular basis for stability in the adenovirus hexon trimer. Submitted for publication.

Laver, W. G., Wrigley, N. G., and Pereira, H. G. (1969). Removal of pentons from particles of adenovirus type 2. *Virology* **39**, 599–605.

Lemay, P., and Boulanger, P. (1980). Physicochemical characteristics of structural and nonstructural proteins of human adenovirus-2. *Ann. Virol. Inst. Pasteur* **131**, 259–275.

Maizel, J. V., Jr., White, D. O., and Scharff, M. D. (1968). The polypeptides of adenovirus. I. Evidence for multiple protein components in the virion and a comparison of types 2, 7A, and 12. *Virology* **36**, 115–125.

Olson, A. J., Bricogne, G., and Harrison, S. C. (1983). Structure of tomato bushy stunt virus. IV. The virus particle at 2.9 Å resolution. *J. Mol. Biol.* **171**, 61–93.

Pereira, H. G., and Wrigley, N. G. (1974). *In vitro* reconstruction, hexon bonding and handedness of incomplete adenovirus capsid. *J. Mol. Biol.* **85**, 617–631.

Philipson, L. (1983). Structure and assembly of adenoviruses. *In* "The Molecular Biology of Adenoviruses" (W. Doerfler, ed.), Vol. 1, pp. 1–52 (*Curr. Top. Microbiol. Immunol.* **109**). Springer-Verlag, Berlin.

Prage, L., Pettersson, U., Höglund, S., Lonberg-Holm, K., and Philipson, L. (1970). Structural proteins of adenoviruses. IV. Sequential degradation of the adenovirus type 2 virion. *Virology* **42**, 341–358.

Roberts, M. M., and Burnett, R. M. (1987). Adenovirus hexon: A novel use of the viral beta-barrel. *In* "Biological Organization: Macromolecular Interactions at High Resolution" (R. M. Burnett and H. J. Vogel, eds.), pp. 113–124. Academic Press, Orlando, Florida.

Roberts, M. M., White, J. L., Grütter, M. G., and Burnett, R. M. (1986). Three-dimensional structure of the adenovirus major coat protein hexon. *Science* **232**, 1148–1151.

Roberts, M. M., van Oostrum, J., and Burnett, R. M. (1987). X-ray diffraction studies on adenovirus type 5 hexon. *Annal. N.Y. Acad. Sci.* **494**, 416–418.

Roberts, R. J., O'Neill, K. E., and Yen, C. T. (1984). DNA sequences from the adenovirus 2 genome. *J. Biol. Chem.* **259**, 13968–13975.

Rossmann, M. G., Abad-Zapatero, C., Hermodson, M. A., and Erickson, J. W. (1983). Subunit interactions in southern bean mosaic virus. *J. Mol. Biol.* **166**, 37–83.

Rossmann, M. G., Arnold, E., Erickson, J. W., Frankenberger, E. A., Griffith, J. P., Hecht, H.-J., Johnson, J. E., Kamer, G., Luo, M., Mosser, A. G., Rueckert, R. R., Sherry, B., and Vriend, G. (1985). Structure of a human common cold virus and functional relationship to other picornaviruses. *Nature* (*London*) **317**, 145–153.

Rowe, W. P., Huebner, R. J., Gillmore, L. K., Parrott, R. H., and Ward, T. G. (1953). Isolation of a cytopathogenic agent from human adenoids undergoing spontaneous degeneration in tissue culture. *Proc. Soc. Exp. Biol. Med.* **84**, 570–573.

Smith, K. O., Gehle, W. D., and Trousdale, M. D. (1965). Architecture of the adenovirus capsid. *J. Bacteriol.* **90**, 254–261.

Toogood, C. I. A., and Hay, R. T. (1988). DNA sequence of the adenovirus type 41 hexon gene and predicted structure of the protein. *J. Gen Virol.* **69**, 2291–2301.

Toogood, C. I. A., Murali, R., Burnett, R. M., and Hay, R. T. (1989). The adenovirus type 40 hexon: Sequence, predicted structure and relationship to other adenovirus hexons. *J. Gen. Virol.,* in press.

van Oostrum, J., and Burnett, R. M. (1985). Molecular composition of the adenovirus type 2 virion. *J. Virol.* **56**, 439–448.

van Oostrum, J., Smith, P. R., Mohraz, M., and Burnett, R. M. (1986). Interpretation of electron micrographs of adenovirus hexon arrays using a crystallographic molecular model. *J. Ultrastruct. Mol. Struct. Res.* **96**, 77–90.

van Oostrum, J., Smith, P. R., Mohraz, M., and Burnett, R. M. (1987a). Morphology of the vertex region of adenovirus. *Ann. N.Y. Acad. Sci.* **494**, 423–426.

van Oostrum, J., Smith, P. R., Mohraz, M., and Burnett, R. M. (1987b). The structure of the adenovirus capsid. III. Hexon packing determined from electron micrographs of capsid fragments. *J. Mol. Biol.* **198**, 73–89.

Wall, J. S., and Hainfield, J. F. (1986). Mass mapping with the scanning transmission electron microscope. *Annu. Rev. Biophys. Biophys. Chem.* **15**, 355–376.

White, D. O., Scharff, M. D., and Maizel, J. V., Jr. (1969). The polypeptides of adenovirus. III. Synthesis in infected cells. *Virology* **38**, 395–406.

5 Crystal Structures of Influenza Virus Neuraminidase Complexed with Monoclonal Antibody Fab Fragments

W. Graeme Laver
Influenza Research Unit, John Curtin School of Medical Research, The
Australian National University, Canberra ACT 2601, Australia

Gillian M. Air and Ming Luo
Department of Microbiology, University of Alabama at Birmingham,
Birmingham, Alabama 35294

A. Portner, S. D. Thompson, and Robert G. Webster
Department of Virology and Molecular Biology, St. Jude Children's Research
Hospital, Memphis, Tennessee 38101

I. Introduction

Antibodies are the first line of defense in animals against invading viruses but, until recently, very little was known about how antibodies recognize and bind to virus particles or how this binding renders the virus particle noninfectious.

A great deal of information has been gathered concerning the structure of combining sites (complementarity determining regions, or CDRs) on antibodies and the interactions involved in the binding of small molecular weight haptens, but there is a considerable debate about the way in which antibodies recognize and bind to the surfaces of foreign proteins. In the past, many methods have been used in attempts to define antigenic determinants (epitopes) on protein molecules (reviewed in Benjamin *et al.*, 1984). The methods have included the use of protein fragments to absorb antisera, the production of anti-peptide antisera and their reactions with intact proteins, proteolytic digestion of antigen in complexes with antibodies as a probe of protected peptide bonds within the epitope, and the protection of amino acids in the epitope from specific chemical derivitization. Other methods include the characterization of "escape mutants" (variants which do not bind neutralizing monoclonal antibodies) and competitive binding studies of monoclonal antibodies for an antigen. Although data from some experiments have implicated particular amino acid residues as participants in particular epitopes, none has allowed a complete structural description of an epitope.

The first description of the structure of an epitope on a protein molecule

Use of X-Ray Crystallography in the Design of Antiviral Agents
49

was obtained by determining the three-dimensional crystal structure of a complex between lysozyme and the Fab fragment of a monoclonal antibody (Amit *et al.*, 1986). In this complex, 16 amino acids on the lysozyme and 17 on the antibody form a tightly packed interface from which water molecules are excluded. The interaction has been described as conforming to a "lock and key" picture of antibody–antigen interaction, in which, apart from some amino acid side chain movements, no structural changes occur in either the antibody or the antigen. Two more lysozyme–Fab complex structures have since been determined at high resolution (Sheriff *et al.*, 1987; Davies *et al.*, 1988; see also this volume).

We have grown crystals of monoclonal antibody Fab fragments complexed with influenza virus neuraminidase (NA) of the N9 subtype. A number of crystalline complexes, two of which diffract X-rays to beyond 3 Å, have been obtained and a low-resolution structure of one of these complexes, (N9 NA–NC41 Fab) has been reported (Colman *et al.*, 1987).

II. The Influenza Virus Neuraminidase

The NA accounts for about 5–10% of the virus protein and exists as a mushroom-shaped spike on the surface of the virion. It is a tetramer with a box-shaped head, $100 \times 100 \times 60$ Å, made out of four coplanar and roughly spherical subunits, and a centrally attached stalk containing a hydrophobic region by which it is embedded in the viral membrane (Fig. 1). Several roles have been suggested for the neuraminidase. The enzyme catalyzes cleavage of the α-ketosidic linkage between terminal sialic acid and an adjacent sugar residue. This destroys the hemagglutinin receptor on the

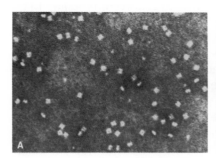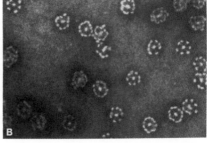

Figure 1. Electron micrographs of (A) tetrameric neuraminidase heads released by pronase and (B) neuraminidase molecules released with detergent, then the detergent removed; in (B) the molecules are associated by the hydrophobic N-terminal sequence as diagrammed in Fig. 2. The protease-released NA heads retain all the enzymatic and antigenic activity. Electron micrographs were kindly provided by Nick Wrigley.

host cell, thus allowing elution of progeny virus particles from infected cells. The removal of sialic acid from the carbohydrate moiety of newly synthesized hemagglutinin and neuraminidase is also necessary to prevent self-aggregation of the virus. Neuraminidase may also facilitate mobility of the virus both to and from the site of infection. (For a review see Colman and Ward, 1985.)

The neuraminidase molecule is composed of a single polypeptide chain, coded by RNA segment 6, oriented in the virus membrane by its N-terminus, which is the opposite orientation to hemagglutinin anchoring (Fig. 2). No posttranslational cleavage of the NA polypeptide occurs, no signal peptide is split off, and even the initiating methionine is retained. Nor is there processing at the C-terminus; the C-terminal sequence -Met-Pro-Ile predicted from the gene sequence for N2 NA is found in intact NA molecules isolated from the virus and in the pronase-released NA heads. A sequence of six polar amino acids at the N-terminus of the NA polypeptide, which are totally conserved in each of the nine different influenza A NA subtypes but not in influenza B, is followed by a sequence of hydrophobic amino acids which must represent the transmembrane regions of the NA. This sequence is not conserved at all between subtypes (apart from conservation of hydrophobicity). Proteases cleave the stalk releasing

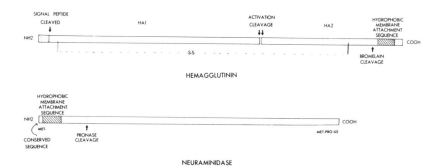

Figure 2. Diagrammatic representation of some features of the HA and NA polypeptides on influenza virus. Following synthesis of the HA precursor, the signal peptide is cleaved and an additional cleavage generates HA_1 and HA_2. This cleavage is essential for fusion activity. The HA is anchored in the viral membrane by a hydrophobic sequence near the C-terminus. The NA is oriented in the opposite way to the HA, by a hydrophobic sequence near the N-terminus. No posttranslational cleavage of the polypeptide chain occurs; even the initiating methionine is retained in the mature protein. The hydrophobic sequence is preceded by six polar amino acids (which are conserved in all nine subtypes of influenza A), and followed by the amino acids of the NA stalk, which are extremely variable in sequence. Pronase cleaves the NA near the top of the stalk, releasing the enzymatically and antigenically active heads, which can be crystallized.

the enzymatically and antigenically active head of the NA, which in some cases can be crystallized.

A. Structure of N2 Neuraminidase

The three-dimensional structure of N2 neuraminidase heads, determined from an electron density map at 2.9 Å resolution, shows that each monomer is composed of six topologically identical β-sheets arranged in a propeller formation. The tetrameric enzyme has circular fourfold symmetry stabilized in part by metal ions bound on the symmetry axis. The catalytic sites are located in deep pockets which occur on the upper corners of the box-shaped tetramer (Varghese *et al.*, 1983; Colman *et al.*, 1983).

B. Structure of N9 Neuraminidase

The three-dimensional structure of the neuraminidase antigen of subtype N9 from an avian influenza virus (A/tern/Australia/G70c/75) which shares 50% sequence identity but no antigenic cross-reactivity with the human N2 influenza virus neuraminidase (Laver *et al.*, 1984) has also been determined by X-ray crystallography and shown to be folded similarly to neuraminidase of subtype N2 isolated from a human influenza virus (Baker *et al.*, 1987). Small differences in the way in which the subunits are organized around the molecular fourfold axis are observed. Insertions and deletions with respect to subtype N2 neuraminidase occur in four regions, only one of which is located within the major antigenic determinants around the enzyme active site.

C. N9 Neuraminidase–NC41 Fab Structure

The crystal structure of the N9–NC41 complex is reported to 2.9 Å resolution (Colman *et al.*, 1987).

The areas of the NA which are in contact with antibody NC41 are shown in Fig. 3 and indicate that the epitope is discontinuous, comprising five separate peptide segments on four loops including about 17 amino acid residues. In addition it appears that complementarity determining region (CDR) H1 of the NC41 Fab may interact with carbohydrate covalently attached to Asn 200 of a neighboring subunit of the tetrameric antigen.

III. Escape Mutants

A key factor in the continued epidemiology of influenza in the human population is antigenic variation, caused by mutations in the genes coding

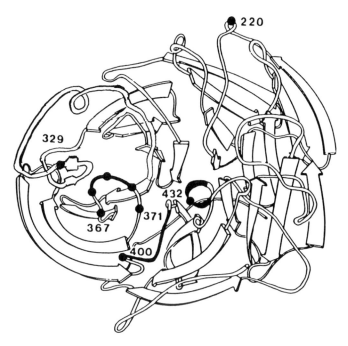

Figure 3. Schematic diagram showing the chain fold in the neuraminidase monomer viewed down the fourfold axis. This diagram has been revised from that previously published (Varghese *et al.*, 1983) in the region involving amino acid 329. The revised 329 loop was redrawn by Ming Luo using a photo of the α-carbon chain trace from Bullough (1988). The epitope recognized by NC41 antibody involves the three loops marked in black. It must also involve part of the 329 loop since a sequence change at this position reduces antibody binding. (Laver *et al.*, 1987.)

N2 numbering is used. The side chains of amino acids 368–370 point towards the viewer, while that of Arg 371 (an active site residue) points away and into the catalytic site located above and to the right of C_α371. Mutations at positions 367, 369, 370, 372, 400, and 432 abolish the binding of NC41 antibody to neuraminidase, whereas mutations at 368 and 329 reduce binding. A mutation at residue 220 (outside the NC41 epitope) has no effect on binding of NC41 to neuraminidase.

for HA and NA which cause amino acid substitutions in the proteins which result in altered antigenicity and escape from circulating antibodies (Webster *et al.*, 1982). Although numerous data are available on the sequence variation in epidemic strains of flu, it is not possible to identify which particular changes in field strains are important in escape from antibodies. Monoclonal antibodies have proved to be powerful tools in the study of epitopes of the NA.

Escape mutants of N2 and N9 were obtained by growing the viruses in the presence of high concentration of monoclonal antibodies to the NA

(Webster *et al.*, 1984; 1987). These antibodies did not directly neutralize the infectivity of the virus, since careful experiments showed that antibodies to the NA did not prevent the virus from infecting cells. However, the presence of antibodies in the fluid surrounding the cells prevented these from yielding virus, and these monoclonal antibodies could be used to select antigenic variants (escape mutants) of the NA. These escape mutants (like those of other virus proteins studied) had single amino acid sequence changes which sufficed to abolish binding of the antibody used for their selection.

In the epitope recognized by NC41 antibody, amino acid sequence changes in escape mutants at positions 367 (Ser to Asn or Gly or Arg), 369 (Ala to Asp), and 370 (Ser to Leu), 372 (Ser to Phe or Tyr), 400 (Asn to Lys), and 432 (Lys to Asn) abolish binding of NC41 to neuraminidase while changes at positions 329 (Asn to Asp) and 368 (Ile to Arg) reduce binding (Webster *et al.*, 1987, and unpublished observations). All of these amino acids contribute to the surface area of the antigen that is buried when NC41 antibody binds (Fig. 3).

IV. Structure of Escape Mutants

How do single amino acid sequence changes in the loops abolish antibody binding? Remember, only 1 out of about 17 contact residues is altered in each mutant. One possible mechanism is that a single sequence change can drastically alter the conformation of the whole epitope, so that it is no longer recognized by antibody. The evidence suggests, however, that this is unlikely.

The structure of an escape mutant of influenza virus N2 NA (A/Tokyo/3/67), which was selected with monoclonal antibody S10/1 (Laver *et al.*, 1982), has been determined. Residue 368, which is lysine in the wild type, changed to glutamic acid in the mutant. This change abolished S10/1 antibody binding.

The difference Fourier map (Varghese *et al.*, 1988) shows difference peaks resulting from a shift of electrons from the lysine side chain to a new position of glutamic acid in the mutant (Fig. 4). Lysine 368 in the wild-type A/Tokyo/3/67 neuraminidase structure is linked by a salt-bridge with aspartic acid 369, and the altered position of glutamic acid 368 in the mutant may result from a charge repulsion by Asp 369. No other differences between the structures are indicated in the difference electron density map.

This experiment shows that the S10/1 epitope does indeed include amino acid residue 368 and that the change in the sidechain has sufficed effectively to abolish binding of the antibody to the epitope. The single

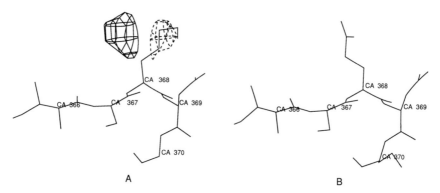

Figure 4. An atomic model of amino acids in the neighborhood of residue 368 in neuraminidase of A/Tokyo/3/67. (A) Wild type NA with Lys at position 368 superimposed on an electron density difference map. (B) The refined atomic model of the same region in the mutant which has Glu at 368. (From Varghese *et al.*, 1988.)

amino acid substitution of lysine to glutamic acid in the mutant introduces two free charged acid groups (Glu 368 and Asp 389) in the interface of the antibody and the antigen. The change in the epitope structure in this case is substantial: two electronic charges plus attending shape changes. In some cases, however, escape mutants that have lost the capacity to bind antibody show more subtle chemical changes that are observed here, such as arginine for lysine (Laver *et al.*, 1982; Air *et al.*, 1985).

A survey of single amino acid changes in escape mutants of influenza antigens (Caton *et al.*, 1982; Wiley *et al.*, 1981; Lentz *et al.*, 1984; Air *et al.*, 1985; Webster *et al.*, 1987) does not show any preference for size or character of the replacing or the replaced residue in the antigen and indicates that even very subtle changes abolish the binding of an antibody for an antigen. The structure has been reported of only one other viral escape mutant, an influenza hemagglutinin (Knossow *et al.*, 1984). In this case, the substitution of an aspartic acid for a glycine residue also caused only local structural alterations on the surface of the molecule.

Although the possibility of some escape mutants influencing antibody binding at a distance cannot be ruled out, the two influenza virus surface antigens studied so far both show that local structural changes suffice to abolish antibody binding to an antigen.

V. Structure of a Second N9 Neuraminidase–Fab Complex (NC10 Fab)

The N9 neuraminidase used here was isolated from an influenza virus found in a whale in Maine in 1984. It differs from the neuraminidase of the virus from terns (isolated in Australia in 1975 and used in the NC41 complex study) by 14 amino acids (Air *et al.*, 1987), mostly on the underside of the globular neuraminidase head (Varghese *et al.*, 1983). The NC10 antibody was raised against the tern virus neuraminidase (Webster *et al.*, 1987) but binds an epitope in which all peptide segments of the tern and whale virus neuraminidases are identical in sequence. Crystals of tern N9–NC10 Fab have also been grown but are disordered and unsuitable for structure analysis.

The structural data reported are not yet of high quality (Colman *et al.*, 1989), but an α-carbon chain trace has been made for the V region of Fab NC10. No electron density is seen for the C-region of the Fab. The general shape of the tetrameric N9–NC10 Fab protomer is similar to that of the NC41 complex (and other complexes examined by electron microscopy) (Tulloch *et al.*, 1986). Packing of the protomers is as described for N9 NA–NC35 Fab crystals (Tulloch *et al.*, 1986), which share the same space group and similar cell dimensions with the N9–NC10 complex crystal. Two neuraminidase heads make close contact via the membrane-proximal surface. Two sets of Fab fragments interdigitate, interacting through the V-modules of the Fab. Crystal growth does not require the participation of the disordered C-module.

The epitope recognized by NC10 antibody is, to a large degree, common to that seen by the NC41 antibody, and escape mutants with sequence changes at positions 329, 369, 370, and 432 do not bind to NC10 antibody. Changes at positions 367, 372, and 400 have little effect on binding of NC10, in contrast to a marked decrease in binding of antibody NC41 (Webster *et al.*, 1987). In the structure (Colman *et al.*, 1989), contacts with surface loops around residues 330, 342, 369, 401, and 432 appear possible.

Although the two antibodies NC10 and NC41 are binding to the same region of the NA surface, the arrangements of the two antibodies on that surface are strikingly different (Colman *et al.*, 1989), each being rotated about 90° from the position of the other (Color Plate 4). Thus, whereas of all the NC41 CDRs, L1 is most remote from the antigen, L2 and H1 of NC10 are most distant from the epitope. Since the antibodies NC10 and NC41 have very different amino acid sequences, belonging to different families of heavy chains and different classes of light chains, this difference in binding configuration of the variable CDRs is not surprising. It will be interesting to compare the interactions of NA with each of the antibodies when high-resolution structural data are available.

VI. Mechanism of Inhibition of Neuraminidase Activity by Antibody

Both NC41 and NC10 antibodies completely inhibit NA activity for the large substrate, fetuin (50,000 Da), presumably by blocking access of the substrate to the active site. When a small substrate, neuraminyl-lactose (NAL) (500 Da), is used, NC41 Fab inhibits neuraminidase activity by about 75%, while NC10 does not inhibit the enzyme at all for this small substrate (Fig. 5).

The crystal structure of the NC41 complex shows that while the Fab does not appear to block access of neuraminyl-lactose to the active site, its binding has altered the conformation of one of the surface loops of the enzyme, distorting the position of an active site arginine, residue 371. This may explain the low neuraminidase activity of this complex, in which there is no steric hindrance to entry of small substrates into the active site pocket. Other explanations which cannot be ruled out include steric hindrance (by the Fab) of the lactose moiety of neuraminyl-lactose and long-range forces originating from the antibody which influence diffusion of the substrate into or out of the active site. Another possibility is that the small shift in the 370 loop of the NA on antibody binding is irrelevant and that antibody instead stabilized the structure so that conformational changes essential to substrate binding and/or catalysis cannot occur.

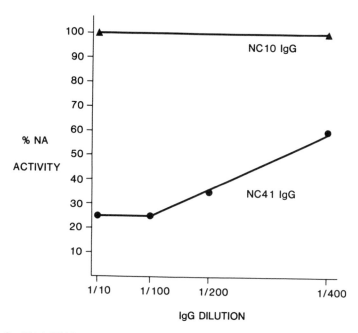

Figure 5. NA inhibition curves for monoclonal antibodies NC41 and NC10 on N9 neuraminidase using *N*-acetyl neuraminyl-lactose (500 Da) as substrate. Both antibodies inhibited N9 NA activity almost totally (95% inhibition) when a large substrate, fetuin (50,000 Da) was used.

VII. Change in Antibody Structure

In the absence of a structure of the uncomplexed NC41 Fab fragment, no statements can be made on this subject.

VIII. Influenza Type B Neuraminidase

Influenza B/Lee/40 neuraminidase has less than 25% sequence identity compared with either N2 or N9 neuraminidase in the head region. Conservation of several of the cysteine residues which form disulfide bonds in N2 and N9 neuraminidases and of several amino acid side chains which line the active site pocket suggest that the polypeptides may be similarly folded, but this can only be confirmed by a full structure determination.

Crystals of neuraminidase heads from two different influenza B virus strains have been grown (Bossart *et al.,* 1988). Neuraminidase crystals of influenza B/Hong Kong/8/73 were grown from solutions of potassium phosphate. The crystals are tetragonal prisms, space group I422; the axes are $a = 123$ Å and $c = 165$ Å. Influenza B/Lee/40 neuraminidase crystals were grown from solutions of polyethylene glycol 4000. The crystals are tetragonal pyramids, space group $P4_12_12$ or its enantiomorph $P4_32_12$; the axes are $a = 125$ Å and $c = 282$ Å.

Fab fragments from four different monoclonal antibodies have also been complexed with influenza B virus neuraminidase (B/Lee/40) and the complexes have been crystallized (Laver *et al.,* 1988). Three of the complex crystals are, so far, not suitable for X-ray diffraction studies, but the fourth (B/Lee/40 NA-B1 Fab) forms large crystals which diffract X-rays to 3.0 Å resolution. The structure of the B complex is being determined.

Interestingly, while the NA crystals from B/Lee and B/HK/8/73 viruses show disorder which makes structural determination difficult, the complex crystals show no such disorder and may provide a route to the three dimensional structure of influenza B NA.

IX. Sendai HN

The hemagglutinin–neuraminidase (HN) protein of Sendai virus has been isolated from virus particles in a biologically active soluble form after removal by proteolytic digestion of the hydrophobic amino-terminal anchor sequence (Thompson *et al.,* 1988). The soluble HN exists as both dimers and tetramers and crystallization trials with each of these forms have so far yielded amorphous material.

Dimers complexed with Fab fragments of a monoclonal antibody formed long needle crystals (Laver *et al.*, 1989). These so far are not suitable for X-ray diffraction analysis but the results suggest that HN molecules from paramyxoviruses, even if not crystallizable, may when complexed with Fab fragments in some cases yield crystals suitable for X-ray diffraction analysis.

Acknowledgments

This work was supported by grants AI 19084, AI 26718, and AI 18203 from NIAID and was greatly helped by international telephone facilities made available by the Australian Overseas Telecommunications Commission. All the X-ray crystallographic analyses were done by Peter Colman, CSIRO Division of Biotechnology. We sincerely regret that Colman and his colleagues did not wish to be co-authors; the results quoted are published.

References

Air, G. M., Ritchie, L. R., Laver, W. G., and Colman, P. M. (1985). Gene and protein sequence of an influenza neuraminidase with hemagglutinin activity. *Virology* **145,** 117–122.

Air, G. M., Webster, R. G., Colman, P. M., and Laver, W. G. (1987). Distribution of sequence differences in influenza N9 neuraminidase of tern and whale viruses and crystallization of the whale neuraminidase complexed with antibodies. *Virology* **160,** 346–354.

Amit, A. G., Mariuzza, R. A., Phillips, S. E. V., and Poljak, R. J. (1986). Three dimensional structure of an antigen–antibody complex at 2.8Å resolution. *Science* **233,** 747–753.

Baker, A. T., Varghese, J. N., Laver, W. G., Air, G. M., and Colman, P. M. (1987). Three-dimensional structure of neuraminidase of subtype N9 from an avian virus. *Proteins* **2,** 111–117.

Benjamin, D. C., Berzofsky, J. A., East, I. J., Gurd, F. R. N., Hannum, C., Leach, S. J., Margoliash, E., Michael, J. G., Miller, A., Prager, E. M., Reichlen, M., Sercarz, E. E., Smith-Gill, S. J., Todd, P. E., and Wilson, A. C. (1984). The antigenic structure of proteins: A reappraisal. *Annu. Rev. Immunol.* **2,** 67–101.

Bossart, P. J., Babu, Y. S., Cook, W. J., Air, G. M., and Laver, W. G. (1988). Crystallization and preliminary X-ray analyses of two neurominidases from influenza B virus strains B/Hong Kong/8/73 and B/Lee/40. *J. Biol. Chem.* **263,** 6421–6423.

Bullough, P. (1988). High resolution imaging of a neuraminidase–antibody fragment complex. Ph.D. Thesis, Univ. of Cambridge.

Caton, A. J., Brownlee, G. G., Yewdell, J. W., and Gerhard, W. (1982). The antigenic structure of the influenza virus A/PR/8/34 hemagglutinin (H1) subtype. *Cell* **31,** 417–427.

Colman, P. M., and Ward, C. W. (1985). Structure and diversity of the influenza virus neuraminidase. *Curr. Top. Microbiol. Immunol.* **114,** 177–255.

Colman, P. M., Varghese, J. N., and Laver, W. G. (1983). Structure of the catalytic and antigenic sites in influenza virus neuraminidase. *Nature (London)* **303,** 41–44.

Colman, P. M., Laver, W. G., Varghese, J. N., Baker, A. T., Tulloch, P. A., Air, G. M., and Webster, R. G. (1987). The three-dimensional structure of a complex of antibody with influenza virus neuraminidase. *Nature (London)* **326,** 358–363.

Colman, P. M., Tulip, W., Varghese, J. N., Tulloch, P. A., Baker, A. T., Laver, W. G., Air, G. M., and Webster, R. G. (1989). Three-dimensional structures of influenza virus neruaminidase–antibody complexes. *Philos. Trans. R. Soc. London, B* **323**, n1217, 511.

Davies, D. R., Sheriff, S., and Padlan, E. A. (1988). Antibody–antigen complexes. *J. Biol. Chem.* **263**, 10541–10544.

Knossow, M., Daniels, R. S., Douglas, A. R., Skehel, J. J., and Wiley, D. C. (1984). Three-dimensional structure of an antigenic mutant of the influenza virus hemagglutinin. *Nature (London)* **311**, 678–680.

Laver, W. G., Air, G. M., Webster, R. G., and Markoff, L. J. (1982). Amino acid sequence changes in antigenic variants of type A influenza virus N2 neuraminidase. *Virology* **122**, 450–460.

Laver, W. G., Colman, P. M., Webster, R. G., Hinshaw, V. S., and Air, G. M., (1984). Influenza virus neuraminidase with hemagglutinin activity. *Virology* **137**, 314–323.

Laver, W. G., Webster, R. G., and Colman, P. M. (1987). Crystals of antibodies complexed with influenza virus neuraminidase show isosteric binding of antibody to wild-type and variant antigens. *Virology* **156**, 181–184.

Laver, W. G., Luo, M., Bossart, P. J., Babu, Y. S., Smith, C., Accavitti, M. A., Tulloch, P. A., and Air, G. M. (1988). Crystallization and preliminary X-ray analysis of type B influenza virus neuraminidase complexed with antibody Fab fragments. *Virology* **167**, 621–624.

Laver, W. G., Thompson, S. D., Murti, K. G., and Portner, A. (1989). Crystallization of Sendai virus HN protein complexed with monoclonal antibody Fab fragments. *Virology* **171**, 291–293.

Lentz, M. R., Air, G. M., Laver, W. G., and Webster, R. G. (1984). Sequence of the neuraminidase gene of influenza virus A/Tokyo/3/67 and previously uncharacterized monoclonal variants. *Virology* **135**, 257–265.

Sheriff, S., Silverton, E. W., Padlan, E. A., Cohen, G. H., Smith-Gill, S. J., Finzel, B. C., and Davies, D. R. (1987). Three-dimensional structure of an antibody–antigen complex. *Proc. Natl. Acad. Sci. USA* **84**, 8075–8079.

Thompson, S. D., Laver, W. G., Murti, K. G., and Portner, A. (1988). Isolation of a biologically active soluble form of the hemagglutinin–neuraminidase protein of Sendai virus. *J. Virol.* **62**, 4653–4660.

Tulloch, P. A., Colman, P. M., Davis, P. C., Laver, W. G., Webster, R. G., and Air, G. M. (1986). Electron and X-ray diffraction studies of influenza neuraminidase complexed with monoclonal antibodies. *J. Mol. Biol.* **190**, 215–225.

Varghese, J. N., Laver, W. G., and Colman, P. M. (1983). Structure of the influenza virus glycoprotein antigen neuraminidase at 2.9 Å resolution. *Nature (London)* **303**, 35–40.

Varghese, J. N., Webster, R. G., Laver, W. G., and Colman, P. M. (1988). Structure of an escape mutant of glycoprotein N2 neuraminidase of influenza virus A/Tokyo/3/67 at 3A. *J. Mol. Biol.* **200**, 201–203.

Webster, R. G., Laver, W. G., Air, G. M., and Schild, G. C. (1982). Molecular mechanisms of variation in influenza viruses. *Nature (London)* **296**, 115–121.

Webster, R. G., Brown, L. E., and Laver, W. G. (1984). Antigenic and biological characterization of influenza virus neuraminidase N2 with monoclonal antibodies. *Virology* **135**, 30–42.

Webster, R. G., Air, G. M., Metzger, D. W., Colman, P. M., Varghese, J. N., Baker, A. T., and Laver, W. G. (1987). Antigenic structure and variation in an influenza virus N9 neuraminidase. *J. Virol.* **61**, 2910–2916.

Wiley, D. C., Wilson, I. A., and Skehel, J. J. (1981). Structural identification of the antibody-binding sites of Hong Kong influenza hemagglutinin and their involvement in antigenic variation. *Nature (London)* **289**, 373–378.

6 An Approach to the Design of Anti-Influenza Agents

Rebecca C. Wade

Department of Chemistry, University of Houston, Houston, Texas 77204-5641

I. Introduction

The influenza hemagglutinin coat glycoprotein is responsible for the attachment of the virus to the host-cell receptors by binding to a terminal sialic acid, and also for mediating a membrane fusion event leading to the release of the viral nucleocapsid into the cytoplasm (Wiley and Skehel, 1987). Hemagglutinin may therefore be a suitable target for the design of anti-influenza agents which would inhibit its action by binding at a functionally important region of the molecule.

The influenza virus proteins undergo a high rate of mutation, enabling the virus to evade the body's immune defense system and giving rise to periodic epidemics and pandemics of the disease. However, it may be possible to design an anti-influenza agent that is effective against a wide range of strains of influenza, regardless of mutation of the virus, by designing it to bind specifically to a highly conserved region of hemagglutinin.

The three-dimensional structure of hemagglutinin has been solved by X-ray crystallography (Wilson et al., 1981) and therefore, the method of receptor fit (Goodford, 1984; Beddell, 1984) may be used in order to design novel anti-influenza agents targeted against this glycoprotein.

There are four conserved regions in hemagglutinin. Of these, the host-cell receptor binding pocket appears to be the most appropriate target binding site for inhibitors designed by the method of receptor fit. Not only is it of known structure, functionally important, and highly conserved, but it also satisfies other criteria necessary for the application of this method (see Section II). In addition, it has recently been proposed as a suitable site for the binding of a drug (Weis et al., 1988).

Therefore, the strategy adopted in this work was to design ligands which would bind tightly to the host-cell receptor binding site, preventing the terminal sialic acid of the cell receptors from binding and so preventing the attachment of influenza viruses to cells and consequent viral infection. In order to design such ligands, the GRID method (Goodford, 1985; Boobbyer et al., 1989) was used to determine the type and position of chemical

Use of X-Ray Crystallography in the Design of Antiviral Agents

61

groups which might bind strongly in the hemagglutinin receptor binding pocket (see Section III). Methods by which these chemical groups could be incorporated into possible anti-influenza agents are discussed in Section IV.

II. Hemagglutinin Host-Cell Receptor Binding Site

There is one host-cell receptor binding site on each monomer of the hemagglutinin trimer. It is situated in the globular head of the external domain of the glycoprotein and it is surrounded by a rim of very variable residues with antigenic character (Weis *et al.*, 1988).

The surface residues of the host-cell receptor binding site in H3N2 A/Hong Kong/1968 hemagglutinin are shown in Fig. 1. The binding site is viewed from on top with the hydroxyl group of Tyr 98 and the aromatic ring of Trp 153 at the base of a depression. Glu 190 and Leu 194 extend from a short α helix which is part of one of the antigenic sites of the protein (Wiley *et al.*, 1981). The shape of the binding pocket may be thought of as a "well" with two "valleys" and a "gully" adjoining it and a slight depression or "platform" to one side as shown in Fig. 1b (Alderson, 1987).

The conservation of the residues of the binding pocket is shown in Fig. 1a. The totally conserved residues participate in intramolecular hydrogen bonds and are involved in making aromatic ring-stacking interactions with other conserved residues behind the surface of the binding site (Weis *et al.*, 1988). The conserved residues may thus serve to stabilize the architecture of the pocket.

There are a number of variable serine residues in the "serine-rich valley" (see Fig. 1). These serines mutate to hydrophobic residues in other strains of influenza A, thus losing their functional hydroxyl groups. Therefore, even though this region may be able to bind strongly to polar or charged ligands, an anti-influenza agent should not be designed to bind to the residues in this region because the extent of binding may be highly susceptible to mutation of the virus.

Residue 226 is also highly variable and the specificity of hemagglutinin for certain sialosides is dependent on the identity of this residue (Rogers *et al.*, 1983).

As well as being highly conserved and functionally important, the hemagglutinin host-cell receptor binding pocket possesses other properties which make it suitable for investigation by the method of receptor fit. It contains a number of conserved polar and charged groups which are capable of binding to a ligand, and this facilitates the design of ligands which will bind specifically to the target site. It also consists of residues

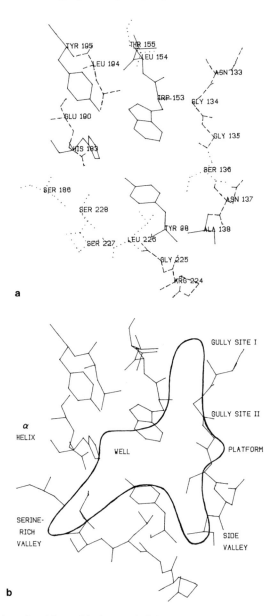

Figure 1. (a) Selected residues of the hemagglutinin host-cell receptor binding site with their conservation in the known sequences of hemagglutinin shown. Totally conserved residues (—); residues undergoing some mutations which should not affect ligand binding (— —); residues undergoing significant mutations in some strains of the virus (····). (b) Schematic diagram showing the structure and "geography" of the host-cell receptor binding site.

that are predominantly of moderate mobility with temperature factors mostly in the range 12–23 Å^2 while the average for the whole structure is 17 Å^2 (Knossow *et al.*, 1986). It is important that the target binding site should not consist of residues of high thermal mobility, because if it did, its crystal structure might not provide sufficient information on the real nature of the binding site for the design of ligands to be possible.

III. Determination of Favorable Ligand Binding Regions in the Hemagglutinin Host-Cell Receptor Binding Site

A. Methods

The GRID method was used to study the hemagglutinin host-cell receptor site. This method is only outlined briefly here as it is described in detail elsewhere (Goodford, 1985; Boobbyer *et al.*, 1989). It is a procedure for determining energetically favorable binding regions on molecules of known structure. It may be applied to macromolecules such as proteins, nucleic acids, and polysaccharides; and to small organic molecules such as drugs. These may all be considered in an aqueous environment.

In this method, a small chemical probe group, for example, a carbonyl oxygen atom or a water molecule, is passed through a regular spatial array around a target molecule, and the interaction energy between the probe and the target is calculated at each point in the array. These energies may then be contoured and displayed using a suitable computer graphics program such as FRODO (Jones, 1985). This permits energy minima to be located and these correspond to favorable binding regions for the probe. For calculations with different probes, different energy minima may be computed, and so regions may be identified where specific binding to the target molecule may occur.

The GRID energy function consists of Lennard–Jones, electrostatic, and hydrogen-bonding terms. It is computed as the sum of the pairwise interactions of the probe group with each atom in the target. The Lennard–Jones term is a 12–6 repulsion–dispersion function. The electrostatic term is given by a Coulombic function with an effective dielectric constant which varies according to the extent to which the interacting atoms are buried in the target molecule (Goodford, 1985). The hydrogen-bonding term is dependent on the length of the hydrogen bond, its orientation at the hydrogen bond donor and acceptor atoms, and the chemical nature of these atoms; it has been formulated so as to reproduce experimental observations of hydrogen bond geometries (Boobbyer *et al.*, 1989). At each point in the array, the probe is oriented so as to make the most favorable interactions with the target molecule.

The interactions of the following probes with the hemagglutinin target were calculated with program GRID on an array around the host-cell receptor binding site: methyl carbon (—CH₃), chlorine (—Cl), fluorine (—F); ammonium (—NH₃⁺), amide (—NH₂ and —NH), and aromatic (>N) nitrogen; carbonyl (=O), carboxyl (—O), ether (>O), and hydroxyl (—OH) oxygen; and water (H₂O).

Regions of the receptor binding site were identified at which a particular probe was found to have a minimum in its interaction energy and was predicted to bind more strongly than any of the other probes. The positions of these energy minima were named GRID ligand points. They indicate the positions at which certain chemical groups might be advantageously incorporated in a ligand designed to bind strongly to the protein.

The GRID energy map for a methyl probe was also used to show the shape of the hemagglutinin binding pocket. By contouring the map at a slightly repulsive energy, a surface similar to a solvent-accessible surface (Lee and Richards, 1971), although calculated with a different probe, can be obtained. This indicates the shape within which a designed ligand should fit on binding at the target site.

B. Results

The positions and binding energies for the GRID ligand points found in different regions of the host-cell receptor binding site are given in Table I and displayed in Fig. 2a. Some of the probes were predicted to interact with hemagglutinin more weakly than other probes at all points in the binding site and hence, no GRID ligand points were identified for these probes. The energy maps for the GRID ligand points are shown in Fig. 2 and are discussed below for each region of the binding pocket (as defined in Fig. 1b).

1. The Well

A binding site for an NH₃⁺ group was predicted at the base of the hemagglutinin binding pocket and is shown by the contours in Fig. 2b. The probe was predicted to make two hydrogen bonds of about 2.8 Å length to the protein at this position. A third hydrogen bond to NE His 183 at a distance of 3.3 Å was not predicted because it would be unfavorably oriented at the probe in relation to the other two predicted hydrogen bonds and because the histidine nitrogen can make an intramolecular hydrogen bond to OH Tyr 98, which is 2.85 Å away. The NH₃⁺ probe was predicted to bind more strongly in the well region than any other probe studied.

Table I. GRID Ligand Points in the Hemagglutinin Host-Cell Receptor Binding Site

Region of binding pocket[a]	Probe	Probe energy (kcal/mol)[b]	Hydrogen bond partners in hemagglutinin	Hydrogen bond energy (kcal/mol)
Well	Nitrogen ($-NH_3^+$)	-11.8	OH Tyr 98	-1.6
			OE1 Glu 190	-1.9
Side valley	Carboxyl oxygen ($-O$)	-11.6	OG Ser 136	-3.4
	Amide nitrogen ($-NH_2$)	-6.7	O Arg 224	-2.7
Platform	Chlorine ($-Cl$)	-6.1		
Gully site II	Hydroxyl oxygen ($-OH$)	-6.8	O Gly 135	-4.0
			N Gly 135	-0.5
Gully site I	Carboxyl oxygen ($-O$)	-11.3	OG1 Thr 155	-3.9
	Carboxyl oxygen ($-O$)	-11.0	N Asn 133	-2.6

[a]The location of these regions in the hemagglutinin binding pocket is shown in Fig. 1b.

[b]This is the minimum energy calculated by program GRID for the probe at the region of the binding pocket considered.

A water molecule has been observed experimentally in this region forming a bridge between Gln 190 and Tyr 98 (Weis *et al.,* 1988). Program GRID predicted a water binding site in the well which corresponded to this observed water position. However, program GRID predicted that an NH_3^+ probe group would displace this water molecule because it was calculated to have a more favorable binding energy of -11.8 kcal/mol than a water probe, for which a binding energy of -9.2 kcal/mol was calculated.

2. The Side Valley

A negatively charged carboxyl oxygen was predicted to bind in the side valley more strongly than any other probe. Energy contours for its interaction with hemagglutinin are shown in Fig. 2c. At the energy minimum, it was predicted to accept a hydrogen bond from OG Ser 136. The GRID carboxyl oxygen probe has the size of only one atom. However, in a ligand, it would be part of a larger, planar carboxylate or amide group. Therefore, for this GRID ligand point to be occupied by a carboxyl oxygen, another carboxyl oxygen or amide nitrogen atom must also be able to interact favorably in the side valley at the same time. In addition, the side valley must be large enough to accommodate a whole carboxylate or amide group in an orientation which allows favorable hydrogen bonds to be made to the protein.

An amide nitrogen probe was calculated to interact less favorably than a carboxyl oxygen in the side valley. It had an energy minimum at -7.6 kcal/mol near to that of the carboxyl oxygen probe, where it was predicted to donate hydrogen bonds to OG Ser 136 and O Arg 224. However, it is not possible for OG Ser 136 to form strong hydrogen bonds to both the oxygen and the nitrogen of an amide group positioned in the side valley at the same time because the angle subtended by these two atoms at OG Ser 136 would only be about 30–40°. Therefore, calculations were performed with program GRID for both a carboxyl oxygen probe and an amide nitrogen probe in the side valley region with no hydrogen bonds permitted between the probe and OG Ser 136.

The carboxyl oxygen probe was now found to have an energy minimum at -8.7 kcal/mol. However, the position of this energy minimum was only 0.7 Å from that of the energy minimum predicted for a carboxyl oxygen probe which could hydrogen bond to OG Ser 136, and so both these energy minima could not be occupied at the same time. An amide nitrogen probe now had an energy minimum of -6.7 kcal/mol, which was less favorable than for a carboxyl oxygen, but this energy minimum was further from the first carboxyl oxygen ligand point at a distance of 1.7 Å and at a position at which the probe could donate a hydrogen bond to O Arg 224 (see Fig. 2d). A carboxyl oxygen probe was predicted to interact with a similar energy at this position but it was less likely to make such a specific interaction with the protein because it could not form a hydrogen bond to O Arg 224.

Therefore, a ligand amide ($CONH_2$) group positioned in the side valley was predicted to interact favorably with the protein. A ligand carboxylate group could also fit in this valley but it was predicted to make a less specific interaction with the protein.

3. The Platform

An energy minimum for a chlorine probe was found in this region and is shown in Fig. 2e. A more favorable interaction was predicted for a carboxyl oxygen but there was insufficient space to fit a whole carboxylate group into this region of the binding site. Chlorine was calculated to bind favorably at this region because it was able to make large attractive dispersive interactions.

4. The Gully Site II

A small favorable binding region (see Fig. 2f) was found here for a hydroxyl group, which could make hydrogen bonds to the nitrogen and the oxygen of Gly 135. Water was calculated to have almost the same binding

Rebecca C. Wade

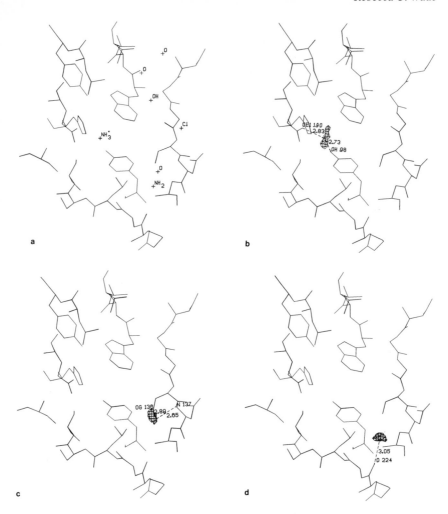

Figure 2. (a) The hemagglutinin host-cell receptor binding site with crosses marking the seven GRID ligand points identified for the particular probe groups shown labeled. (b–h) The hemagglutinin host-cell receptor binding site with energy contours around the GRID ligand points. Hydrogen bonds calculated between the protein and the probes positioned at the GRID ligand points are shown by dashed lines and their lengths are given in Angstroms. (b) Energy contours at −10 kcal/mol around the energy minimum for an NH_3^+ probe in the well. (c) Energy contours at −10 kcal/mol for the interaction of a carboxyl oxygen probe in the side valley. (d) Energy contours at −6 kcal/mol for an amide nitrogen probe in the side valley. In this map, OG Ser 136 is not permitted to make a hydrogen bond to the probe

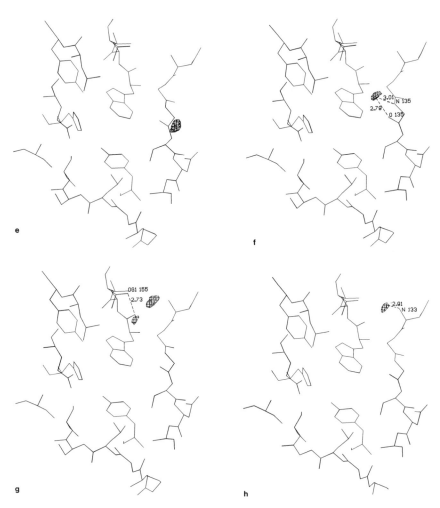

because it is assumed that it is already donating one hydrogen bond to a carboxyl oxygen positioned at the GRID ligand point shown in (c). (e) Energy contours at −5 kcal/mol for a chlorine probe in the platform region. (f) Energy contours at −6 kcal/mol for a hydroxyl probe at gully site II. (g) Energy contours at −10 kcal/mol for a carboxyl oxygen probe at gully site I. (h) Energy contours at −10 kcal/mol for a carboxyl oxygen probe at gully site I. In this map, OG1 Thr 155 is not permitted to donate a hydrogen bond as it is assumed that it is already donating one hydrogen bond to a carboxyl oxygen positioned at the GRID ligand point shown in (g).

energy (-6.5 kcal/mol) as a hydroxyl group (-6.8 kcal/mol) at this position because it could make hydrogen bonds to the same protein atoms.

5. The Gully site I

A carboxyl oxygen probe was calculated to interact more favorably than any other probe at gully site I. Two minima of very similar binding energy were detected, as shown in Fig. 2g, but at both of these, the probe was predicted to make a hydrogen bond to OG1 Thr 155. However, the two oxygens of a carboxylate group cannot both accept hydrogen bonds from OG1 Thr 155 at once. Therefore, a GRID ligand point was assigned at the energy minimum at which the shortest and strongest hydrogen bond could be made to OG1 Thr 155. Then the interaction of an oxygen probe in this region of the gully was calculated again, but this time, OG1 Thr 155 was not permitted to donate a hydrogen bond to the probe. This resulted in the predicted binding site shown in Fig. 2h, at which the carboxyl oxygen could accept a strong hydrogen bond from N Asn 133. This energy minimum was 3.0 Å from the other GRID ligand point in this gully site. The separation of the oxygens in a carboxylate group is about 2.2 Å, so the ligand carboxylate oxygens cannot both be positioned so that they make the most favorable interactions with the protein simultaneously. However, they could be positioned to interact favorably compared to other probes including water (for which the energy minimum calculated in this part of the gully was -9.6 kcal/mol). An amide nitrogen was calculated to have a binding energy of about -6 kcal/mol in this region, indicating that it would interact far less favorably at this gully site than a carboxyl oxygen. Therefore, these calculations suggest that it would not be appropriate to include an amide group at this position in a strongly binding ligand.

The principal results of probing the host-cell receptor binding site with program GRID may be summarized by the seven GRID ligand points identified. These, along with the energy maps which indicate how the binding energy of the probe groups varies with their displacement from the GRID ligand points, may be used for modeling anti-influenza agents.

IV. Design of Ligands to Block the Attachment of Influenza Viruses to Cells

Two different approaches to constructing ligands based on the GRID energy maps for the receptor binding site may be considered.

Completely novel compounds may be modeled containing the appropriate chemical groups positioned so as to satisfy the GRID ligand points. This may be done by using molecular graphics to model a carbon skeleton which connects the appropriate probe groups positioned at the GRID ligand points and does not penetrate the protein surface. Energy minimization may be used in order to relieve conformational strain in the model-built compounds. While modeling these compounds, it is important to consider not only the optimization of their binding energy, but also that of other properties which are important in the therapeutic action of a drug, such as lipophilicity, pK_a, and the rate of metabolism and excretion.

Graphical model-building is a somewhat subjective procedure, and it is likely that many possible molecules that could satisfy the GRID ligand points will be overlooked. Automated methods of constructing a molecule to satisfy a set of ligand points are under development. One method is to search databases of known compounds for those which contain the required chemical groups in a particular geometry (see, e.g., Sheridan and Venkataraghavan, 1987; DesJarlais *et al.,* 1988). Another is the use of expert systems to generate a planar template of aromatic rings and to adjust its size, shape, and ring substituents to satisfy the ligand points (Lewis and Dean, 1989).

A problem in designing compounds *de novo* is that they may be difficult to synthesize. An alternative approach is to design compounds that may be synthesized by standard techniques, thus facilitating the experimental testing of their binding and activity. These compounds may also be constructed by using molecular graphics modeling and energy minimization techniques.

For example, peptides may be made by automated solid phase synthesis (Erickson and Merrifield, 1976). The type and position of the functional groups that may be incorporated into peptides is limited and so their binding energy may be less favorable than that for a ligand designed *de novo*. However, in order to satisfy the GRID ligand points and to optimize complementarity to the binding site, natural peptides may be modified, for example, by the incorporation of nonstandard or D-amino acids, N-methylation, cross-linking, or cyclization. Such modifications are also advantageous because natural peptides are readily metabolized *in vivo* and their modification may result in a reduction in the rate at which they are degraded. Chemical modification may also enhance binding by reducing the flexibility of the peptides, thus causing a decrease in the unfavorable entropic loss that occurs on the binding of short linear peptides, which are very flexible in solution but are constrained on binding to a target molecule.

The ligands should be designed to maximize the amount of water displaced from the hemagglutinin binding pocket because undisplaced water might shield the interaction of a ligand with the protein and because the release of bound water makes a favorable entropic contribution to the free energy of ligand binding. The extent to which a modeled ligand excludes water from the binding pocket may be assessed using program GRID by examining energy contours for a water probe interacting with a target consisting of both hemagglutinin and the docked ligand. Attractive energy contours between the ligand and the protein indicate regions where water could remain undisplaced in the receptor binding pocket when the ligand binds; these should be eliminated during the modeling of the ligands.

Ligands could be designed to bind to the protein via water bridges by including water molecules that were predicted to bind strongly to the protein in the target. However, at all the binding sites for water calculated by program GRID, other probes were found which were predicted to bind more strongly than water and therefore, to displace water from the binding site. In addition, the three water molecules observed by crystallography are all in positions occupied by sialic acid in the crystal structure of a complex of hemagglutinin and sialyllactose (Weis *et al.*, 1988). This suggests that there are no water molecules in the host-cell receptor binding site which can be considered as an integral part of the target and that a strongly binding ligand should displace all water from the receptor binding site.

Following the methods outlined here, some *de novo* compounds and some peptides have been designed to block attachment of influenza viruses to cells and these are undergoing experimental investigation (Wade, 1988).

V. Conclusions

In summary, the hemagglutinin host-cell receptor binding site has been identified as a suitable target binding site for anti-influenza agents designed to be active against a range of strains of the virus. This site has been mapped energetically by using program GRID to calculate its interaction with a variety of chemical groups that might be incorporated in potential antiviral compounds. Favorable binding sites for particular chemical groups have been found in the host-cell receptor binding pocket and these may be used in the design of anti-influenza agents.

This work demonstrates one way in which the method of receptor fit might be used to design anti-influenza agents which inhibit the action of hemagglutinin. The structure of the other influenza coat glycoprotein, neuraminidase, has also been solved (Varghese *et al.*, 1983), and a similar procedure to that described here may be applicable to it. This approach may also be of value in the design of therapeutic agents active against other diseases and it should find increasing application as the number of experimentally determined structures of target molecules of pharmacological importance multiplies.

Program GRID is available on request from Peter Goodford, Laboratory of Molecular Biophysics, The Rex Richards Building, South Parks Rd, Oxford OX1 3QU, UK.

Acknowledgments

I would like to thank Peter Goodford for many helpful discussions and suggestions. Support from the Science and Engineering Research Council and the Medical Research Council, UK, is gratefully acknowledged.

References

Alderson, G. M. (1987). Ligands for glycoproteins. Part II. Thesis, Univ. of Oxford.

Beddell, C. R. (1984). Designing drugs to fit a macromolecular receptor. *Chem. Soc. Rev.* **13**, 279–319.

Boobbyer, D. N. A., Goodford, P. J., McWhinnie, P. W., and Wade, R. C. (1989). New hydrogen-bond potentials for use in determining favorable binding sites on molecules of known structure. *J. Med. Chem.* **32**, 1083–1094.

DesJarlais, R. L., Sheridan, R. P., Seibel, G. L., Dixon, J. S., Kuntz, I. D., and Venkataraghavan, R. (1988). Using shape complementarity as an initial screen in designing ligands for a receptor binding site of known three-dimensional structure. *J. Med. Chem.* **31**, 722–729.

Erickson, B. W., and Merrifield, R. B. (1976). Solid phase peptide synthesis. *In* "The Proteins" (H. Neurath and R. L. Hill, eds.), 3rd Ed., Vol. 2, pp. 255–527. Academic Press, New York.

Goodford, P. J. (1984). Drug design by the method of receptor fit. *J. Med. Chem.* **27**, 557–564.

Goodford, P. J. (1985). A computational procedure for determining energetically favorable binding sites on biologically important macromolecules. *J. Med. Chem.* **28**, 849–857.

Jones, T. A. (1985). Interactive computer graphics: FRODO. *Methods Enzymol.* **115**, 157–171.

Knossow, M., Lewis, M., Rees, D., Wilson, I. A., Skehel, J. J., and Wiley, D. C. (1986). The refinement of the hemagglutinin membrane glycoprotein of influenza virus. *Acta Crystallogr., Sect. B* **B42**, 627–632.

Lee, B., and Richards, F. M. (1971). The interpretation of protein structure: Estimation of static accessibility. *J. Mol. Biol.* **55**, 379–400.

Lewis, R. A., and Dean, P. M. (1989). Automated site-directed drug design: The formation of molecular templates in primary structure generation. *Proc. Roy. Soc.* **B236**, 141–162.

Rogers, G. N., Paulson, J. C., Daniels, R. S., Skehel, J. J., Wilson, I. A., and Wiley, D. C. (1983). Single amino acid substitutions in influenza haemagglutinin change receptor binding specificity. *Nature (London)* **304**, 76–78.

Sheridan, R. P., and Venkataraghavan, R. (1987). Designing novel nicotinic agonists by searching a database of molecular shapes. *J. Comput. Aided Mol. Des.* **1**, 243–256.

Varghese, J. N., Laver, W. G., and Colman, P. M. (1983). Structure of the influenza virus glycoprotein antigen neuraminidase at 2.9Å resolution. *Nature (London)* **303**, 35–40.

Wade, R. C. (1988). Ligand–macromolecule interactions. D.Phil. Thesis, Univ. of Oxford.

Weis, W., Brown, J. H., Cusack, S., Paulson, J. C., Skehel, J. J., and Wiley, D. C. (1988). Structure of the influenza virus haemagglutinin complexed with its receptor, sialic acid. *Nature (London)* **333,** 426–431.

Wiley, D. C., and Skehel, J. J. (1987). The structure and function of the haemagglutinin membrane glycoprotein of influenza virus. *Annu. Rev. Biochem.* **56,** 365–394.

Wiley, D. C., Wilson, I. A., and Skehel, J. J. (1981). Structural identification of the antibody-binding sites of Hong Kong influenza haemagglutinin and their involvement in antigenic variation. *Nature (London)* **289,** 373–378.

Wilson, I. A., Skehel, J. J., and Wiley, D. C. (1981). Structure of the haemagglutinin membrane glycoprotein of influenza virus at 3Å resolution. *Nature (London)* **289,** 366–373.

7 Immunochemical and Crystallographic Studies on the Interaction between Antibody and a Synthetic Peptide of Influenza Hemagglutinin

David C. Jackson
Department of Microbiology, University of Melbourne, Parkville 3052,
Australia

Vilma M. Zubak and Sonia M. Dayan
C.S.I.R.O. Division of Biotechnology, Parkville 3052, Australia

Lorena E. Brown
Department of Microbiology, University of Melbourne, Parkville 3052,
Australia

I. Antigenic Properties of a Synthetic Peptide of Influenza Virus Hemagglutinin

The synthetic peptide representing the C-terminal 24 amino acid residues of the heavy chain (HA_1) of the influenza virus hemagglutinin (HA) is able to elicit an antibody response in the absence of a carrier protein and therefore possesses epitopes for B cells and also for helper T cells (Nestorowicz et al., 1985). In mice, a strong antibody response is restricted to animals of the H-2^d haplotype which bear the I-E^d gene, indicating that the immune response to this immunogen is controlled by the major histocompatibility complex (Brown et al., 1988).

Antisera and the monoclonal antibodies (MAbs) 1/1 and 2/1, elicited by peptide 305–328 in BALB/c mice, bind equally well to influenza virus which has been exposed to pH 5 treatment and to the peptide immunogen itself (Fig. 1). The conformational change in HA that results from exposure to acid pH (Skehel et al., 1982) is thought to result in exposure of the N-terminus of the HA light chain (HA_2), which allows it to become available for membrane fusion. This conformational change is also accompanied by alterations in the expression of some of the major antigenic determinants of HA which are remote from the N terminus of HA_2 (Jackson and Nestorowicz, 1985). The increased antigenicity of acid-treated HA when examined with antibodies raised against synthetic peptide 305–328 indicate that this region of the molecule, removed some 21 Å from the N terminus of HA_2 (Wilson et al., 1981), is also exposed by acid treatment.

Although antibodies elicited by this peptide bind to viruses of the influenza H3 subtype, they do not bind to viruses belonging to the H2 subtype

Use of X-Ray Crystallography in the Design of Antiviral Agents

75

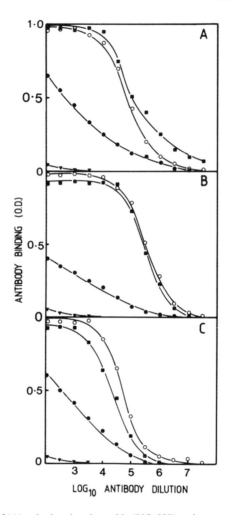

Figure 1. Binding of (A) polyclonal anti-peptide (305–328) antiserum; (B) monoclonal antibody 1/1 and (C) monoclonal antibody 2/1 to peptide 305–328 (■) and to untreated (●) or pH 5-treated (○) H3 virus. Binding of antibody to bovine serum albumin (▼). (From Jackson *et al.*, 1986.)

(Nestorowicz *et al.*, 1985). Comparison of the amino acid sequences of this region of the HA$_1$ molecule from strains of the H3 and the H2 subtype (Fig. 2) reveals that differences in sequence are located within two clusters of amino acids. The lack of binding to H2 virus of antibodies raised to the H3 sequence suggests that amino acid residues in one or both of the H3 specific regions contribute to the epitopes for antibody binding.

Two synthetic peptide analogs, each of which contains only one of the

C P K Y V K **Q N T** L **K** L A T G **M** R N V P **E K Q T**   H3 sequence, residues (305-328)

C P K Y V K S E K L V L A T G L R N V P Q I E S   H2 sequence, residues (305-328)

C P K Y V K S E K L V L A T G L R N V P **E K Q T**   Analog A
C P K Y V K **Q N T** L **K** L A T G L R N V P Q I E S   Analog B

Figure 2. Sequences of the regions of the HA$_1$ chain of the influenza hemagglutinin corresponding to residues 305–328 of the H3 and H2 subtypes. Amino acid differences between the two subtypes are shown in bold face and underlined. Also shown are the two analogs in which the H3-specific clusters of amino acids have been substituted in the H2 sequence.

distinctive H3 segments (Fig 2), are each equally capable of binding antiserum raised against the parent peptide. The two monoclonal antibodies 1/1 and 2/1 raised against the parent sequence, however, discriminate between the two analogs: MAb 1/1 binds only to analog A and MAb 2/1 binds only to analog B (Fig. 3A and B, respectively). These results indicate

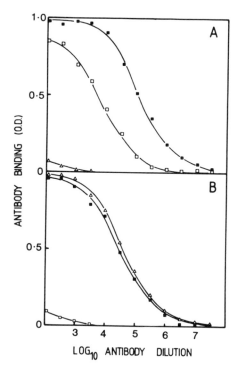

Figure 3. Binding of (A) monoclonal antibody 1/1 and (B) monoclonal antibody 2/1 to peptide 305–328 (■), analog A (□) and analog B (△). (From Jackson *et al.*, 1986.)

the presence of at least two distinct epitopes for B cells within the parent peptide, at least some of the residues of which are contributed by each of the areas 311QNTLK315 and 325EKQT328 (Jackson et al., 1986).

II. Identification of the Amino Acid Residues Involved in Antibody Binding

Further resolution of those amino acids actually involved in the binding of antibodies was carried out by synthesizing all of the possible penta-, hexa-, hepta-, and octapeptides homologous with the native sequence of peptide (305–328) and then examining the ability of each of these homologs to bind to antiserum raised against the parent peptide. The results (Schoofs et al., 1988) show that polyclonal antisera elicited by the parent peptide contain antibodies that bind to three sites within the parent sequence. Two of these sites are also recognized by the monoclonal antibodies: MAb 1/1 binds to peptides which contain residues 322NVPEKQT328 and MAb 2/1 was found to bind to all of the homolog peptides that include residues 314LKLAT318.

Identification of those residues within the epitopes defined by the two MAbs which are essential for antibody binding was carried out by examining the binding of the MAbs to a set of peptide analogs. These analogs were constructed by replacing each residue of the native sequence encompassing the antigenic site under investigation in turn by each of the 19 alternative amino acids. In the case of the epitope recognized by MAb 1/1 and encompassed by residues 322NVPEKQT328, only three residues N322, E325, and Q327 were irreplaceable (Fig. 4, upper panel). When the ability of monoclonal antibody 2/1 to bind to each of the analogs of the sequence 313TLKLAT318 was determined, it was found that each of the adjacent residues L314, K315, L316, A317, and T318 were essential for binding. In contrast, residue T313 could be replaced by a number of different amino acids without significant loss of antibody binding (Fig. 4, lower panel). It is possible that these irreplaceable amino acids represent the residues which form essential contacts with the antigen binding site, or paratope, of the corresponding antibody. Furthermore, because MAb 1/1 exhibits an absolute requirement for only three amino acid residues compared with the five contiguous residues which are essential for the binding of MAb 2/1, it might be expected that the fit of antibody 2/1 for its epitope is better than the fit of antibody 1/1. This is supported by the finding that the affinity of 2/1, 1.8×10^{10} l/mol, is higher than the affinity of MAb 1/1, 5.1×10^9 l/mol, for the parent peptide (Jackson et al., 1988).

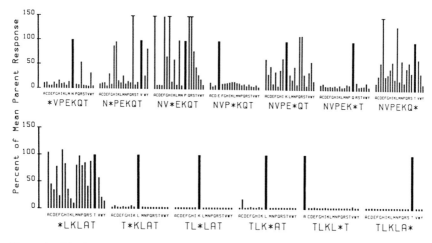

Figure 4. Binding of antibody to analogs of peptide 305–328. The bars represent the percentage binding obtained compared with binding to the parent peptide. Upper panel: The analogs are each identical in sequence to the heptapeptide 322NVPEKQT328 except that in each, one of the residues has been replaced by one of the other 19 amino acids. The position in the parent sequence that has been substituted is identified by an asterisk and the replacement amino acid is shown by its single letter code. Lower panel: The analogs are each identical in sequence to the hexapeptide 313TLKLAT318 except that in each, one of the residues has been replaced by one of the other 19 amino acids. (Modified from Schoofs *et al.*, 1988.)

III. Stoichiometry of the Interaction between Antibody and Peptide Antigen

Inspection of the sequence of the parent peptide shows that the boundaries of the two epitopes defined by MAbs 1/1 and 2/1 are separated by only three residues, that is, about 10 Å if the peptide chain is in extended conformation. When the stoichiometry of the interaction between MAbs 1/1 and 2/1 and the parent peptide is examined by competitive binding assays using virus as antigen, it is found that complete inhibition is achieved only with the homologous antibody (Fig. 5). The lack of inhibition when the heterologous antibody is used as the competitor indicates that both antibodies can bind simultaneously to a single antigen molecule. Neither MAb 1/1 nor MAb 2/1 was inhibited from attaching to virus by MAb 508/2, which is directed to a remote site on the hemagglutinin involving residues 188 and 189.

The equilibria established between different combinations of peptide and antibody are clearly demonstrated by gel permeation chromatography and ultracentrifugation studies. When radiolabeled parent peptide and both monoclonal antibodies are present, most of the radiolabel elutes

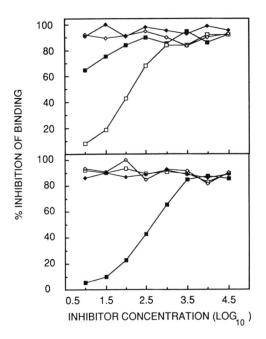

Figure 5. Inhibition of binding of radioiodinated MAb 1/1 IgG (upper panel) or radioiodinated MAb 2/1 IgG (lower panel) to influenza virus. Each of the labeled antibodies was examined for its binding to virus in the presence of ascitic fluid containing MAb 1/1 (□), 2/1 (■), or the irrelevant MAb 508/2 (◇). The levels of binding obtained in the absence of unlabeled antibody are also shown (◆). (From Jackson *et al.,* 1988.)

from a column of Sephacryl S-200 at a position corresponding to a molecular species of much higher molecular mass than that formed in the presence of either antibody alone (Fig. 6). Values derived for the Stokes radius and for the Svedberg constant allow a molecular mass of approximately 240,000 Da to be assigned to the immune complex formed between peptide and both MAbs. This value, although somewhat lower than the sum of the molecular masses of the components (~300,000 Da), is consistent with the dissociation and reassociation which occur during the course of these experiments. The elution positions of the complexes formed in the presence of the individual monoclonal antibodies and also the presence of the minority species which appear as shoulders in these profiles again indicate that the affinity of MAb 1/1 is less than the affinity of MAb 2/1. Taken together, these results also support the idea that the immune complexes formed between the peptide and MAbs 1/1 and 2/1 are composed of two antibody molecules in association with one or two peptide molecules, in other words, two antibody molecules can bind simultaneously to a single

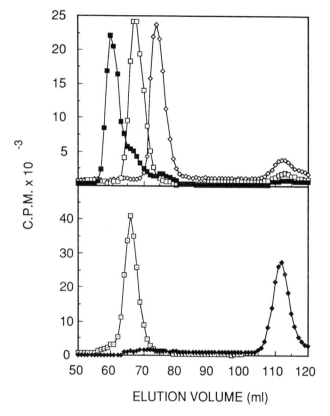

Figure 6. Examination of the chromatographic behaviour of peptide 305–328 in the presence of antibodies. Upper panel: Radioiodinated peptide was introduced to a column (160 cm × 1 cm) of Sephacryl S-200 equilibrated in phosphate-buffered saline containing 1 mg/ml of bovine serum albumin and monoclonal antibody 1/1 (◇), monoclonal antibody 2/1 (□), or a mixture of both 1/1 and 2/1 (■) The column was developed at a flow rate of 15 ml/hr and 1 ml fractions collected and assayed for radioactivity. Lower panel: Radioiodinated monoclonal 1/1 IgG (□) or radioiodinated peptide 305–328 (◆) was introduced to the Sephacryl column and the chromatogram developed as above. (From Jackson *et al.*, 1988.)

peptide molecule (Jackson *et al.*, 1988). Using anti-dinitrophenyl (DNP) antibodies and a series of dinitrophenyl polymethylenediamines, Valentine and Green (1967) showed that the shortest compound which allowed the two haptenic groups to be occupied by two separate antibody molecules had an overall length of 25 Å. The configuration of the haptenic system allows end-insertion of the terminal DNP groups into the antibody paratopes. This is in contrast to the present system, in which end-insertion of both epitopes into their respective antibody binding sites (paratopes) is known not to occur because monoclonal antibody 1/1 binds to the heptapeptide 322NVPEKQT328, even attached at T328 to a solid support, (Schoofs *et al.*, 1988), and monoclonal antibody 2/1 recognizes the sequence 314LKLAT318, which is in the center of the molecule. It is possible that to minimize steric inhibition these antibodies approach their respective epitopes from opposite sides of a single antigen molecule, forming a trans configuration of antibody and antigen. Alternatively, if there is a β turn in the peptide chain between the two epitopes, simultaneous binding of two antibodies could also occur with minimum steric hindrance. The simultaneous binding of two antibody molecules within such a short distance of each other suggests that the paratopes of MAbs 1/1 and 2/1 are not situated in deep clefts that are traditionally associated with antibody binding sites. In this case, molecular association would be better accommodated by the epitope and paratope of the antigen and antibody molecules interacting through planar surfaces.

IV. Properties of Anti-Idiotypic Antibodies Raised Against MAbs 1/1 and 2/1

Two anti-idiotypic MAbs, 4C1 and 51/3, have been prepared against MAbs 1/1 and 2/1 respectively. The binding specificity of these anti-idiotypic reagents to the respective antipeptide MAbs was examined in binding assays in the presence of synthetic peptide homologs of the parent peptide sequence. The results (data not shown) show that a sequence as short as the pentapeptide 324PEKQT328 was able to inhibit the binding between MAbs 1/1 and 4C1. Complete inhibition is achieved with the peptide 322NVPEKQT328, which contains all of the residues necessary for optimum binding between antibody and antigen (Fig. 4, upper panel). These results imply that occupation of the paratope of MAb 1/1 by a peptide which may represent the "naked" epitope defined by this antibody is sufficient to prevent the binding of the anti-idiotypic monoclonal antibody. Because the small size of such peptides eliminates the probability of significant steric inhibition, these results indicate that MAb 4C1 is a paratope-directed anti-idiotypic antibody. In similar studies with MAb 51/3, no inhibition was observed with synthetic peptides up to and including analog B and the parent peptide 305–328, suggesting that this anti-idiotypic antibody is directed to a framework region of MAb 2/1.

V. Conclusions

These studies have allowed some of the properties of this antigen–antibody system to be defined in detail. The regions to which the two

Figure 7. Photomicrograph of crystals of Fab′ fragments derived from monoclonal antibody 2/1.

monoclonal antibodies bind have been located and important amino acid residues involved in the binding process have been identified. Each antibody binds strongly to the antigenic peptide with MAb 2/1 exhibiting somewhat higher affinity than MAb 1/1, possibly due to the larger number of contact residues involved in the binding interaction. Both antibodies bind to hemagglutinin, particularly if first exposed to acid pH, and are able to bind simultaneously to the same antigen molecule even though the boundaries of the epitopes are separated by only three residues. These findings may indicate that the antibody binding sites are not deep clefts and that each antibody could approach its epitope from different sides of the peptide to overcome steric hindrance. A similar sandwich arrangement has been proposed for the recognition of an antigenic peptide by class II histocompatibility antigen and a T cell receptor (Sette *et al.,* 1987).

Because the properties of this antigen–antibody system may be amenable to X-ray crystallographic studies, we have been encouraged to carry out crystallization trials on Fab′ fragments, the Fab′–peptide immune complexes and also the idiotype–anti-idiotype complexes. Using a pulse diffusion method, trigonal crystals of MAb 2/1 (Fig. 7) with unit cell dimensions $a = b = 160$ Å, $c = 77$ Å have been obtained and crystals of this Fab′ fragment complexed with synthetic peptide have also been prepared. Similar crystallization trials are underway with MAb 1/1 as well as with the anti-idiotypic antibodies. If the structure of these antigen–antibody complexes and also of the idiotype–anti-idiotype complexes can be solved by X-ray diffraction methods, we may be closer to understanding how such small peptides bind to antibodies when only 3–5 residues are involved, compared with 16 residues implicated in the lysozyme–anti-lysozyme system (Amit *et al.,* 1986). Also, we may be able to shed some light on the molecular mechanisms underlying the phenomenon of antigen mimicry in idiotype–anti-idiotype systems.

References

Amit, A. G., Mariuzza, R. A., Phillips, S. E. V., and Poljak, R. J. (1986). Three-dimensional structure of an antigen–antibody complex at 2.8Å resolution. *Science* **233,** 747–753.

Brown, L. E., Murrary, J. M., Anders, E. M., Tang, X.-L., White, D. O., Tregear, G. W., and Jackson, D. C. (1988). Genetic control and fine specificity of the immune response to a synthetic peptide of influenza hemagglutinin. *J. Virol.* **62,** 1746–1752.

Jackson, D. C., and Nestorowicz, A. (1985). Antigenic determinants of influenza virus hemagglutinin. XI. Conformational changes detected by monoclonal antibodies. *Virology* **145,** 72–83.

Jackson, D. C., Tang, X.-L., Brown, L. E., Murrary, J. M., White, D. O., and Tregear, G. M. (1986). Antigenic determinants of influenza virus hemagglutinin. XII. The epitopes of a synthetic peptide representing the C-terminus of HA_1. *Virology* **155,** 625–632.

Jackson, D. C., Poumbourios, P., and White, D. O. (1988). Simultaneous binding of two monoclonal antibodies to epitopes separated in sequence by only three amino acid residues. *Mol. Immunol.* **25,** 465–471.

Nestorowicz, A., Tregear, G. W., Southwell, C. N., Martyn, J., Murray, J. M., White, D. O., and Jackson, D. C. (1985). Antibodies elicited by influenza virus hemagglutinin fail to bind to synthetic peptides representing putative antigenic sites. *Mol. Immunol.* **22,** 145–154.

Schoofs, P. G., Geysen, H. M., Jackson, D. C., Brown, L. E., Tang, X.-L., and White, D. O. (1988). Epitopes of an influenza viral peptide recognized by antibody at single amino acid resolution. *J. Immunol.* **140,** 611–616.

Sette, A., Buus, S., Colon, S., Smith, J. A., Miles, C., and Grey, H. M. (1987). Structural characteristics of an antigen required for its interaction with Ia and recognition by T cells. *Nature (London)* **328,** 395–399.

Skehel, J. J., Bayley, P. M., Brown, E. B., Martin, S. R., Waterfield, M. D., White, J. M., Wilson, I. A., and Wiley, D. C. (1982). Changes in the conformation of influenza virus hemagglutinin at the pH optimum of virus mediated membrane fusion. *Proc. Natl. Acad. Sci. USA* **79,** 968–972.

Valentine, R. C., and Green, M. M. (1967). Electron microscopy of an antibody–hapten complex. *J. Mol. Biol.* **27,** 615–617.

Wilson, I. A., Skehel, J. J., and Wiley, D. C. (1981). Structure of the hemagglutinin membrane glycoprotein of influenza virus at 3Å resolution. *Nature (London)* **289,** 366–372.

8 Structural Studies of Antipeptide Antibodies

James M. Rini, Robyn L. Stanfield, Enrico A. Stura,
Ursula Schulze-Gahmen, and Ian A. Wilson
Department of Molecular Biology, Research Institute of Scripps Clinic,
La Jolla, California 92307

I. Introduction

The question of how an antipeptide antibody can recognize both a peptide immunogen and its corresponding sequence in the intact antigen is still to be answered in structural terms. An example of a well-characterized peptide determinant from the influenza hemagglutinin (HA) molecule is illustrated in Color Plate 5. The peptide sequence is embedded on the surface of the protein and does not particularly protrude from the protein. However, the residues assigned to the determinant are all in fact on the protein exterior with a compact solvent-accessible surface. Even if the immunizing peptide were in fact to adopt a solution conformation similar to that in the intact protein, the surrounding environment presumably affects binding to the intact protein. We have indeed quantitated the relative affinities of the antibodies for both the peptide and the protein and found that the relative binding is 100 times less to the protein than to the peptide (Wilson, 1985; Wilson et al., 1986; K. F. Bergmann and I. A. Wilson, unpublished observations). Thus we have decided to study by X-ray crystallographic methods the structures of antipeptide antibodies in the presence and absence of peptide and protein antigens. Analysis of both the complexed and uncomplexed Fabs will allow us to determine the extent to which structural changes in the Fab accompany antigen binding.

Major advances in our understanding of how proteins interact with antibodies and the extent to which this interaction causes conformational changes in the antigen have been advanced significantly with the structure determination of four Fab–protein complexes. These structures, which include three lysozyme complexes, D1.3 (Amit et al., 1986a,b); HyHEL-5 (Sheriff et al., 1987a); and HyHEL-10 (Davies et al., 1988, also contains a general review of the three complexes) and a neuraminidase–Fab complex, NC41 (Colman and Webster, 1987; Colman et al., 1987), all show the intimate association of antibody and antigen. In these cases the antigen

Use of X-Ray Crystallography in the Design of Antiviral Agents
87

combining surfaces are flatter than the clefts seen in the earlier studies of Fabs and their hapten complexes from multiple myeloma immunoglobulins (see reviews in Amzel and Poljak, 1979; Davies and Metzger, 1983). The surfaces on the antibody may also involve ridges and grooves, however, (see Davies *et al.*, this volume) and all have a surface area of approximately 690–777 $Å^2$ buried by the antigen with very few if any water molecules between the antibody and antigen (Davies *et al.*, 1988).

Our goal, then, is to determine the structure of antipeptide antibodies and show how they are able to interact with both peptides and proteins. This question is addressed by our study of two separate antipeptide antibody systems. Monoclonal antipeptide antibodies have been generated against residues 75–110 of the influenza virus hemagglutinin (HA) (Niman *et al.*, 1983; Wilson *et al.*, 1984) and against the C-helix (residues 69–87) (Sheriff *et al.*, 1987b) of myohemerythrin (Mhr) (Fieser *et al.*, 1987). We have obtained crystals of the free Fabs and several Fab–peptide complexes and are in the process of solving these structures by X-ray crystallography. In addition, the structures of two of these peptides have been determined by two-dimensional nuclear magnetic resonance spectroscopy; both, surprisingly, have different but well-defined conformations in aqueous solution (Dyson *et al.*, 1985, 1988a,b).

II. Anti-Hemagglutinin Peptide Fabs

Mouse monoclonal antibodies were raised against an immunodominant peptide (HA_1 75–110) (Niman *et al.*, 1983; Wilson *et al.*, 1984) that lies in the trimer interface of the intact HA molecule. The antibodies have also been shown to bind to the low pH fusion active form of the hemagglutinin, in which the HA presumably changes conformation so that the determinant becomes exposed (White and Wilson, 1987). As part of our efforts to determine the structural details of the antibody–peptide interaction we have crystallized the Fab from a monoclonal antibody (17/9) both in its native form and as a complex containing the peptide determinant HA_1 100–108 (Schulze-Gahmen *et al.*, 1988). The 50% inhibition value for this peptide in a competition ELISA assay is 3×10^{-7} *M*. The native Fab molecule has been crystallized from PEG 600 in imidazole–malate buffer pH 5.6 (Color Plate 6a). The crystals are monoclinic, space group $P2_1$, with unit cell dimensions of $a = 90.2$ Å, $b = 82.9$ Å, $c = 73.4$ Å, $\beta = 122.5°$ (Schulze-Gahmen *et al.*, 1988). There are two molecules in the asymmetric unit positioned at *x, y, z* and approximately *x* + 1/2, *y* + 1/2, *z*, such that the space group can be considered C121 at low resolution. The crystals are very well ordered, diffracting to better than 1.9 Å resolution. The Fab–

peptide complex also crystallizes in space group P2₁, from PEG 600 solutions, in this case with unit cell dimensions of $a = 63.9$ Å, $b = 73.0$ Å, $c = 49.1$ Å, $\beta = 120.6°$ (Schulze-Gahmen *et al.,* 1988) (Color Plate 6b). Again the crystals diffract to high resolution (<2.1 Å) and we expect that a comparison of these two structures will allow us not only to determine what interactions are involved in peptide binding but whether structural changes in the antibody also occur.

We have recently solved the native Fab 17/9 structure by molecular replacement (J. M. Rini and I. A. Wilson, unpublished observations). Due to the wide range of elbow angles seen for the Fab structures which have been solved to date we made no assumptions about the relative orientations of the variable and constant domains and determined the rotation and translation parameters for both domains independently, as had been done previously for the determination of other Fab structures (Cygler *et al.,* 1987; Cygler and Anderson, 1988a,b; Vitali *et al.,* 1987; Prasad *et al.,* 1988; Sheriff *et al.,* 1987a; Davies *et al.,* 1988). Both the variable and constant domains from each of the five Fab structures in the Brookhaven data bank were used in the Crowther fast rotation function. All the variable domain models yielded a consistent determination for the rotational orientation with a high standard deviation and a good discrimination over the second highest value. However, with the constant domain models the results were much more variable. Although the three κ light chain models gave a consistent value for the rotational orientation, the two λ light chain models did not yield interpretable results. Significantly, the best results came from HyHEL-5, which like Fab 17/9 is an IgG with the κ type light chain (Sheriff *et al.,* 1987a). Using the variable domain of McPC603 and the constant domain from HyHEL-5 as models, the Crowther and Blow translation function from the Merlot package of Fitzgerald (1988) was solved for each domain independently for both molecules in the asymmetric unit. The vector relating the two molecules in the asymmetric unit confirmed the pseudocentered relationship evident in the diffraction patterns. The model was then improved by rigid body domain refinement using CORELS (Sussman, 1985). At this stage the hypervariable loops were removed from the model and those side chains which were not conserved were truncated to alanine or glycine. Several cycles of restrained atomic refinement were then performed (Hendrickson, 1985), resulting in a model which was then used in calculating 2Fo–Fc electron density maps. Since the hypervariable loops were not included in the model used in the refinement or phase calculation, the resulting maps are as unbiased as possible in these regions. The hypervariable loops were then manually built into the map on an interactive graphics device using FRODO (Jones, 1982) followed by the molecular dynamics X-ray refinement protocol of XPLOR (Brunger *et al.,*

1987, 1989). The current R factor is 0.25 for all data between 8.0 and 2.5 Å with an overall temperature factor of 12.0.

The hypervariable loops of Fab 17/9 define a pronounced cleft, which presumably forms the peptide binding site. As mentioned above, for those protein–antibody complexes for which structures are known the contact region on the antibody covers approximately 750 Å^2 and excludes nearly all water molecules from the interface. Conceivably, the cleft in this structure serves to maximize the contact area between the peptide and the antibody. However, since this antibody can also bind to the low-pH form of the hemagglutinin it must also be able to interact with the peptide embedded in that form of the protein.

We have soaked these native crystals in solutions of various peptides containing the epitope with the hope of diffusing the peptide into the binding site. However, with all peptides tried, the crystals crack and become disordered. Although we have been able to collect data from some of these crystals, we see at present no evidence of bound peptide. This observation is consistent with that fact that the molecules are packed in a head-to-tail fashion in the crystal lattice. Conceivably the binding site is blocked and peptide binding results in the disruption of the lattice. Thus we are pursuing the cocrystals to determine the details of the antibody–antigen interaction.

In addition to Fab 17/9 we have also obtained various peptide cocrystals with an Fab from another clone (26/9) in the antibody panel (Niman *et al.*, 1983; Wilson *et al.*, 1984). The crystals grow in space group P2$_1$2$_1$2$_1$ with unit cell dimensions $a = 72.0$ Å, $b = 103.5$ Å, $c = 66.0$ Å. We have just solved the rotation and translation functions for this structure and are in the process of refining the model to 3Å. This structure should soon yield direct structural information on the mode of peptide binding to the antibody. Like Fab 17/9, Fab 26/9 is an IgG2a:κ monoclonal antibody subcloned from the original fusions performed by Niman *et al.* (1983).

III. Anti-Mhr Peptide Fabs

Five monoclonal antibodies (A,C,F,I,L) were generated against the C-helix peptide (residues 69–87) of Mhr and one (M) against intact Mhr which recognizes the C-helix peptide (Fieser *et al.*, 1987). Antibodies A and F recognize a determinant (69–73) at the amino end of the peptide immunogen while antibodies C, I, and L recognize a determinant (79–84) at the other end of the peptide. Like the HA peptide, the C-helix peptide has a conformation in water and adopts a "nascent helix" structure as shown by high-resolution two-dimensional NMR (Dyson *et al.*, 1988b). As no helical structure is observed by circular dichroism, the NMR results suggest that in solution the peptide populates a flickering set of turnlike structures. We have begun structure determinations by X-ray crystallography on all of these anti-Mhr peptide Fabs either as free Fabs or as Fab–peptide complexes. Putative Fab–Mhr complex crystals, which are brown in color, are under review to determine whether both Fab L2 and Mhr are present.

At present the structure determination of the native anti-Mhr Fab I2 and its Fab–peptide complex is well advanced (R. L. Stanfield, E. A. Stura, and I. A. Wilson, unpublished observations). The native Fab crystallizes in the orthorhombic space group $P2_12_12_1$ with the unit cell dimensions $a = 98.0$ Å, $b = 151.7$ Å, $c = 80.8$ Å and diffracts to better than 2.7 Å resolution (Color Plate 7a). The Fab–peptide complex crystals are in the hexagonal space group $P6_322$ with $a = b = 142.5$ Å and $c = 101.5$ Å and diffract to better than 2.6 Å resolution (Plate 7b) (Stura *et al.*, 1989). Complete three-dimensional data have been collected on the Siemens–Nicolet–Xentronics area detector to 2.6 Å for the complex and 2.8 Å for the native. Rotation and translation solutions have been found for both the native Fab (2 molecules/asymmetric unit) and the Fab–peptide complex (1 molecule/asymmetric unit) which currently has an R-factor of 26% after one cycle of simulated annealing and refinement using the XPLOR package (Brunger, 1988; Brunger *et al.*, 1989).

In addition, X-ray quality crystals of the native anti-Mhr Fab A2 have been grown, and the structure determination is in progress (D. H. Fremont, E. A. Stura, and I. A. Wilson, unpublished observations). Putative A2–peptide complex crystals also exist and are under analysis. Solution of these two antibody Fab structures, which bind to different determinants on the same peptide, will then permit an evaluation of how different antibodies recognize the same peptide and whether any conformation is induced in the other regions of the peptide by Fab interaction with the short peptide determinant.

Acknowledgments

We would like to thank Terry Fieser and Rod Samodal for the Mhr Fabs and Gail Fieser for excellent technical assistance in the preparation of the anti-HA peptide Fabs. Support is gratefully acknowledged from NIH grants AI19499 and GM38794 (to IAW), a Canadian MRC Fellowship (to JMR), and a Fellowship from DFG (to USG). This is publication #5876MB from the Research Institute of Scripps Clinic.

References

Amit, A. G., Mariuzza, R. A., Phillips, S. E. V., and Poljak, R. J. (1986a). Three-dimensional structure of an antigen–antibody complex at 6 Å resolution. *Science* **233**, 747–753.

Amit, A. G., Mariuzza, R. A., Phillips, S. E. V., and Poljak, R. J. (1986b). Three-dimensional structure of an antigen–antibody complex at 2.8 Å resolution. *Nature (London)* **313**, 156–158.

Amzel, L. M., and Poljack, R. J. (1979) Three-dimensional structure of immunoglobulins. *Annu. Rev. Biochem.* **48**, 961–967.

Brunger, A. T. (1988). X-PLOR Manual, Version 1.5, Yale University, New Haven, Connecticut.

Brunger, A. T., Kuriyan, J., and Karplus, M. (1987). Crystallographic R-factor refinement by molecular dynamics. *Science* **235**, 458–460.

Brunger, A. T., Karplus, M., and Petsko, G. A. (1989). Crystallographic refinement by simulated annealing: Application to crambin. *Acta Crystallogr., Sect. A* **A45**, 50–61.

Colman, P. M., and Webster, R. G. (1987). *In* "Biological Organization: Macromolecular Interactions at High Resolution" (R. Burnett, ed.), pp. 125–133. Academic Press, New York.

Colman, P. M., Laver, W. G., Varghese, J. N., Baker, A. T., Tulloch, P. A., Air, G. M., and Webster, R. G. (1987). Three-dimensional structure of a complex of antibody with influenza virus neuraminidase. *Nature (London)* **326**, 358–363.

Cygler, M., and Anderson, W. F. (1988a). Application of the molecular replacement method to multidomain proteins. 1. Determination of the orientation of an immunoglobulin Fab fragment. *Acta Crystallogr., Sect. A* **A44**, 38–45.

Cygler, M., and Anderson, W. F. (1988b). Application of the molecular replacement method of multidomain proteins. 2. Comparison of various methods for positioning an oriented fragment in the unit cell. *Acta Crystallogr., Sect. A* **A44**, 300–308.

Cygler, M., Boodhoo, A., Lee, J. S., and Anderson, W. F. (1987). Crystallization and structure determination of an autoimmune antipoly (dT) immunoglobulin Fab fragment at 3.0 Å resolution. *J. Biol Chem.* **262**, 643–648.

Davies, D. R., and Metzger, H. (1983). Structural basis of antibody function. *Annu. Rev. Immunol.* **1**, 87–117.

Davies, D. R., Sheriff, S., and Padlan, E. A. (1988). Antibody–antigen complexes. *J. Biol. Chem.* **263**, 10541–10544.

Dyson, H. J., Cross, K., Houghten, R., Wilson, I. A., Wright, P. E., and Lerner, R. A. (1985). The immunodominant site of a synthetic immunogen as a conformational preference in water for a Type II reverse turn. *Nature (London)* **318**, 480–483.

Dyson, H. J., Rance, M., Houghten, R. A., Lerner, R. A., and Wright, P. E. (1988a). Folding of immunogenic peptide fragments of proteins in water solution. I. Sequence requirements for the formation of a reverse turn. *J. Mol. Biol.* **201**, 161–200.

Dyson, H. J., Rance, M., Houghten, R. A., Lerner, R. A., and Wright, P. E. (1988b). Folding of immunogenic peptide fragments of proteins in water solution. II. The nascent helix. *J. Mol. Biol.* **201**, 201–217.

Fieser, T. M., Tainer, J. A., Geysen, H. M., Houghten, R. A., and Lerner, R. A. (1987). Influence of protein flexibility and peptide conformation on reactivity of monoclonal anti-peptide antibodies with a protein α-helix. *Proc. Natl. Acad. Sci. USA* **84**, 8568–8572.

Fitzgerald, P. M. D. (1988). MERLOT, an integrated package of computer programs for the determination of crystal structures by molecular replacement. *J. Appl. Crystallogr.* **21**, 273–278.

Hendrickson, W. A. (1985). Stereochemically restrained refinement of macromolecular structures. *Methods Enzymol.* **115**, 252–270.

Jones, T. A. (1982). FRODO: A graphics fitting program for macromolecules. *In* "Computational Crystallography" (D. Sayre, ed.), pp. 303–317. Clarendon Press, Oxford.

Niman, H. L., Houghten, R. A., Walker, L. A., Reisfeld, R. A., Wilson, I. A., Hogle, J. M., and Lerner, R. A. (1983). Generation of protein-reactive antibody by short peptides is an event of high frequency: Implications for the structure basis of immune recognition. *Proc. Natl. Acad. Sci. USA* **80**, 4949–4953.

Prasad, L., Vandonselaar, M., Lee, J. S., and Delbaere, L. T. J. (1988). Structure determination of a monoclonal Fab fragment specific for histidine-containing protein of the phosphoenolpyruvate: Sugar phosphotransferase system of *E. coli. J. Biol. Chem.* **263**, 2571–3574.

Schulze-Gahmen, U., Rini, J. M., Arevalo, J. H., Stura, E. A., Kenten, J. H., and Wilson, I. A. (1988). Preliminary crystal data, primary sequence, and binding data for an anti-peptide from influenza virus hemagglutin. *J. Biol. Chem.* **263**, 17100–17105.

Sheriff, S., Silverten, E. W., Padlan, E. A., Cohen, G. H., Smith-Gill, S. J., Finzel, B. C., and Davies, D. R. (1987a). Three-dimensional structure of an antibody–antigen complex. *Proc. Natl. Acad. Sci. USA* **84**, 8075–8079.

Sheriff, S., Hendrickson, W. A., and Smith, J. L., (1987b). Structure of myohemerythrin in the Azidomff state at 1.7/1.3 Å resolution *J. Mol. Biol.* **197**, 273–296.

Stura, E. A., Stanfield, R. L., Fieser, T. M., Balderas, R. S., Smith, L. R., Lerner, R. A., and Wilson, I. A. (1989). Preliminary crystallographic data and preliminary sequence for anti-peptide Fab' B13IZ and its complex with the C-helix peptide from myohemerythrin. *J. Biol. Chem.* **264**, 15721–15725.

Sussman, J. L. (1985). Constrained-restrained least-square (CORELS) refinement of proteins and nucleic acids. *Methods Enzymol.* **115**, 271–303.

Vitali, J., Young, W. W., Schatz, V. B., Subottka, S. E., and Kretsinger, R. H. (1987). Crystal structure of an anti-Lewis alpha Fab determined by molecular replacement methods *J. Mol. Biol.* **198**, 351–355.

White, J. M., and Wilson, I. A. (1987). Anti-peptide antibodies detect steps in a protein conformational change: Low pH activation of the influenza virus hemagglutinin. *J. Cell Biol.* **105**, 2887–2896.

Wilson, I. A. (1985). Probing the structure and antigenic determinants of influenza virus hemagglutinin using antipeptide monoclonal antibodies. *In* "Immune Recognition of Protein Antigens" (W. G. Laver and G. M. Air, eds.), pp. 30–34. Cold Spring Harbor Lab., Cold Spring Harbor, New York.

Wilson, I. A., Niman, H. L., Houghten, R. A., Cherenson, A. R., Connolly, M. L., and Lerner, R. A. (1984). The structure of an antigenic determinant in a protein. *Cell* **37**, 767–778.

Wilson, I. A., Bergmann, K. F., and Stura, E. A. (1986). Structure analysis of anti-peptide antibodies against influenza virus hemagglutinin. *In* "Vaccines '86: New Approaches to Immunization: Developing Vaccines Against Parasitic, Bacterial and Viral Disease" (R. A. Lerner, R. Channock, and F. Brown, eds.), pp. 33–37. Cold Spring Harbor Lab., Cold Spring Harbor, New York.

9 Complexes of Peptides, Nucleotides, and Fluorescein with Immunoglobulin Fragments: Effects of Solvent on Crystal Structures and Ligand Binding

Allen B. Edmundson, James N. Herron, Kathryn R. Ely*, and
Debra L. Harris
Department of Biology, University of Utah, Salt Lake City, Utah 84112

Edward W. Voss, Jr.
Department of Microbiology, University of Illinois, Urbana, Illinois 61801

Gordon Tribbick and H. Mario Geysen
Coselco Mimotopes Party Limited, Parkville, Victoria 3052, Australia

I. Introduction

The structures and binding properties of three proteins will be discussed in this article: (1) a human Bence–Jones dimer (Mcg); (2) the Fab from a murine monoclonal autoantibody (BV04-01) which binds single-stranded DNA; and (3) the Fab from a murine monoclonal antibody (4-4-20) with high affinity for fluorescein in aqueous solutions. Structures of the Mcg dimer crystallized in deionized water and in ammonium sulfate have recently been refined at 2.0 Å resolution (Ely *et al.*, 1989; see also Abola *et al.*, 1980). Complexes produced by cocrystallization or by diffusion of ligands into preexisting crystals (Edmundson *et al.*, 1987, 1989a,b) have all been conducted in the ammonium sulfate form, which is more suitable for the peptide binding experiments to be described.

The unliganded BV04-01 Fab and complexes of Fab with tri- and pentadeoxynucleotides of thymidylic acid were crystallized in ammonium sulfate (Gibson *et al.*, 1985). A refined structure will be presented for the unliganded Fab at 2.7 Å resolution (Herron *et al.*, 1987, 1989), and the principal features of the putative binding site will be related to binding studies with oligodeoxynucleotides.

Cocrystals of the fluorescein hapten and the 4-4-20 Fab (Gibson *et al.*, 1988) were obtained in 2-methyl-2,4-pentanediol (MPD) and in polyethylene glycol (PEG). The structure of the complex in MPD was determined first because the crystals were isomorphous with those of the unliganded BV04-01 Fab. This system was particularly interesting because the affinity for fluorescein and the thermal stability of the protein were both

---

* Present address: La Jolla Cancer Research Foundation, La Jolla, California 92037

Use of X-Ray Crystallography in the Design of Antiviral Agents
95

decreased in MPD. The structure of the hapten combining site and the effect of solvent on its high affinity will be emphasized in the present report.

Binding interactions of the small ligands mentioned above, as well as the peptides described by Rini *et al.* (Chapter 8, this volume), occur in more protected environments than the block end types of surfaces of Fabs cocrystallized with larger antigens, like those discussed in Chapters 5, 15, 16, and 17. Among proteins that we have investigated, the combining sites include a conical cavity and an ellipsoidal pocket (Mcg), a groove (BV04-01), and a slot (4-4-20). In all three of these systems there is direct or indirect evidence for conformational changes on ligand binding. Conformational changes were not observed in the binding of vitamin K_1OH to the human New Fab' (Amzel *et al.*, 1974), in the binding of phosphorylcholine to the murine McPC603 Fab (Segal *et al.*, 1974), or in the studies of cocrystals of lysozyme and the murine Fab D1.3 (Amit *et al.*, 1986). Conformational changes were detected in the epitope region of lysozyme in the complex with the murine HyHEL-5 Fab (Sheriff *et al.*, 1987). Changes in both antigen and antibody structures were also believed to have occurred in the formation of neuraminidase–Fab complexes (Colman *et al.*, 1987).

Mutual conformational adjustments in the ligand and protein have been found in many examples of compounds diffused into crystals of the Mcg dimer (Edmundson *et al.*, 1974, 1984, 1987; Edmundson and Ely, 1985). In cocrystals of bis(DNP) lysine and the dimer, two molecules of ligand were bound in tandem along the interface of the V domain pair (Edmundson *et al.*, 1989a). The ligand in the main binding cavity assumed an extended conformation and seemed molded to the surface contours of the native (unaltered) cavity. In the adjacent deep binding pocket, the ligand adopted a highly compact conformation, expanded the walls of the binding site, and caused significant shifts in the orientations of protein side chains. Thus, within one complex there were examples to support both the lock and key and induced fit models of ligand binding. In designing peptides optimized for complementarity with the Mcg dimer (see Section II), we found additional cases that could be included in the lock and key category, as well as examples to be added to the list of induced fit models.

The large number of ligands binding to the Mcg dimer is incompatible with a monospecificity concept for ligand binding in this versatile, but primitive, immunoglobulin system. It is interesting that Lavoie *et al.* (Chapter 17, this volume) also have evidence for multireactivity in the lysozyme–antilysozyme system. With the 4-4-20 antibody we can consider the structural features that influence the binding of fluorescein and the exclusion of rhodamine derivatives in a highly specialized (and less versatile) combining site.

II. Binding Properties of the Mcg Bence–Jones Dimer

A. Comparison of the Structures of the Dimer Crystallized in Ammonium Sulfate and in Deionized Water

Photographs of the C_α tracings of the dimer crystallized in the two solvents are shown side by side in Color Plate 8, with the trigonal form (ammonium sulfate) on the left and the orthorhombic form (deionized water) on the right (Ely *et al.*, 1989). Each light chain dimer is composed of two conformational isomers derived from the same amino acid sequences. In both structural features and binding properties, the dimer displays the characteristics of a primitive Fab. Surface dot representations (Connolly, 1983) of the main binding cavities of the two dimers are presented as end-on-views in Color Plate 9.

It is clear from Plate 8 that the dimer adopts significantly different structures in the two solvents. If pseudotwofold axes are drawn between pairs of constant (C) and variable (V) domains, the "elbow bend" angles between the pseudodiads are 115° in the trigonal form and 132° in the orthorhombic form (Abola *et al.*, 1980; Ely *et al.*, 1983).

Patterns of ordered water molecules, assigned conservatively and included in the 2.0 Å crystallographic refinements (Ely *et al.*, 1989), were quite distinct in the two types of crystals. In some cases the water structures could be correlated with conformational changes in the protein. For example, the close contacts between the V and C domains of monomer 1 of the trigonal form were not retained in orthorhombic crystals. Water molecules filled the space created when these domains moved apart.

The modes of association of the V domains and to a lesser extent the pairing interactions of the C domains were also different in the trigonal and orthorhombic forms. Differences in the V domain pairing were reflected in the main binding cavities, as depicted in Plate 9. The shapes of the cavities, as well as the orientations of the side chains lining the walls, were altered when water was substituted for ammonium sulfate in the crystallization mixture.

We can begin to build a greater understanding of solvent effects now that refined structures are being obtained for proteins in different media. It is curious that the same protein (Mcg) should assume such different structures in two solvents, while seemingly dissimilar molecules like the unliganded BV04-01 Fab and the liganded 4-4-20 Fab should produce almost identical crystals in ammonium sulfate (BV04-01) and MPD (4-4-20).

B. Design of Site-Directed Peptides for the Mcg Bence–Jones Dimer

1. Testing and Synthesis of Peptides

In the initial stages of this project the Mcg dimer was systematically tested for binding activity against peptides coupled to solid supports by the method of Geysen *et al.* (1987). Minimal binding units were first identified from the 400 possible dipeptides of the 20 common amino acid residues (Tribbick *et al.*, 1989). Peptides increasing in infinity for the protein were then synthesized by successive additions of amino acids to the minimal binding units. Additions to both the amino and carboxyl ends, as well as combinations of optical isomers (D- and L-), were screened to optimize binding activity at each stage of the syntheses. Peptides containing from two to six residues were selected from these sets of ligands and synthesized in milligram quantities for diffusion into trigonal crystals of the Mcg dimer. The α-amino groups were acetylated to simulate conditions when the peptides were tethered to rods. At the other end the peptides were initially synthesized in the amide form. To increase the solubilities of longer or more hydrophobic peptides, carboxyl groups were sometimes substituted for the amide groups.

2. X-Ray Analyses of Peptide–Protein Complexes

All crystals were exposed to saturated solutions of peptides in 1.9 M ammonium sulfate buffered with 0.2 M phosphate, pH 6.2 (the crystallizing medium). Soaking periods were extended to at least 1 month to ensure limiting binding conditions for all samples. X-ray diffraction data were collected to d spacings of 2.7 Å, and the locations and relative occupancies of the peptides were determined by difference Fourier analyses. Protein phase angles were calculated from the atomic coordinates of the crystallographically refined structure of the Mcg dimer (Ely *et al.*, 1989). After the peptides were fitted to the "difference" electron density by interactive computer graphics (programs written by R. J. Athay), the structures of the ligand–protein complexes were crystallographically refined with the programs of Hendrickson and Konnert (1981; Hendrickson, 1985). Alternating procedures of model building and refinement were continued until the sterochemistry appeared satisfactory and the crystallographic R-factors plateaued at 20–22%.

At the first level of the X-ray analyses we verified that peptides with binding activity when linked to rods also formed complexes with the Mcg dimer in trigonal crystals. All 20 peptides tested lodged in the main binding cavity of the dimer and nowhere else in the protein molecule. Relative occupancies of peptides within a series increased with size in the order of di<<tri<tetrapeptides. No appreciable changes in occupancies were noted for peptides containing more than four residues unless there were special interactions (e.g., Ac-QfHp$\beta\beta$-OH, as discussed below).[1]

3. General Binding Patterns of Peptides

The following peptides were included in a series based on the incremental additions of amino acids to the minimal binding unit of Ac-Hp-NH$_2$: Ac-fHp-NH$_2$; Ac-fβHp-NH$_2$; Ac-QfHp-OH; Ac-QfHpβ-OH; and Ac-QfHp$\beta\beta$-OH. Binding patterns for Ac-Hp-NH$_2$ were difficult to interpret because this dipeptide was accommodated in two overlapping sites with low occupancies. For the remaining members of the series there was only one site and the electron density was well defined. The "cage" electron density modules for three of these peptides are shown in Fig. 1. Skeletal models of the binding regions of the Mcg dimer are codisplayed to indicate the sites occupied by the ligands. In Fig. 2 superimposed models of Ac-fHp-NH$_2$ and Ac-fβHp-NH$_2$ are presented in the conformations adopted in the main cavity. These peptides differ only by a "spacer" β-alanine residue between the D-phenylalanine and L-histidine residues. In Fig. 3 the skeletal model of the ligand Ac-QfHp-OH is superimposed on models of penta- and hexapeptide extensions (Ac-QfHpβ-OH and Ac-QfHp$\beta\beta$-OH).

The *N*-acetyl group of each peptide was inserted into the cavity first. In all cases this group made a significant contribution to the binding energy, as evidenced by the number of contacts with protein substituents (the number ranged from 12 to 22 in tri-, tetra-, and pentapeptides). Pairs of atoms were considered potential contacts if they were separated by less than 4 Å.

4. Binding of Ac-fHp-NH$_2$ and Ac-fβHp-NH$_2$

In Ac-fHp-NH$_2$ D-phenylalanine occupied a hydrophobic pocket and made 20 of the total of 45 contacts with the protein. The histidine side chain provided 10 and the proline 3 contacts. When the β-alanine spacer was included in Ac-fβHp-NH$_2$, the acetyl group moved into closer apposition with monomer 2, with an increase in the number of contacts from 12 to 22.

---

[1] One letter codes are used for the amino acid residues. Capital letters designate L-amino acids and lower case letters represent D-optical isomers. Other abbreviations are Ac for acetyl, β for β-alanine, OH for carboxyl group, and NH$_2$ for amide group.

N-ACETYL-L-GLN-D-PHE-L-HIS-D-PRO-COO⁻

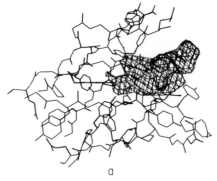

a

N-ACETYL-L-GLN-D-PHE-L-HIS-D-PRO-β-ALA-COO⁻

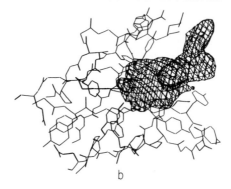

b

N-ACETYL-L-GLN-D-PHE-L-HIS-D-PRO-β-ALA-β-ALA-COO⁻

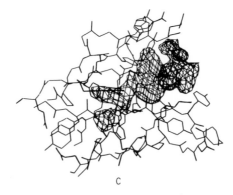

c

Figure 1. "Cage" electron density modules for three peptides in 2.7 Å difference Fourier maps, calculated after crystallographic refinement of the ligand–protein complexes. Skeletal models of the binding regions of the Mcg Bence–Jones dimer (trigonal form) are codisplayed with the cage density to illustrate the general modes of binding of the peptides. The peptide in

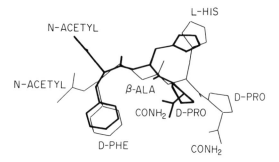

Figure 2. Superimposed skeletal models of two related peptides in the conformations assumed when they were bound in the main cavity of the Mcg dimer. These peptides differ by the insertion of a β-alanine residue between D-phenylalanine and L-histidine. Note that the phenyl rings and the histidine imidazolium groups of both peptides occupy the same general subsites despite the sequence differences.

Phenylalanine remained in the same pocket and in practically the same orientation (26 contacts, 18 with monomer 1), but proline was shifted out into the solvent. Although displaced along the peptide by β-alanine, the histidine side chain still extended into the subsite occupied by the imidazole ring in the tripeptide (see Fig. 2). The flexibility introduced into the peptide by β-alanine permitted this unusual binding event to occur.

These results suggest that there are preferred sites for the D-phenylalanine and L-histidine side chains. D-proline probably increases the rigidity of one end of the peptide and may help to stabilize the histidine orientation. In the binding of β-casomorphin-4 to the Mcg dimer, proline also tends to lodge at the entry of the binding cavity without providing a significant number of contacts with the protein (Edmundson *et al.*, 1987). Proline apparently has the appropriate size, shape, and chemical properties to act as a useful constituent of peptide ligands for the dimer.

5. Binding of Ac-QfHp-OH and Ac-QfHpβ-OH

Addition of L-glutamine to the amino end was clearly favored over other alternatives in the ELISA assays of tethered peptides (Tribbick *et al.*,

(a) was designed to bind to the Mcg dimer by the method of Geysen *et al.* (1987); (b and c) show the binding patterns obtained when one (b) and two (c) β-alanine residues were added to the carboxyl end of the "designer" peptide in (a). Peptides (a and b) were complementary to the topography of the main cavity and caused relatively few distortions in the protein structure. The binding patterns are therefore consistent with a lock and key model of binding. Conformational adjustments occurred in both the peptide and the protein in the complex depicted in (c). This pattern serves as a good example of the induced fit type of binding.

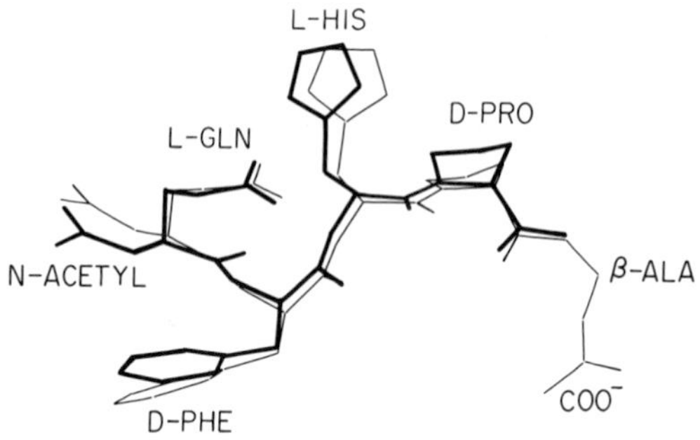

Figure 3. Superimposed skeletal models representing the conformations of two peptides when bound to the Mcg dimer. These peptides are identical to those depicted in (a) and (b) of Fig. 1. In each peptide the internal structure is stabilized by the formation of a hydrogen bond between the amide carbonyl group of L-glutamine and an imidazolium NH of L-histidine. The tetrapeptide was designed to act as an optimized site-filling ligand for the main cavity. Note that the addition of one β-alanine residue at the carboxyl end did not appreciably alter the binding pattern of the optimized portion of the ligand.

1989). Reasons for this increase in affinity were apparent from the binding pattern in the crystal (see Fig. 3). The *N*-acetyl group was held in a fixed position at the bottom of the cavity, with its carbonyl oxygen atom located within hydrogen bonding distance of the phenolic hydroxyl group of tyrosine 38, monomer 2. The glutamine side chain extended toward the entrance to the cavity. Its methylene groups interacted mainly with the side chains of phenylalanine 99 and tyrosine 93 side chains of monomer 1. Internal stabilization of the bound peptide occurred through the formation of a hydrogen bond between the carbonyl oxygen of glutamine and an imidazolium group of histidine. This hydrogen bonding helped to maintain the histidine side chain in an orientation suitable for 11 interactions with atoms of tyrosine 34 and glutamic acid 52 of monomer 2. D-Phenylalanine occupied the same hydrophobic pocket as before, but the orientation of the phenyl ring and the direction of approach to the pocket were different because of the elongation of the ligand with glutamine. The phenylalanine side chain participated in an exceptional number (33) of interactions with the atoms of a network of six aromatic and aliphatic apolar residues of the protein (especially tyrosine 34, valine 48, and tyrosine 51 of monomer 1). As in Ac-fβHp-NH₂ the carboxyl terminal proline residue of Ac-QfHp-OH was not in contact with the protein.

The packing of the ligand in the cavity was not appreciably changed when the sequence of Ac-QfHp-OH was extended to Ac-QfHpβ-OH (see Fig. 3). Interactions with the protein inside the combining site were marginally increased from 80 to 87, but the β-alanine remained outside the cavity between protein molecules in the crystal lattice.

6. Binding of Ac-QfHpββ-OH

Since β-alanine residues can be used as spacers connecting the carboxyl ends of peptides to the rods, they are not considered as part of the ligand in the ELISA screening procedures of Geysen *et al.* (1987). As a control, however, the hexapeptide Ac-QfHpββ-OH was included in the crystallographic series. It was surprising to find a severely modified binding pattern compared to that expected from previous results. Two views of the "cage" electron density for the peptide are shown in Fig. 4 (also see Fig. 1). The view in Fig. 4a is useful in understanding the chemical events

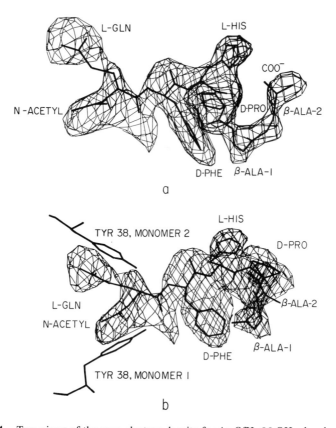

Figure 4. Two views of the cage electron density for Ac-QfHpββ-OH when bound to the Mcg dimer (this corresponds to the binding pattern in Fig. 1c). (a) On the right note that the carboxyl group of β-alanine 2 forms an ion pair with the imidazolium group of L-histidine. A hydrogen bond can no longer be formed between the side chains of L-glutamine and L-histidine, as in the peptides shown in Fig. 3. L-Glutamine and its associated *N*-acetyl groups penetrate the barrier between the main cavity and the deep binding pocket and interact with the inner surfaces of the two tyrosine 38 side chains of the protein. These interactions are illustrated at the left of the view in (b).

which triggered the change in the binding pattern. Apparently, the tail of the peptide was sufficiently flexible for the carboxyl group of β-alanine to form an ion pair with one of the imidazolium nitrogen atoms of histidine (the histidine is expected to have a partial positive charge at the pH of the crystals). This interaction was probably more favorable than the hydrogen bonding between the second imidazolium group and the carbonyl oxygen of glutamine. With this bond broken, the glutamine side chain and N-acetyl group participated in a concerted action to become wedged between the two tyrosine 38 side chains on the floor of the cavity of the dimer (see Fig. 4b).

In the native protein these two tyrosine residues are in van der Waals contact with a pair of phenylalanine (101) side chains to form a barrier between the main cavity and a deep binding pocket. In breaching this barrier the glutamine and acetyl groups caused local conformational changes in the protein and doubled the number of interactions (182 versus 87 with Ac-QfHpβ-OH). Stacking interactions of the planar amide group of glutamine with the phenolic ring of tyrosine have been observed previously in the crystal structure of the dimer. A similar interaction with the acetyl group is also probably to be expected under these unusual circumstances.

7. Summary and Implications of the Peptide Binding Studies

The relative affinities in the series of tethered peptides were commensurate with the progressive increase in the number of ligand–protein interactions in crystals: Ac-fHp-OH, 45 possible contacts; Ac-fβHp-NH$_2$, 54, Ac-QfHp-OH, 80; AcQfHpβ-OH, 87; and AcQfHp$\beta\beta$-OH, 182. Criteria for a site-filling ligand were met at the tetrapeptide level. The shape adopted by the Ac-QfHp-OH peptide closely matched the topography of the main cavity and there were many interactions with protein atoms. Conformational changes were minimal in comparison with those observed in the binding of chemotactic and opioid peptides (Edmundson and Ely, 1985; Edmundson et al., 1987).

The incremental approach to ligand design worked very well in this series. Site-filling peptides with secondary structure and relatively high affinity could be custom-fitted without major distortions of the combining site. Such interactions conformed to a lock and key model. However, the binding patterns were radically changed when the secondary structure in the site-filling pentapeptide Ac-QfHpβ-OH was inadvertently disrupted by the addition of a second β-alanine residue with a charged carboxyl group. This type of chemistry would not occur in epitope mapping or other applications of the methods of Geysen et al. (1987), but it is very instructive for exploring the binding capabilities of such molecules as the Mcg dimer. With the imidazolium group diverted to form an ion pair with carboxylate, the liberated glutamine side chain and its associated acetyl group participated in two concerted sets of interactions with the protein. In this case, which is commensurate with an induced fit model, the occupancy of auxiliary binding sites in the deep pocket was accompanied by extensive conformational adjustments in both ligand and protein. Although we suspect that the activation energy for the formation of such a ligand–protein complex may have been high, the final structure was stabilized by more than twice the number of interactions observed when binding was restricted to the main cavity.

III. Monoclonal Autoantibody with Activity against Single-Stranded DNA

Mice spontaneously developing a syndrome analogous to human systemic lupus erythematosus synthesize antibodies with anti-DNA activity. Splenocytes from one of these mice (Female F_1, New Zealand Black × New Zealand White) were fused with a nonsecreting myeloma cell line (SP 2/0-Ag 14) to produce a monoclonal IgG2b (κ) autoantibody which bound single-stranded DNA (Ballard *et al.*, 1984; Ballard and Voss, 1985). Pyrimidine bases were bound by this antibody (designated BV04-01) in preference to purines. Thymine was favored over uracil. Electrostatic interactions with phosphate groups appeared to be of limited importance, since variations in ionic strength did not appreciably affect the binding of synthetic nucleotides.

In solution the binding of 5'-phosphorylated oligodeoxythymidilic acid reached detectable levels when the length of the ligand exceeded five nucleotides (Ballard and Voss, 1985). Relative binding constants were augmented 5- to 10-fold if the size was increased from 10 to 18 nucleotides (Smith *et al.*, 1988). The Fab has been cocrystallized with ligands as small as trinucleotides of deoxythymidine (Edmundson *et al.*, 1989b). Murine Fabs from other anti-DNA antibodies have also been cocrystallized with oligonucleotides (Cygler *et al.*, 1987; Anderson *et al.*, 1988), and it will be very interesting to compare their structures with that of the BV04-01 Fab.

The three-dimensional structure of the unliganded BV04-01 Fab was determined at 2.7 Å resolution (Herron *et al.*, 1987). Atomic coordinates of the murine McPC 603 Fab were kindly made available to us by Gerson Cohen, Eduardo Padlan and David Davies. With the V_H–V_L and C_{H1}–C_L pairs of domains of this protein as starting models, the structure of the BV04-01 Fab was solved by molecular replacement methods (Rossmann and Blow, 1962; Crowther, 1972; Lattman and Love, 1970; Fitzgerald,

1988) in combination with crystallographic refinement techniques (Herz-berg and Sussman, 1983; Hendrickson and Konnert, 1981; Hendrickson, 1985). Interactive model building was interspersed with cycles of refinement with the program FRODO (Jones, 1978; Pflugrath *et al.*, 1984), implemented on an Evans and Sutherland PS 330 color graphics system. Amino acid sequences of the V_H and V_L domains were deduced from nucleotide sequences by D. W. Ballard, P. R. Blier, P. E. Pace, and A. Bothwell at Yale University (see Smith *et al.*, 1988). These sequences were substituted for those of the McPC 603 Fab at an early stage of the refinement process.

An C_α skeletal model of the BV04-01 Fab is presented as a stereo diagram in Color Plate 10. Constituents of the putative binding site for DNA are illustrated in Color Plate 11.

This Fab is an extended molecule with a large irregular groove between the tips of the V_H and V_L domains. The pseudotwofold axis of rotation between the V domains was nearly confluent with the pseudodiad be-tween the C domains. There were few contacts between V and C domains in the central solvent cavity. As in other IgG2b antibodies, the heavy and light chains were covalently linked by a disulfide bond between residues 136 (heavy) and 219 (light). This bond is depicted at the lower end of the model in Plate 8. Both C domains terminated at residue 219. In the light chain this residue was the C-terminus of the gene product, and in the heavy chain glutamic acid 219 furnished the peptide bond hydrolyzed by papain.

The groove was designated as the probable binding site for DNA. All three hypervariable loops of the light chain and the second and third hypervariable regions of the heavy chain contributed to the lining of this groove. A cluster of the following aromatic side chains was one of the most prominent structural features of the putative binding site: histidines 31 and 98, and tyrosines 37 and 54 of the light chain; tryptophan 107 of the heavy chain (see Plate 11). Considering the low sensitivity of the binding of synthetic oligonucleotides to changes in ionic strength (Ballard and Voss, 1985), we did not expect to find such a high concentration of positively charged residues in and around the groove (e.g., lysine 55, light chain, and arginines 50, 52, and 100 in the heavy chain, plus histidines 31 and 98 in the light chain, at pH values <7).

In such closely related molecules as the immunoglobulins, it is interest-ing to consider what structural features are responsible for the formation of a groove rather than a cavity for ligand binding. For example, the V domains of the BV04-01 Fab do not cross each other at a severe angle like that observed in the Bence–Jones dimer, which has a cavity-type binding site. The third hypervariable region of the heavy chain is especially impor-tant in determining the size and shape of the binding site of the BV04-01

Fab. This loop is bent away from the more cylindrical part of the domain and directed toward the first hypervariable loop of the light chain (the third hypervariable region is the large red loop extending into the V domain interface at the top of Plate 10).

The presence of a potential binding cavity in the unliganded BV04-01 Fab was obscured by a set of side chains acting like a false floor over the entry. These residues included serine 96 and leucine 101 of the light chain and arginine 50 and aspartic acid 101 of the heavy chain (see Plate 11). The arginine and aspartic acid side chains formed a salt bridge over the entry to the cavity. Access to the cavity was also partially blocked by segments of polypeptide backbone of the third hypervariable loop of the heavy chain, particularly around alanine 106.

A. X-ray Analyses of Complexes of the BV04-01 Fab with Oligodeoxynucleotides

Diffusion of a trinucleotide of deoxythymidylate into a crystal of unliganded BV04-01 Fab was accompanied by overt damage to the crystal. Under mild conditions (2 days soak with 2 molar quantities of nucleotide) it was possible to limit the damage and collect X-ray diffraction data to 2.7 Å resolution. The difference Fourier map indicated changes in many regions of the protein, although one module of electron density near tryptophan 107 of the heavy chain may have represented a ligand molecule in the putative binding site. Use of a hexanucleotide did not lead to crystal damage, but again the difference Fourier map showed a wide distribution of structural perturbations.

To circumvent these disruptive effects, attempts were made to cocrystallize the BV04-01 Fab with ligands ranging in size from three to eight nucleotides. Cocrystals with tri- and pentanucleotides have thus far proved suitable for X-ray analyses. In ammonium sulfate, these ligand–protein complexes crystallized in the same space group (monoclinic P2$_1$), and the unit cell dimensions were also nearly identical. Apparently, the binding of relatively small ligands by the BV04-01 Fab led to significant changes in crystal properties (the unliganded Fab crystallized in the triclinic space group P1).

Trial solutions of the structures of the complexes were obtained by molecular replacement methods, with the unliganded BV04-01 Fab as the starting model. The structures of the protein components are being refined with the program PROLSQ (Hendrickson, 1985), used in conjunction with interactive model building. Ligands will be included in the refinement procedures in the near future. At the present it appears that the binding region has been altered in the formation of the ligand–Fab complex. The V domains have moved apart, and the third hypervariable loop of the heavy chain has shifted from its orientation in the unliganded protein. The nature and magnitude of these shifts will be more comprehensively defined as the electron density maps improve during the refinement process. It is already apparent that aspartic acid 101 and arginine 50 are not sufficiently close to form the ion pair that partially blocked the entry to the cavity in the unliganded Fab.

IV. Three-Dimensional Structure of a Fluorescein–Fab Complex Crystallized in MPD

It was extremely good fortune that a complex of fluorescein with an Fab from a high-affinity monoclonal antibody (4-4-20) crystallized from MPD in a form nearly isomorphous with the triclinic crystals of the unliganded BV04-01 Fab (Gibson *et al.*, 1988). With the latter protein as starting model, the structure of the fluorescein–Fab complex was readily solved at 2.7 Å resolution by molecular replacement methods (Herron *et al.*, 1989).

Anti-fluorescein monoclonal antibodies have provided an excellent system for investigating the molecular basis of antigenic specificity. As a group these antibodies exhibit a wide range of binding affinities (10^5–10^{10} M^{-1}). They offer an additional advantage of being amenable to a variety of experimental fluorescence techniques which can be used in correlating binding affinities, kinetics, and thermodynamics (Kranz *et al.*, 1982; Herron *et al.*, 1986). Bates *et al.* (1985) developed a family of idiotypically cross-reactive antibodies which collectively exhibited binding affinities ranging over several orders of magnitude. Moreover, at least six of the monoclonals were encoded by the same germline "variable" genes (Bedzyk and Voss, 1989). This family of antibodies can thus be used to further the understanding of both idiotypy and affinity maturation. The crystal structure of the hapten–Fab complex discussed here is of special interest because the 4-4-20 monoclonal shows the highest affinity for fluorescein among members of the idiotype family.

In Color Plate 12 the C_α tracings of the 4-4-20 and BV04-01 Fabs are superimposed to illustrate the marked similarities in the structures of the two proteins. A stereo diagram of the model of the complex of the 4-4-20 Fab with fluorescein is presented in Color Plate 13. The structure of the combining site with fluorescein in place is shown in Color Plate 14.

Similarities in the tertiary structures of the light chains of BV04-01 and 4-4-20 were probably to be expected, since the amino acid sequences differ in only six positions (Bedzyk *et al.*, 1989). In both cases the light chains predominate in providing the major crystal packing interactions (Herron *et al.*, 1989). A close correspondence in surface topography would help explain the appearance of isomorphous crystal forms, but we do not yet understand why such packing should occur in media as different as MPD and ammonium sulfate.

Despite 42 differences in the sequences of the V_H domains of 4-4-20 and BV04-01 (Bedzyk *et al.*, 1989), the polypeptide chains adopt nearly identical conformations (see Plate 12). However, one notable difference is observed in the third hypervariable loop, in which the 4-4-20 is three residues shorter than BV04-01.

Relatively few substitutions appear to be critical for the formation of a combining site suitable for fluorescein rather than DNA. In the light chain these substitutions include arginine 39 for histidine and tryptophan 101 for leucine; tryptophan 33 and glutamine 50 replace alanine and arginine in the heavy chain. Tyrosines 102 and 103 have no obvious equivalents in BV04-01, presumably because of the differences in lengths of the third hypervariable loop. Arginine 39 is replaced by histidine in other monoclonals of the antifluorescyl idiotype family and by glutamine in the Mcg Bence–Jones dimer. This glutamine is not present in the regions involved in ligand binding (Edmundson *et al.*, 1974).

The hapten combining site in 4-4-20 is a relatively narrow (9 Å across) slot between the V_H and V_L domains (see Plate 13 and 14). In the complex the planar xanthonyl group occupies the bottom of the slot and the phenylcarboxyl moiety faces the solvent. The slot is produced mainly by a network of the tyrosines and tryptophans mentioned above. Tyrosine L37 and tryptophan H33 flank the xanthonyl ring system of fluorescein and tryptophan L101 provides the floor of the binding site. Tyrosines H102 and H103 form part of the upper boundary of the slot, with residue 103 being near the phenyl ring of fluorescein.

The phenyl ring of tyrosine L37 participates in a stacking interaction with the xanthonyl moiety of fluorescein and its phenolic hydroxyl forms a hydrogen bond with the phenylcarboxyl group of the ligand. Histidine L31, uncharged at the pH (7.3) of the crystallization mixture, is located within hydrogen bonding distance of one of the two enolic oxygen atoms on the xanthonyl ring. On the opposite side of this ring the second enolic group forms an ion pair with arginine L39. Thus one of the two formal charges on the fluorescein dianion is neutralized and the second is partially compensated by formation of a hydrogen bond. The presence of an enol–arginine ion pair in a medium of low dielectric constant is probably the key factor in the large incremental increase in affinity for fluorescein (two to three orders of magnitude) relative to other monoclonals in the idiotype

family. These proteins contained residues appropriate for aromatic interactions and hydrogen bonding with the ligand, but lacked the critical
arginine residue.

Two rhodamine derivatives (110 and B), which are structural analogs of
fluorescein, are not bound by the 4-4-20 antibody (Kranz *et al.*, 1982). In
these compounds the enolic groups of fluorescein are replaced by amino
(110) and diethylamino groups (B). At lower pH values (<5) these aryla-
mino groups would be positively charged and therefore would be repelled
by like charges on the histidine L31 and arginine L39 side chains. At pH 8
these rhodamines would be monoanionic, but the pH profile for fluorescein
binding shows a strong dependence on the presence of a dianion. Moreover, the crystallographic results suggest that the interactions with the
enolic groups are of prime importance in the binding of fluorescein. With
rhodamine B, an additional steric factor should be considered. In crystals
of the Mcg dimer, the tetraethyl derivative (B) could not be accommodated
in the binding site, whereas 6-carboxytetramethylrhodamine readily
formed a tight complex with the protein (Edmundson *et al.*, 1984).

A. Possible Effects of the Carrier Protein on Hapten Binding

In the preparation of the immunogen used to elicit the production of
antifluorescyl antibodies, isothiocyanate derivatives of fluorescein amine
are coupled to the ε-amino groups of lysine residues of keyhole limpet
hemocyanin (KLH). The structure of the fluorescein–Fab complex was
examined to assess what effects the KLH carrier protein might have had
on the permissible orientations of the free hapten in the combining site of
the elicited antibody. Plate 14 shows that there is a clear channel leading
from the *para* position of the phenyl ring of fluorescein to the outside
surface of the heavy chain. This channel has suitable dimensions (3 Å
wide, 7 Å long) for a lysine side chain from KLH to attach to bound
fluorescein. It could be argued that the restraints associated with this
linkage, together with the necessity for compatible docking surfaces for
KLH, may have been important factors in the selection or induction of a
slot-shaped binding site for fluorescein.

B. Sequestering of MPD below the Fluorescein Combining Site

An unassigned module of electron density was found in the interface of the
V domains 4.6 Å below the hapten binding site. This electron density,
which was similar in size and shape to that expected for an MPD molecule,
persisted in Fourier maps calculated at various intervals throughout the
crystallographic refinement process. The putative solvent molecule was

sequestered in a pocket of polar and apolar side chains, including those of threonine 99, serine 101, methionine 105, and tryptophan 108 of the heavy chain and arginine 39 and tyrosine 41 of the light chain.

MPD was introduced into the crystallization mixture after formation of the fluorescein–Fab complex. It therefore seems likely that the hapten would have to be dissociated from the protein before such a large solvent molecule could be admitted into a deeper region of the V domain interface.

Previous studies (Gibson *et al.*, 1988) indicated that the affinity of 4-4-20 for fluorescein decreased 300-fold in 47% (v/v) MPD (the association constant in aqueous solution was $3.4 \times 10^{10} \, M^{-1}$). The temperature stabilities of both the intact antibody and the Fab were also significantly lower in the presence of MPD. An analysis of thermodynamic results further suggested that conformational changes in the protein were largely responsible for the decrease of affinity. In the future the structure in MPD will be compared with that of the fluorescein–Fab complex crystallized in polyethylene glycol (Gibson *et al.*, 1988). Hopefully, these studies will provide greater insight into the general effects of these solvents. At our present level of understanding, it seems likely that local perturbations accompanying the direct binding of an MPD molecule in the V domain interface would have a significant influence on the affinity constant of a hapten bound in a neighboring site.

Acknowledgments

We are deeply grateful to Stuart Rodda, Thomas Mason, and the remainder of the Coselco group in Melbourne for the design, testing, and synthesis of the peptides used in this work; to Brad Nelson for his photography, Barbara Staker for her art work, and Judy Baker for typing the manuscript. This work was supported by Grant CA19616 to A. B. E. from the National Cancer Institute, Department of Health and Human Services; Grant AI22898 to J. N. H.; and Grant AI20960 to E. W. V. Coselco Mimotopes Ltd. provided the support for G. T. and H. M. G., as well as facilities and care for A. B. E. on several working visits to the laboratory. We thank the managing director of Coselco, N. J. McCarthy, for his kindness and advice.

References

Abola, E. E., Ely, K. R., and Edmundson, A. B. (1980). Marked structural differences of the Mcg Bence–Jones dimer in two crystal systems. *Biochemistry* **19**, 432–439.

Amit, A. G., Mariuzza, R. A., Phillips, S. E. V., and Poljak, R. J. (1986). Three-dimensional structure of an antigen–antibody complex at 2.8-Å resolution. *Science* **233**, 747–753.

Amzel, L. M., Poljak, R. J., Saul, F., Varga, J. M., and Richards, F. F. (1974). The three-dimensional structure of a combining-region–ligand complex of immunoglobulin New at 3.5 Å resolution. *Proc. Natl. Acad. Sci. USA* **71**, 1427–1430.

Anderson, W. F., Cygler, M., Braun, R. P., and Lee, J. S. (1988). Antibodies to DNA. *BioEssays* **8**, 69–74.

Ballard, D. W., and Voss, E. W., Jr. (1985). Base specificity and idiotypy of anti-DNA autoantibodies with synthetic nucleic acids. *J. Immunol.* **135**, 3372–3386.

Ballard, D. W., Lynn, S. P., Gardner, J. E., and Voss, E. W., Jr. (1984). Specificity and kinetics defining the interaction between a murine monoclonal autoantibody and DNA. *J. Biol. Chem.* **259**, 3492–3498.

Bates, R. M., Ballard, D. W., and Voss, E. W., Jr. (1985). Comparative properties of monoclonal antibodies comprising a high affinity anti-fluorescyl idiotype family. *Mol. Immunol.* **22**, 871–877.

Bedzyk, W. D., and Voss, E. W., Jr. (1989). To be published.

Bedzyk, W. D., Johnson, L. S., Riordan, G. S., and Voss, E. W., Jr. (1989). Variable region primary structure of a high affinity monoclonal anti-fluorescein antibody. *J. Biol. Chem.* (in press).

Colman, P. M., Laver, W. G., Varghese, J. N., Baker, A. T., Tulloch, P. A., Air, G. M., and Webster, R. G. (1987). Three-dimensional structure of a complex of antibody with influenza virus neuraminidase. *Nature (London)* **326**, 358–363.

Connolly, M. L. (1983). Solvent-accessible surfaces of proteins and nucleic acids. *Science* **221**, 709–713.

Crowther, R. A. (1972). Fast rotation function. *In* "The Molecular Replacement Method: A Collection of Papers on the Use of Non-Crystallographic Symmetry" (M. G. Rossmann, ed.), pp. 173–178. Gordon & Breach, New York.

Cygler, M., Boodhoo, A., Lee, J. S., and Anderson, W. F. (1987). Crystallization and structure determination of an autoimmune anti-poly (dT) immunoglobulin Fab fragment at 3.0 Å resolution. *J. Biol. Chem.* **262**, 643–648.

Edmundson, A. B., and Ely, K. R. (1985). Binding of *N*-formylated chemotactic peptides in crystals of the Mcg light chain dimer: Similarities with neutrophil receptors. *Mol. Immunol.* **22**, 463–475.

Edmundson, A. B., Ely, K. R., Girling, R. L., Abola, E. E., Schiffer, M., Westholm, F. A., Fausch, M. D., and Deutsch, H. F. (1974). Binding of 2,4-dinitrophenyl compounds and other small molecules to a crystalline λ-type Bence–Jones dimer. *Biochemistry* **13**, 3816–3827.

Edmundson, A. B., Ely, K. R., and Herron, J. N. (1984). A search for site-filling ligands in the Mcg Bence–Jones dimer: Crystal-binding studies of fluorescent compounds. *Mol. Immunol.* **21**, 561–576.

Edmundson, A. B., Ely, K. R., Herron, J. N., and Cheson, B. D. (1987). The binding of opioid peptides to the Mcg light chain dimer: Flexible keys and adjustable locks. *Mol. Immunol.* **24**, 915–935.

Edmundson, A. B., Ely, K. R., He, X.-M., and Herron, J. N. (1989a). Co-crystallization of an immunoglobulin light chain dimer with bis(dinitrophenyl)lysine: Tandem binding of two ligands, one with and one without accompanying conformational changes in the protein. *Mol. Immunol.* **26**, 207–220.

Edmundson, A. B., Herron, J. N., Ely, K. R., He, X.-M., Harris, D. L., and Voss, E. W., Jr. (1989b). Synthetic site-directed ligands. *Philos. Trans. R. Soc. London, Ser. B* **323**, 495–509.

Ely, K. R., Herron, J. N., and Edmundson, A. B. (1983). Three-dimensional structure of the orthorhombic form of the Mcg Bence–Jones dimer. *Prog. Immunol.* **5**, 61–66.

Ely, K. R., Herron, J. N., Harker, M., and Edmundson, A. B. (1989). Three-dimensional structure of a light chain dimer crystallized in water; Conformational flexibility of a molecule in two crystal forms. *Mol. Biol.* (in press).

Fitzgerald, P. M. D. (1988). Merlot, an integrated package of computer programs for the determination of crystal structures by molecular replacement. *J. Appl. Crystallogr.* **21**, 273–278.

Geysen, H. M., Rodda, S. J., Mason, T. J., Tribbick, G., and Schoofs, P. G. (1987). Strategies for epitope analysis using peptide synthesis. *J. Immunol. Methods* **102**, 259–274.

Gibson, A. L., Herron, J. N., Ballard, D. W., Voss, E. W., Jr., He, X.-M., Patrick, V. A., and Edmundson, A. B. (1985). Crystallographic characterization of the Fab fragment of a monoclonal anti-ss-DNA antibody. *Mol. Immunol.* **22**, 499–502.

Gibson, A. L., Herron, J. N., He, X.-M., Patrick, V. A., Mason, M. L., Lin, J.-N., Kranz, D. M., Voss, E. W., Jr., and Edmundson, A. B. (1988). Differences in crystal properties and ligand affinities of an antifluorescyl Fab (4-4-20) in two solvent systems. *Proteins* **3**, 155–160.

Hendrickson, W. A. (1985). Stereochemically restrained refinement of macromolecular structures. *Methods Enzymol.* **115**, 252–270.

Hendrickson, W. A., and Konnert, J. H. (1981). Stereochemically restrained crystallographic least-squares refinement of macromolecule structures. *In* "Biomolecular Structure, Conformation, Function and Evolution" (R. Srinivasan, E. Subramanian, and N. Yathindra, eds.), Vol. 1, pp. 43–57. Pergamon, New York.

Herron, J. N., Kranz, D. M., Jameson, D. M., and Voss, E. W., Jr. (1986). Thermodynamic properties of ligand binding by monoclonal anti-fluorescyl antibodies. *Biochemistry* **25**, 4602–4609.

Herron, J. N., He, X.-M., Gibson, A. L., Voss, E. W., Jr., and Edmundson, A. B. (1987). Crystal structure of a murine Fab fragment with specificity for single-stranded DNA. *Fed. Proc.* **46**, 1626.

Herron, J. N., He, X.-M., Mason, M. L., Voss, E. W., Jr., and Edmundson, A. B. (1989). Three-dimensional structure of a fluorescein-Fab complex crystallized in 2-methyl-2,4-pentanediol. *Proteins* **5**, 271–280.

Herzberg, O., and Sussman, J. L. (1983). Protein model building by the use of a constrained-restrained least-squares procedure. *J. Appl. Crystallogr.* **16**, 144–150.

Jones, T. A. (1978). A graphics model building and refinement system for macromolecules. *J. Appl. Crystallogr.* **11**, 268–272.

Kabat, E. A., Wu, T. T., and Bilofsky, H. (1977). Unusual distributions of amino acids in complementarity-determining (hypervariable) segments of heavy and light chains of immunoglobulins and their possible roles in specificity of antibody-combining sites. *J. Biol. Chem.* **252**, 6609–6616.

Kranz, D. M., Herron, J. N., and Voss, E. W., Jr. (1982). Ligand binding by monoclonal anti-fluorescyl antibodies. *J. Biol. Chem.* **257**, 6987–6995.

Lattman, E. E., and Love, W. E. (1970). A rotational search procedure for detecting a known molecule in a crystal. *Acta Crystallogr.*, **B26**, 1854–1857.

Pflugrath, J. W., Saper, M. A., and Quiocho, F. A. (1984). New generation graphics system for molecular modeling. *In* "Methods and Applications in Crystallographic Computing" (S. Hall and T. Ashiaka, eds.), pp. 404–407. Clarendon Press, Oxford.

Rossmann, M. G., and Blow, D. M. (1962). Detection of subunits within the crystallographic asymmetric unit. *Acta Crystallogr.* **15**, 24–31.

Segal, D. M., Padlan, E. A., Cohen, G. H., Rudikoff, S., Potter, M., and Davies, D. R. (1974). The three-dimensional structure of a phosphorylcholine-binding mouse immunoglobulin Fab and the nature of the antigen-binding site. *Proc. Natl. Acad. Sci. USA* **71**, 4298–4302.

Sheriff, S., Silverton, E. W., Padlan, E. A., Cohen, G. H., Smith-Gill, S. J., Finzel, B. C., and Davies, D. R. (1987). Three-dimensional structure of an antibody–antigen complex. *Proc. Natl. Acad. Sci. USA* **84,** 8075–8079.

Smith, R. G., Ballard, D. W., Blier, P. R., Pace, P. E., Bothwell, A. L. M., Herron, J. N., Edmundson, A. B., and Voss, E. W., Jr. (1988). Structural features of a murine monoclonal anti-ssDNA autoantibody. *J. Indian Inst. Sci.* (in press).

Tribbick, G., Edmundson, A. B., Mason, T. J., and Geysen, H. M. (1989). Similar binding properties of peptide ligands for a human immunoglobulin and its light chain dimer. *Mol. Immunol.* (in press).

10 Neutralizing Rhinoviruses with Antiviral Agents That Inhibit Attachment and Uncoating

Michael G. Rossmann

Department of Biological Sciences, Purdue University, West Lafayette, Indiana 47907

I. Introduction

Unlike plant viruses, most animal and insect viruses attach to specific cellular receptors partially embedded in the membrane of selected host cells. Receptors are frequently glycoproteins, which may serve a specific cellular function (see, e.g., Co *et al.*, 1985). Viruses have adapted themselves to utilize these cell components for their own life cycles, mostly to the disadvantage of the host. Viral host and tissue specificity may be largely due to the recognition of the cellular receptor by virions. The number of studies that identify the attachment site on a viral surface is very limited; Weis *et al.* (1988) have determined the binding site of sialyllactose in a depression on the hemagglutinin spike of influenza virus and the "canyon" has now been largely confirmed as the receptor attachment site in rhinoviruses (Colonno *et al.*, 1988; Pevear *et al.*, 1989). Here I review the latter finding and generalize as to the requirements of such sites on animal viruses.

Picornaviruses can be divided into at least six receptor families based on results of competition binding and receptor antibody studies (Colonno, 1986; Crowell *et al.*, 1987). Rhinoviruses can be grouped into two distinct receptor groups: the major group, to which belong at least 78 serotypes of HRV (including human rhinovirus 14 or HRV14) and some coxsackie A serotypes, and the minor group, to which belong at least 10 serotypes of HRV (including human rhinovirus 1A or HRV1A) (Abraham and Colonno, 1984; Colonno *et al.*, 1986).

The structure of HRV14 (Rossmann *et al.*, 1985) showed that the polypeptide topology and assembly of the three major proteins (VP1, VP2, VP3) in the viral capsid were remarkably similar to one another and to other plant and animal spherical RNA viruses. Each capsid protein is folded into an eight-stranded antiparallel β-barrel topology (Fig. 1) which can accept numerous insertions at the corners between strands. These insertions mostly decorate the viral exterior and form "puffs" and loops

Use of X-Ray Crystallography in the Design of Antiviral Agents

SBMV

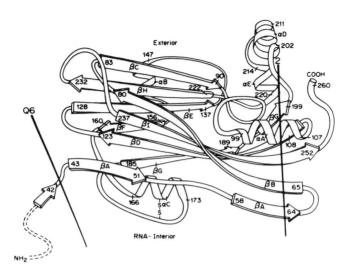

Figure 1. Diagrammatic drawing showing the polypeptide fold of southern bean mosaic virus (SBMV) into an eight-stranded antiparallel β-barrel. Other viral capsid proteins have homologous folds. The nomenclature of the secondary structural elements is given. There are four excursions of the polypeptide chains towards the wedge-shaped end. Each excursion makes a sharp bend or "corner." The most exterior (top left) corner is formed between the β-sheets βB and βC. (Adapted from a drawing by Jane Richardson.) (From Rossmann *et al.*, 1983.)

which are hypervariable and have been shown to be the site of binding of neutralizing antibodies in picornaviruses (Rossmann *et al.*, 1985). The viruses are thus differentiated primarily by variations of surface features. The surface of HRV14 contains a series of remarkably deep crevices or "canyons," unlike anything observed in plant virus structures. Similar observations were subsequently made for Mengo virus (Luo *et al.*, 1987) (in which the depression is in the shape of a pit) and other picornaviruses.

It was hypothesized (Rossmann *et al.*, 1985) that the canyon (one around each fivefold vertex) (Fig. 2) in HRV14 was the site of receptor attachment, largely inaccessible to the broad antigen binding region seen in antibodies (Fig. 3). Thus residues in the lining of the canyon, which should be resistant to accepting mutations that might inhibit receptor attachment, would avoid presenting an unchanging target to neutralizing antibodies. Indeed, the neutralizing immunogenic sites that had been mapped by escape mutants were not in the canyon but on the most exposed and variable parts of the virion both in HRV14 (Rossmann *et al.*,

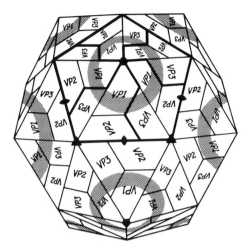

Figure 2. Diagrammatic representation of an icosahedral capsid in picornaviruses. The thickly outlined VP2, VP3, VP1 unit corresponds to the 6S (VP0, VP3, VP1) protomer. The thickly outlined 15-mer cap corresponds to the 14S pentamer observed in assembly experiments. Shown also is the position of the canyon around each fivefold axis. (Adapted from Rossmann *et al.*, 1985.)

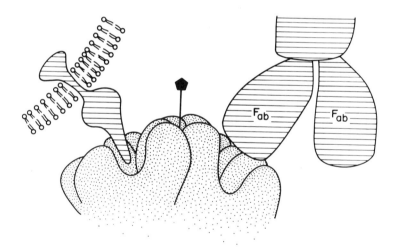

Figure 3. The presence of depressions on the picornavirus surface suggests a strategy for the evasion of immune surveillance. The dimensions of the putative receptor binding sites (the canyon in HRV14, the pit in Mengo virus) sterically hinder an antibody's (top right) recognition of residues at the base of the site, while still allowing recognition and binding by a smaller cellular receptor (top left). This would allow for receptor specificity while at the same time permitting evolution of new serotypes by mutating residues about the rim of the canyon or pit. (From Luo *et al.*, 1987. Copyright 1987 by the AAAS.)

1985; Sherry and Rueckert, 1985; Sherry *et al.*, 1986) and in poliovirus (Hogle *et al.*, 1985; Page *et al.*, 1988). The canyon hypothesis suggests that one strategy for viruses to escape the host's immune surveillance is to protect the receptor attachment site in a surface depression (Fig. 3). Similar depressions related to host cell attachment had been found earlier on the surface of the hemagglutinin spike of influenza virus (Wilson *et al.*, 1981) and at the end of the influenza neuraminidase spike (Varghese *et al.*, 1983). Both sites are now known to bind sialic acid (Weis *et al.*, 1988; Varghese *et al.*, 1983), the terminal carbohydrate moiety of surface glycoproteins on erythrocytes.

The initial evidence supporting the canyon hypothesis was circumstantial:

1. It gave a ready explanation for conservation of viral receptor attachment sites in the face of hypervariable surface properties.
2. It rationalized the remarkable contrast between vertebrate virus structures and plant [e.g., tomato bushy stunt virus (Harrison *et al.*, 1978), southern bean mosaic virus (Abad-Zapatero *et al.*, 1980), satellite tobacco necrosis virus (Liljas *et al.*, 1982)] or insect viruses [black beetle virus (Hosur *et al.*, 1987)]. Namely, animal viruses had depressions while viruses whose hosts did not have immune systems had smooth surfaces or protrusions on their surfaces.
3. Idiotypic antiserum raised against an anti-receptor monoclonal antibody failed to neutralize the virion (Tomassini and Colonno, 1986).
4. Residues in the canyon lining were shown to be more conserved than the other surface residues (Rossmann and Palmenberg, 1988) on comparing different rhinovirus and enterovirus sequences. Similarly, residues lining the deep pit on the surface of Mengo virus were shown to be more conserved than other surfaces on comparing cardioviruses (Rossmann and Palmenberg, 1988).
5. The folding motif of viral capsid proteins is similar to that of concanavalin A (Argos *et al.*, 1980), which can compete with polioviruses for HeLa cell receptors (Lonberg-Holm, 1975). This might be due to similar binding properties of the common fold.
6. The principal amino acid differences between neurovirulent and attenuated polio strains of the same serotype occur on the north rim of the canyon (βB–βC loop) suggesting the differences could in part be due to alteration in receptor specificity. However, that this is not the complete basis of attenuation is seen from the identification of base changes in the noncoding regions of polio RNA that affect virulence (Minor *et al.*, 1986). Nevertheless, some support for this is given by the sequence of a mouse-adapted polio strain (La Monica *et*

al., 1986) and the properties of a polio chimera in which the βB–βC loop has been exchanged between serotypes (Murray *et al.*, 1988; Martin *et al.*, 1988).

Direct verification of the canyon hypothesis was to follow. It is however, necessary to describe first some structural features of icosahedral viruses.

II. Canyon Structure

Picornaviruses and many isometric RNA plant and insect viruses have a diameter of approximately 300 Å with a roughly 50-Å-thick protein coat. The molecular weight of the particles varies from 6.5 to 8.5×10^6 with 20–33% RNA by weight. There are 180 surface protein subunits or domains in the structure, which either are all covalently identical (as in many plant and insect viruses) or consist of three different types of polypeptide folded the same way. Their organization into the capsid (Fig. 2) is always essentially the same and they exhibit quasi-$T = 3$ (where all three types are identical) or pseudo-$T = 3$, as in picornaviruses or the plant comoviruses (Chen *et al.*, 1989), symmetry. In pseudo-$T = 3$ viruses, the three types of domains (VP1, VP2, and VP3) are translated from the RNA genome as a polyprotein and, in some cases, posttranslationally processed into separate protein subunits. The original protomer, before processing, is outlined in Fig. 2, as is the pentameric unit which associates on processing of the protomer prior to assembly. The RNA, which is a single oligonucleotide, in general lacks icosahedral symmetry, but sometimes the protein coat can impose some symmetry on the RNA secondary structure (Chen *et al.*, 1989).

The immunodominant VP1 clusters about the fivefold axes in picornaviruses, while VP2 and VP3 cluster around the pseudo-sixfold axes, corresponding to the icosahedral threefold axes. The canyon is formed roughly at the junction of VP1 (in the north) and VP3 (in the south). The "FMDV loop" of VP1 (βG–βH) forms much of the floor of the canyon and, together with the carboxy termini of VP1 and VP3, also forms the south rim of the canyon.

In studies of HRV14 (Sherry and Rueckert, 1985; Sherry *et al.*, 1986) and poliovirus (Hogle *et al.*, 1985), a series of neutralizing monoclonal antibodies were isolated and used to screen for mutants that were infectious in the presence of the antibodies. These mutants were then sequenced and found to be mostly single-site mutations. Furthermore, both in HRV14 and poliovirus, the monoclonals could be organized into four groups related to different sites on the virus. The escape mutants did not all cluster on the same small peptide. They mapped onto protrusions on either side of the canyon. All the escape mutants corresponded to highly exposed residues. The mutants clustered into regions on the virus corresponding to the four epitopes suggested by the monoclonal groupings. Furthermore, sequence comparisons with other rhinoviruses showed that these neutralizing immunogenic sites (NIm) correspond to the most variable parts of the virus surface. Thus, the deep canyon, lined by conserved residues, is surrounded by hypervariable regions that can bind neutralizing antibodies.

Residues that line the canyon (HRV14 and HRV1A) or pit (Mengo) were selected (Table I) as those residues that have atoms exposed to the outside and are at a viral radius of less than 138 Å, but excluding residues in dimples at the fivefold or twofold axes. The surface residues within one icosahedral asymmetric unit, projected onto a plane perpendicular to an icosahedral twofold axis, are shown in Fig. 4 for HRV14, HRV1A, and Mengo virus. The canyon is identified by a thick contour that demarcates surface atoms at less than 138 Å radius.

Conservation of the canyon and pit residues in comparison with the other surface residues is shown in Table II. The identified residues have been compared on the basis of an alignment of picornaviruses (Palmenberg, 1989) using a scheme in which residues are weighted more in proportion to their distance below the arbitrarily selected 138 Å radius that defines the rim of the canyon in HRV14 or the pit in Mengo virus.

III. Alterations of the Canyon Shape That Inhibit Attachment

Colonno *et al.* (1988) have selected a limited number of residues (K1103, P1155, H1220, S1223 in VP1) that line the canyon (Fig. 4) and specifically mutated these. They showed that in most cases there was a change in the ability of the virus to bind to cell membranes (Fig. 5), the appearance of plaques, and the virus yield. In one case (P155→G) there was an increased affinity of the virus to membranes; in two cases (K103→R, K103→I) there was little change in the binding affinity although plaque shape and virus yield were affected; and in all other cases there was a decreased affinity of virus to membranes. Interestingly, changing H1220 had the greatest effect on the properties of virus attachment. A control experiment was also performed in which it was shown that the mutation I1273→N had no marked effect. While these experiments do not map the complete receptor attachment site, they do show some of the viral residues that participate in the cell attachment process.

Table I. Residues Lining the Protected Receptor Attachment Sites of HRV14, HRV1A, and Mengo Virus

HRV14[a]		HRV1A[b]		Mengo virus[c]	
H1078 βB	K1240 βI	H1078 βB	D1201 FMDV loop	R1085 loop 1	Q3081 βC
V1079 βB	V1278 C terminus	I1079 βB	V1212 FMDV loop	R1088 loop 1	T3083 αA$_0$
T1080 βB	F3086 βC-αA	R1081 βB	D1216 FMDV loop	F1104 loop 2	S3085 αA$_0$
D1101 βC (boxed)	K3093 βC-αA	I1082 βB	T1219 FMDV loop	G1149 βE-puff	C3086 αA$_0$
W1102 βC (boxed)	G3135 βE-αB	F1096 βB-βC	R1223 βH	T1150 βE-puff	S3087 αA$_0$
*K1103 βC	A3136 βE-αB	T1097 βB-βC	E1227 βH	P1151 βE-puff	C3088 αA$_0$
*P1155 βE-αB	D3177 βG-βH	K1098 βB-βC	K1230 βH-βI	K1206 FMDV loop	R3099 αA
G1156 βE-αB	P3178 βG-βH	W1099 βC	A1277 C terminus	R1207 FMDV loop	T3175 βG-βH
N1159 βE-αB	D3179 βG-βH	K1100 βC	R2216 βG-βH	F1208 FMDV loop	V3183 βG-βH
W1163 βE-αB	T3180 βG-βH	E1105 αA$_0$	T3090 αA	I1216 FMDV loop	W3186 βH
D1164 αB	S3183 βG-βH	M1106 αA$_0$	S3091 αA	P1218 FMDV loop	P3225 C terminus
D1165 αB	G3185 βH	G1151 βE-αB	T3092 αA	N1219 FMDV loop	A3226 C terminus
Y1166 αB	Q3226 C terminus	A1152 βE-αB	D3180 βG-βH	F1256 C terminus	P3227 C terminus
I1215 FMDV loop (boxed)	T3227 C terminus	P1153 βE-αB	N3181 βG-βH	N3056 βB knob	W3228 C terminus
V1217 FMDV loop (boxed)	S3229 C terminus	I1154 βE-αB	K3182 βG-βH	A3078 βC	S3229 C terminus
*H1220 FMDV loop	Q3230 C terminus	K1157 βE-αB	M3185 βH	V3079 βC	
*S1223 βH	A3233 C terminus	N1159 βE-αB	D3228 C terminus		
		D1160 βE-αB	L3229 C terminus		
		F1161 αB	I3231 C terminus		

[a] A probe of 3.5 Å radius was used to select the residues. Residues which undergo conformational change are boxed. Residues which have mutated to cause a change in viral attachment properties are marked with an asterisk (*). Residues are arranged in sequence and identified with their secondary structural element. The first digit of the residue number identifies the VP (e.g., residue 1078 is the 78th amino acid in VP1).

[b] A probe of 2.5 Å radius was used to select the residues. Residues are arranged in sequence and identified with their secondary structural element.

[c] A probe of 3.5 Å radius was used to select the residues. Residues are arranged in sequence and identified with their secondary structural element.

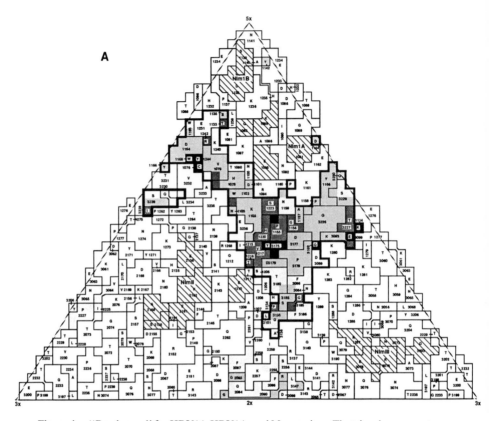

Figure 4. "Road maps" for HRV14, HRV1A, and Mengo virus. The triangle represents one icosahedral asymmetric unit projected onto a plane perpendicular to an icosahedral twofold axis. Shown are the exposed surface residues mapped onto a 2-Å-square grid. A thick black contour at a radius of 138 Å outlines the putative receptor attachment site. The depth of residues below the canyon rim is shown by progressively darker shading. (A) HRV14 showing the escape mutant sites (hatched) to neutralizing residues as well as residues in the canyon that have been mutated, causing altered cell binding properties (probe radius 2.5 Å), (B) HRV1A (probe radius 2.5 Å) and (C) Mengo virus (probe radius 3.5 Å). The first digit identifies the viral protein, while the second, third, and fourth digits give the amino acid sequence number for each surface residue. [Figure (C) from Rossmann and Palmenberg (1988).] *(Figure continues)*

A series of antirhinovirus drugs (WIN compounds) have been under development by the Sterling Research Group (Diana *et al.,* 1977a,b, 1985, 1989; McKinlay and Steinberg, 1986; Otto *et al.,* 1985; Fox *et al.,* 1986). Similar compounds have also been under investigation elsewhere. They probably all bind to the same site on the virion as they are all affected to some extent by escape mutants selected against a specific drug. These

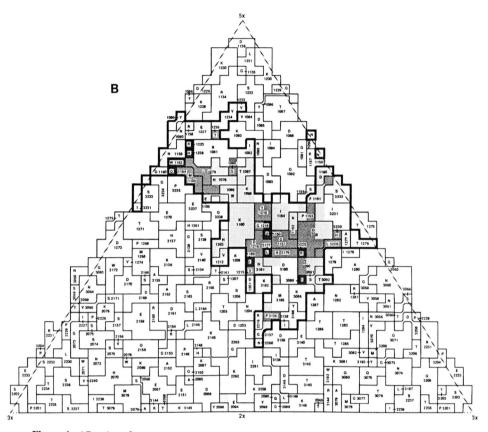

Figure 4   *(Continued)*

compounds were shown to inhibit viral uncoating after host cell membrane penetration (Caliguiri *et al.*, 1980; McSharry *et al.*, 1979). They have been studied structurally (Smith *et al.*, 1986; Badger *et al.*, 1988) and shown to bind into an interior hydrophobic pocket within VP1 underneath the canyon floor. In HRV14 they also cause sizable conformational changes of up to 4 Å in main chain and 7.5 Å in side chain positions. These changes are triggered by a displacement of M1221 when the drug enters the pocket and are mostly independent of which WIN compound is bound. The largest changes are in the FMDV loop where it crosses the canyon floor with H1220 experiencing the biggest conformational change. The structural changes, induced by drug binding to HRV14 into the internal pocket, cause conformational changes to the virus exterior that are largely confined to the canyon (Fig. 6). It was, therefore, speculated that drug binding would

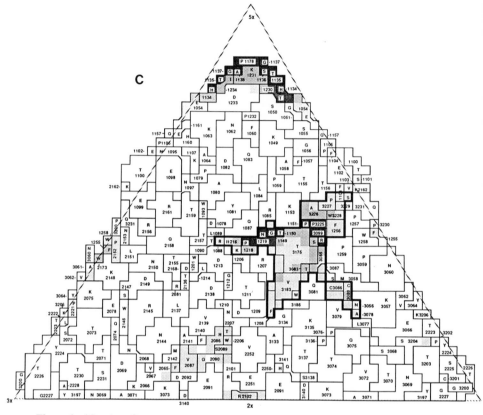

Figure 4 (*Continued*)

also inhibit attachment. This was, indeed, demonstrated in a series of experiments measuring labeled virus attachment to membranes (Pevear *et al.*, 1989) (Fig. 7). The inhibiting concentrations to reduce viral attachment (measured with labeled virus on isolated membranes) to one-half of wild type were of the same order of magnitude as the minimal inhibitory concentrations to reduce plaque counts to one-half of wild type. Thus, in HRV14 these drugs neutralize infectivity both by inhibition of attachment and by inhibition of uncoating. Furthermore, attachment is presumably prevented by the conformational changes on the canyon floor induced by WIN drug binding.

A study of HRV1A showed that the wild-type virus had a conformation closely similar to that of HRV14 when complexed with WIN drugs. When WIN compounds are bound to HRV1A no further conformational changes

Table II. Conservation of Canyon Residues in Rhinoviruses and Enteroviruses

Type	Comparison						
	1	2	3	4	5	6	7
Surface: noncanyon							
1. Polio 1 Mahoney	0.0	0.94	1.54	1.46	1.33	1.33	1.60
2. Polio 3 Sabin		0.0	1.46	1.43	1.51	1.31	1.72
3. Coxsackie B3			0.0	1.78	1.45	1.73	1.79
4. HRV14				0.0	1.28	1.48	1.82
5. HRV89					0.0	1.25	1.21
6. HRV2						0.0	1.18
7. HRV1A[a]							0.0
Surface: canyon	1	2	3	4	5	6	7
1. Polio 1 Mahoney	0.0	0.52	1.01	1.16	0.86	1.10	0.70
2. Polio 3 Sabin		0.0	0.94	1.15	0.81	1.09	0.82
3. Coxsackie B3			0.0	1.00	1.16	1.05	1.03
4. HRV14				0.0	0.85	1.03	0.52
5. HRV89					0.0	0.83	0.51
6. HRV2						0.0	1.15
7. HRV1A[a]							0.0
Internal	1	2	3	4	5	6	7
1. Polio 1 Mahoney	0.0	0.15	0.59	0.62	0.69	0.68	0.91
2. Polio 3 Sabin		0.0	0.61	0.65	0.70	0.70	0.94
3. Coxsackie B3			0.0	0.63	0.64	0.63	0.82
4. HRV14				0.0	0.61	0.62	0.87
5. HRV89					0.0	0.27	0.31
6. HRV2						0.0	0.18
7. HRV1A[a]							0.0

[a] Comparisons with HRV1A apply only to VP1.

occur (Kim *et al.,* 1989a). HRV2, which like HRV1A belongs to the minor rhinovirus receptor group, cannot be inhibited in binding to membranes by the presence of WIN drugs. Tentative results for HRV1A likewise show no inhibition of attachment by WIN compounds (D. C. Pevear, personal communication). These observations strongly correlate conformational changes in the canyon with inhibition of viral attachment (Table III).

Cardioviruses, unlike rhinoviruses, are moderately stable at low pH. On lowering the pH of Mengo virus crystals from 7.3 to 4.6, large con-

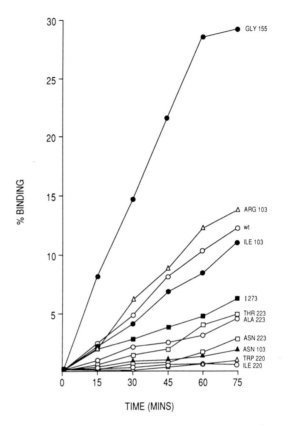

Figure 5.   Binding kinetics of HRV14 mutants. Virus labeled with ^{35}S-labeled Met and representing each of 10 mutant and wild-type virus were used in membrane binding assays. After incubation at 25°C for the times indicated, membranes were pelleted and the percentage of virus specifically associated with the cell pellet was determined. (From Colonno *et al.*, 1988.)

formational changes are observed in residues lining the pit (Fig. 8), the putative receptor attachment site. The conformational changes are extremely localized (Fig. 8), immediately suggesting a functionally important phenomenon (Huber and Bennett, 1983). Similar changes can also be produced by increasing the halide ion concentration to 500 mM at pH 6.2 (Kim *et al.*, 1989b). These conformational changes are observed in the presence of 0.1 M PO$_4$ buffer, in which the crystals were grown. It was found that the virus does not bind to membranes nor is infectious at 0.1 M PO$_4^{-2}$ ion concentration at neutral pH. However, if the pH is lowered or the halide ion concentration increased the virus does attach to membranes and is infectious (Kim *et al.*, 1989b). The infectious

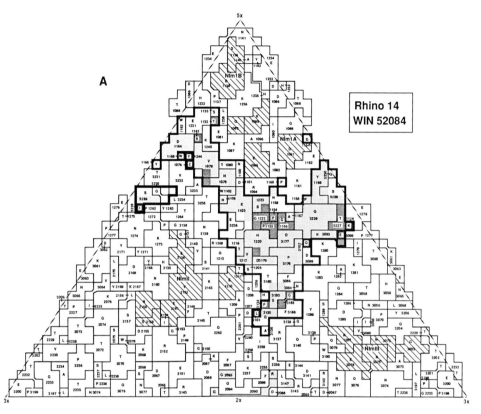

Figure 6. "Road maps" (see Fig. 4 for details) for HRV14 showing residues (shaded) that undergo conformational change on binding WIN compounds. (A) HRV14 complexed with a WIN compound (probe radius 2.5 Å) and (B) difference between (A) and native HRV14 (Fig. 4A). Note that the conformational changes are almost entirely within the canyon region. These changes inhibit cell attachment. (*Figure continues*)

form of the virus loses a phosphate ion bound in a central region close to the pit and is associated with the conformational changes. This again demonstrates that the attachment properties of the virus to cells are altered when residues lining deep surface depressions on a virion undergo a conformational change.

IV. Antiviral Agents That Interfere with Capsid Function

Disoxaril (WIN 51711) is representative of a class of compounds (Diana *et al.*, 1977b, 1989; McSharry *et al.*, 1979; Ninomiya *et al.*, 1984; Lonberg-Holm *et al.*, 1985) that inhibit picornavirus replication in tissue culture

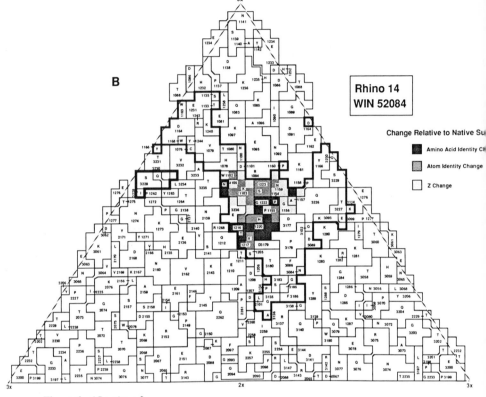

B

Rhino 14
WIN 52084

Change Relative to Native Su

■ Amino Acid Identity Ch
▨ Atom Identity Change
□ Z Change

Figure 6 (*Continued*)

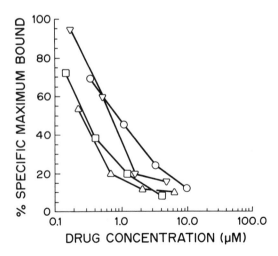

Figure 7. Dose–response curves of adsorption of HRV14 in the presence of four WIN compunds (△,▽,□,○). (From Pevear *et al.*, 1989.)

Table III.   Correlation of Conformational Change in Canyon with
Inhibition of Attachment

	Receptor group	Conformational change on WIN binding	Inhibition of attachment on WIN binding
HRV14	Major	Yes	Yes
HRV1A	Minor	No	No
HRV2	Minor	?	No

(Otto *et al.*, 1985) and in animal models of human enterovirus disease (McKinlay and Steinberg, 1986; McKinlay *et al.*, 1986). These compounds (Fig. 9) bind to the hydrophobic interior of VP1 (Figs. 10 and 11) causing significant conformational changes in HRV14 (Smith *et al.*, 1986), but not in HRV1A (Kim *et al.*, 1989a). In all cases they stabilize the coat protein, causing loss of flexibility, and thus prevent uncoating. As mentioned above, when the compounds cause conformational changes in the floor of the canyon, there is also inhibition of attachment.

The presence of a potential hydrophobic pocket in the typical viral capsid is probably not fortuitous. Some degree of flexibility may be required to accommodate the assembly and disassembly process. This can be provided by the loosely packed internal hydrophobic pocket of the standard viral capsid protein. If this pocket is filled by a molecule of appropriate size and physical characteristics (e.g., WIN 51711), then conformational changes are induced and the protein becomes rigid and fails to perform its normal assembly and disassembly functions. Indeed, the requirement for this function in a protein that can also assemble into an icosahedral particle may, in part, be the cause for the retention of the same protein fold in the evolution of so many viral capsid structures. The shape and size of the hydrophobic pocket varies from one virus to another according to the particular amino acids that line the pocket. The pocket is not necessarily equally accessible in different viruses. For instance, Mengo virus (Luo *et al.*, 1987) has a hydrophobic interior to VP1 but it is not readily accessible and WIN 51711 is not active in this virus. Similarly, the WIN compounds penetrate only into VP1, not into the homologously folded VP2 or VP3 of HRV14. The design of a suitable antiviral agent that inhibits uncoating will thus depend on the knowledge of the precise structure of the targeted virus capsid protein. The agent must be sufficiently flexible to enter the pocket through an available pore on the capsid's exterior, sufficiently hydrophobic to be retained by the pocket, and of suitable size to fit into the pocket. Drug design can further be aided by experiments with well-chosen compounds.

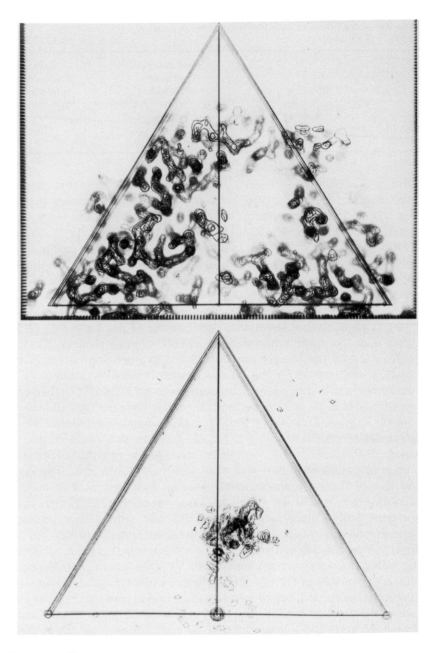

Figure 8. Difference map (below) showing the conformational changes between Mengo virus at pH 4.6 and pH 7.3. These same sections for the native virus are shown above. Note how the conformational changes line the pit, the putative receptor attachment site.

Compound Bound		MIC (µM)

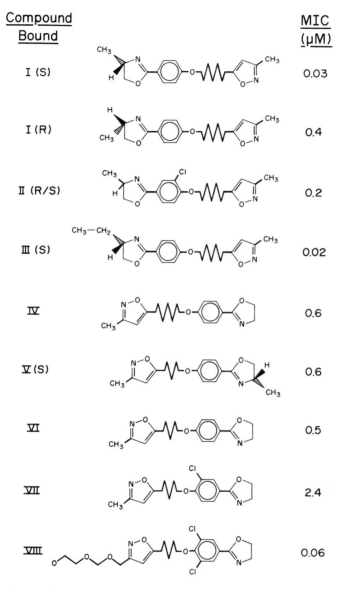

	MIC (µM)
I (S)	0.03
I (R)	0.4
II (R/S)	0.2
III (S)	0.02
IV	0.6
V (S)	0.6
VI	0.5
VII	2.4
VIII	0.06

Figure 9. Some of the compounds studied crystallographically when complexed with HRV14 and HRV1A. The orientation of the compounds shows diagrammatically the orientation of the compound in the VP1 pocket. The left-hand side is closer to the "pore" in the floor of the canyon by which the compounds enter the pocket. Shown also are the compounds' *in vitro* activity against HRV14 measured in terms of the concentration (µM) required to reduce the plaque counts by a factor of two (minimal inhibitory concentration). Compound IV is also known as WIN 51711 or disoxaril.

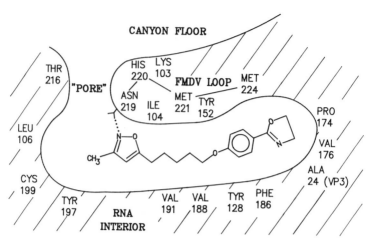

Figure 10. Diagrammatic representation of compound VI bound in the WIN pocket. (From Badger *et al.*, 1988.)

The eight-stranded antiparallel β-barrel motif, with a topology as in viral capsid proteins, has not been found in other classes of proteins (Richardson, 1979). Thus compounds like WIN 51711, which have been particularly adapted to bind with high affinity to a specific viral capsid protein, are unlikely to bind with the same affinity to other types of proteins with different folds. This is a possible explanation for the limited toxicity seen with compounds of this class. In contrast, antiviral agents targeted at, for instance, viral proteases or polymerases have to have greater specificity in order not to interfere with essential metabolic processes that are dependent on proteins with similar function and, therefore, probably also with a similar fold.

The greatest conservation between different picornaviruses occurs in the internal residues, whereas the greatest variability occurs on the antigenic surface (Rossmann and Palmenberg, 1988). The high surface variability (Rossmann *et al.*, 1985) accounts for the large number of serologically distinct viruses, which nevertheless bind to only a few different receptors (Colonno *et al.*, 1986; Mapoles *et al.*, 1985). Thus, antiviral agents such as WIN 51711 have a relatively large range covering not only most rhinoviruses but also many enteroviruses as well.

The hydrophobic character of compounds such as WIN 51711 is essential for their binding to the hydrophobic pocket in VP1 of picornaviruses. However, their hydrophobic character is likely to allow them to be ab-

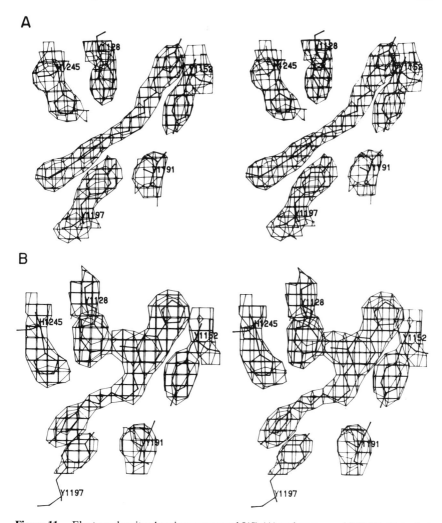

Figure 11. Electron density showing compound I(S) (A) and compound VII (B). Note that the chlorine atom of compound VII readily defines the compound's orientation, although the larger volume of density for the phenyloxazoline group compared to the isoxazole group is also clear in both cases. The two compounds differ in the length of their aliphatic chains. (From Badger *et al.,* 1988.)

sorbed and transported across viral membranes. They could also be ad-sorbed on the capsid during assembly. Thus, variations of these com-pounds might also inhibit uncoating of enveloped viruses such as Sindbis virus or human immunodeficiency virus (Rossmann, 1988).

V. Conclusion

When the structure of HRV14 was solved in 1985 it was hypothesized on the basis of the requirements for conservation, and by contrast to plant virus surfaces, that the canyon was the likely site for receptor attachment. This has now been largely confirmed experimentally. Similar depressions associated with receptor attachment occur on the hemagglutinin and neuraminidase spikes of influenza virus. The receptor attachment site on the major surface antigen, gp120, of human immunodeficiency virus may also be hidden from attack by neutralizing antibodies (Matthews *et al.,* 1987). Thus, the strategy of hiding receptor sites in viral surface depressions may be a frequent occurrence. It follows that vaccines aimed at the more conserved regions of surface antigens (McCray and Werner, 1987) are less likely to be effective (Palmenberg, 1987) since these sites may be inaccessible to neutralizing antibodies.

The structure of HRV14 has also afforded the study of antiviral agents that bind to the viral capsid and inhibit uncoating and, in some cases, attachment. These are the first structural studies of an antiviral agent complexed with a virion. They have stimulated numerous other investigations, many of which are described in other chapters.

Ackowledgments

The "road maps" shown in Figs. 4 and 6 were computed by a program written by myself but were then transcribed by Ann Palmenberg, using a Macintosh computer, into the graphical representations shown here. I am greatly indebted to Ann for her interest, enthusiasm, and dedicated attention to detail. The observations made here are based on numerous results in many laboratories. In particular, I would like to express my appreciation to Edward Arnold, John E. Johnson, Sangsoo Kim, S. Krishnaswamy, Ming Luo, Tom Smith, and Gert Vriend at Purdue University; Beverly Heinz, Roland Rueckert, and Barbara Sherry at the University of Wisconsin; and Guy Diana, Frank Dutko, Mark A. McKinlay, and Dan Pevear at the Sterling Research Group. I am most grateful to Helene Prongay and Sharon Wilder for preparing the manuscript. The work was supported by grants from the National Science Foundation, National Institutes of Health, Sterling Research Group, and Markey Foundation to MGR.

References

Abad-Zapatero, C., Abdel-Meguid, S. S., Johnson, J. E., Leslie, A. G. W., Rayment, I., Rossmann, M. G., Suck, D., and Tsukihara, T. (1980). Structure of southern bean mosaic virus at 2.8 Å resolution. *Nature (London)* **286,** 33–39.
Abraham, G., and Colonno, R. J. (1984). Many rhinovirus serotypes share the same cellular receptor. *J. Virol.* **51,** 340–345.
Argos, P., Tsukihara, T., and Rossmann, M. G. (1980). A structural comparison of concanavalin A and tomato bushy stunt virus protein. *J. Mol. Evol.* **15,** 169–179.

Badger, J., Minor, I., Kremer, M. J., Oliveira, M. A., Smith, T. J., Griffith, J. P., Guerin, D. M. A., Krishnaswamy, S., Luo, M., Rossmann, M. G., McKinlay, M. A., Diana, G. D., Dutko, F. J., Fancher, M., Rueckert, R. R., and Heinz, B. A. (1988). Structural analysis of series of antiviral agents complexed with human rhinovirus 14. *Proc. Natl. Acad. Sci. USA* **85,** 3304–3308.

Caliguiri, L. A., McSharry, J. J., and Lawrence, G. W. (1980). Effect of arildone on modifications of poliovirus *in vitro. Virology* **105,** 86–93.

Chen, Z., Stauffacher, C., Li, Y., Schmidt, T., Bomu, W., Kamer, G., Shanks, M., Lomonossoff, G., and Johnson, J. E. (1989). Protein–RNA interactions in an icosahedral virus at 3.0 Å resolution. *Science* **245,** 154–159.

Co, M. S., Gaulton, G. N., Tominago, A., Homcy, C. J., Fields, B. N., and Greene, M. I. (1985). Structural similarities between the mammalian β-adrenergic and reovirus type 3 receptors. *Proc. Natl. Acad. Sci. USA* **82,** 5315–5318.

Colonno, R. J. (1986). Cell surface receptors for picornaviruses. *BioEssays* **5,** 270–274.

Colonno, R. J., Callahan, P. L., and Long, W. J. (1986). Isolation of a monoclonal antibody that blocks attachment of the major group of human rhinoviruses. *J. Virol.* **57,** 7–12.

Colonno, R. J., Condra, J. H., Mizutani, S., Callahan, P. L., Davies, M. E., and Murcko, M. A. (1988). Evidence for the direct involvement of the rhinovirus canyon in receptor binding. *Proc. Natl. Acad. Sci. USA* **85,** 5449–5453.

Crowell, R. L., Hsu, K. H. L., Schultz, M., and Landau, B. J. (1987). Cellular receptors in coxsackievirus infections. *In* "Positive Strand RNA Viruses" (M. A. Brinton and R. R. Rueckert, eds.), pp. 453–466. Alan R. Liss, New York.

Diana, G. D., Salvador, U. J., Carabateas, P. M., Williams, G. L., Collins, J. C., and Pancic, F. (1977a). Antiviral activity of some β-diketones. 2. Aryloxy alkyl diketones. *In vitro* activity against both RNA and DNA viruses. *J. Med. Chem.* **20,** 757–761.

Diana, G. D., Salvador, U. J., Zalay, E. S., Johnson, R. E., Collins, J. C., Johnson, D., Hinshaw, W. B., Lorenz, R. R., Thielking, W. H., and Pancic, F. (1977b). Antiviral activity of some β-diketones. 1. Aryl alkyl diketones. *In vitro* activity against both RNA and DNA viruses. *J. Med. Chem.* **20,** 750–756.

Diana, G. D., McKinlay, M. A., Otto, M. J., Akullian, V., and Oglesby, C. (1985). [[(4,5-Dihydro-2-oxazolyl)phenoxy]alkyl]isoxazoles. Inhibitors of picornavirus uncoating. *J. Med. Chem.* **28,** 1906–1910.

Diana, G. D., Pevear, D. C., Otto, M. J., McKinlay, M. A., Rossmann, M. G., Smith, T. J., and Badger, J. (1989). Inhibitors of viral uncoating. *Pharmacol. Therapeut.* **42,** 289–305.

Fox, M. P., Otto, M. J., and McKinlay, M. A. (1986). The prevention of rhinovirus and poliovirus uncoating by WIN 51711: A new antiviral drug. *Antimicrob. Agents Chemother.* **30,** 110–116.

Harrison, S. C., Olson, A. J., Schutt, C. E., Winkler, F. K., and Bricogne, G. (1978). Tomato bushy stunt virus at 2.9 Å resolution. *Nature (London)* **276,** 368–373.

Hogle, J. M., Chow, M., and Filman, D. J. (1985). Three-dimensional structure of poliovirus at 2.9 Å resolution. *Science* **229,** 1358–1365.

Hosur, M. V., Schmidt, T., Tucker, R. C., Johnson, J. E., Gallagher, T. M., Selling, B. H., and Rueckert, R. R. (1987). Structure of an insect virus at 3.0 Å resolution. *Proteins* **2,** 167–176.

Huber, R., and Bennett, W. S., Jr. (1983). Functional significance of flexibility in proteins. *Biopolymers* **22,** 261–279.

Kim, S., Smith, T. J., Chapman, M. S., Rossmann, M. G., Pevear, D. C., Dutko, F. J., Felock, P. J., Diana, G. D., and McKinlay, M. A. (1989a). The crystal structure of human rhinovirus serotype 1A (HRV1A). *J. Mol. Biol.* (in press).

Kim, S., Boege, U., Krishnaswamy, S., Minor, I., Smith, T. J., Luo, M., Scraba, D. G., and

Rossmann, M. G. (1989b). Conformational variability of a picornavirus capsid: pH-dependent structural changes of Mengo virus related to its host receptor attachment site and disassembly. Submitted for publication.

La Monica, N., Meriam, C., and Racaniello, V. R. (1986). Mapping of sequences required for mouse neurovirulence of poliovirus type 2 Lansing. *J. Virol.* **57**, 515–525.

Liljas, L., Unge, T., Jones, T. A., Fridborg, K., Lövgren, S., Skoglund, U., and Strandberg, B. (1982). Structure of satellite tobacco necrosis virus at 3.0 Å resolution. *J. Mol. Biol.* **159**, 93–108.

Lonberg-Holm, K. (1975). The effects of concanavalin A on the early events of infection by rhinovirus type 2 and poliovirus type 2. *J. Gen. Virol.* **28**, 313–327.

Lonberg-Holm, K., Gosser, L. G., and Kauer, J. C. (1975). Early alteration of poliovirus in infected cells and its specific inhibition. *J. Gen. Virol.* **27**, 329–342.

Luo, M., Vriend, G., Kamer, G., Minor, I., Arnold, E., Rossmann, M. G., Boege, U., Scraba, D. G., Duke, G. M., and Palmenberg, A. C. (1987). The atomic structure of Mengo virus at 3.0 Å resolution. *Science* **235**, 182–191.

McCray, J., and Werner, G. (1987). Different rhinovirus serotypes neutralized by antipeptide antibodies. *Nature (London)* **329**, 736–738.

McKinlay, M. A., and Steinberg, B. A. (1986). Oral efficacy of WIN 51711 in mice infected with human poliovirus. *Antimicrob. Agents Chemother.* **29**, 30–32.

McKinlay, M. A., Frank, J. A., and Steinberg, B. A. (1986). Use of WIN 51711 to prevent echovirus type 9-induced paralysis in suckling mice. *J. Infect. Dis.* **154**, 676–681.

McSharry, J. J., Caliguiri, L. A., and Eggers, H. J. (1979). Inhibition of uncoating of poliovirus by arildone, a new antiviral drug. *Virology* **97**, 307–315.

Mapoles, J. E., Krah, D. L., and Crowell, R. L. (1985). Purification of a HeLa cell receptor protein for the group B coxsackieviruses. *J. Virol.* **55**, 560–566.

Martin, A., Wychowski, C., Couderc, T., Crainic, R., Hogle, J., and Girard, M. (1988). Engineering a poliovirus type 2 antigenic site on a type 1 capsid results in a chimaeric virus which is neurovirulent for mice. *EMBO J.* **7**, 2839–2847.

Matthews, T. J., Weinhold, K. J., Lyerly, H. K., Langlois, A. J., Wigzell, H., and Bolognesi, D. P. (1987). Interaction between the human T-cell lymphotropic virus type III$_B$ envelope glycoprotein gp120 and the surface antigen CD4: Role of carbohydrate in binding and cell fusion. *Proc. Natl. Acad. Sci. USA* **84**, 5424–5428.

Minor, P. D., John, A., Ferguson, M., and Icenogle, J. P. (1986). Antigenic and molecular evolution of the vaccine strain of type 3 poliovirus during the period of excretion by a primary vaccine. *J. Gen. Virol.* **67**, 693–706.

Murray, M. G., Bradley, J., Yang, X.-F., Wimmer, E., Moss, E. G., and Racaniello, V. R. (1988). Poliovirus host range is determined by a short amino acid sequence in neutralization antigenic site I. *Science* **241**, 213–215.

Ninomiya, Y., Ohsawa, C., Aoyama, M., Umeda, I., Suhara, Y., and Ishitsuka, H. (1984). Antivirus agent, Ro 09-0410, binds to rhinovirus specifically and stabilizes the virus conformation. *Virology* **134**, 269–276.

Otto, M. J., Fox, M. P., Fancher, M. J., Kuhrt, M. F., Diana, G. D., and McKinlay, M. A. (1985). *In vitro* activity of WIN 51711, a new broad-spectrum antipicornavirus drug. *Antimicrob. Agents Chemother.* **27**, 883–886.

Page, G. S., Mosser, A. G., Hogle, J. M., Filman, D. J., Rueckert, R. R., and Chow, M. (1988). Three-dimensional structure of poliovirus serotype 1 neutralizing determinants. *J. Virol.* **62**, 1781–1794.

Palmenberg, A. (1987). A vaccine for the common cold. *Nature (London)* **329**, 668–669.

Palmenberg, A. C. (1989). Sequence alignments of picornaviral capsid proteins. *In* "Molecular Aspects of Picornavirus Infection and Detection" (B. L. Semler and E. Ehrenfeld, eds.), pp. 211–241. Am. Soc. Microbiol., Washington, D.C.

Pevear, D. C., Fancher, M. J., Felock, P. J., Rossmann, M. G., Miller, M. S., Diana, G., Treasurywala, A. M., McKinlay, M. A., and Dutko, F. J. (1989). Conformational change in the floor of the human rhinovirus canyon blocks adsorption to HeLa cell receptors. *J. Virol.* **63,** 2002–2007.

Richardson, J. S. (1979). The anatomy and taxonomy of protein structure. *Adv. Protein Chem.* **34,** 167–339.

Rossmann, M. G. (1988). Antiviral agents targeted to interact with viral capsid proteins and a possible application to human immunodeficiency virus. *Proc. Natl. Acad. Sci. USA* **85,** 4625–4627.

Rossmann, M. G., and Palmenberg, A. C. (1988). Conservation of the putative receptor attachment site in picornaviruses. *Virology* **164,** 373–382.

Rossmann, M. G., Abad-Zapatero, C., Hermodson, M. A., and Erickson, J. W. (1983). Subunit interactions in southern bean mosaic virus. *J. Mol. Biol.* **166,** 37–83.

Rossmann, M. G., Arnold, E., Erickson, J. W., Frankenberger, E. A., Griffith, J. P., Hecht, H. J., Johnson, J. E., Kamer, G., Luo, M., Mosser, A. G., Rueckert, R. R., Sherry, B., and Vriend, G. (1985). Structure of a human common cold virus and functional relationship to other picornaviruses. *Nature (London)* **317,** 145–153.

Sherry, B., and Rueckert, R. (1985). Evidence for at least two dominant neutralization antigens on human rhinovirus 14. *J. Virol.* **53,** 137–143.

Sherry, B., Mosser, A. G., Colonno, R. J., and Rueckert, R. R. (1986). Use of monoclonal antibodies to identify four neutralization immunogens on a common cold picornavirus, human rhinovirus 14. *J. Virol.* **57,** 246–257.

Smith, T. J., Kremer, M. J., Luo, M., Vriend, G., Arnold, E., Kamer, G., Rossmann, M. G., McKinlay, M. A., Diana, G. D., and Otto, M. J. (1986). The site of attachment in human rhinovirus 14 for antiviral agents that inhibit uncoating. *Science* **233,** 1286–1293.

Tomassini, J. P., and Colonno, R. J. (1986). Isolation of a receptor protein involved in attachment of human rhinoviruses. *J. Virol.* **58,** 290–295.

Varghese, J. N., Laver, W. G., and Colman, P. M. (1983). Structure of the influenza virus glycoprotein antigen neuraminidase at 2.9 Å resolution. *Nature (London)* **303,** 35–40.

Weis, W., Brown, J. H., Cusack, S., Paulson, J. C., Skehel, J. J., and Wiley, D. C. (1988). Structure of the influenza virus haemagglutinin complexed with its receptor, sialic acid. *Nature (London)* **333,** 426–431.

Wilson, I. A., Skehel, J. J., and Wiley, D. C. (1981). Structure of the haemagglutinin membrane glycoprotein of influenza virus at 3 Å resolution. *Nature (London)* **289,** 366–373.

11 Structural Determinants of Serotype Specificity, Host Range, and Thermostability in Poliovirus

James M. Hogle, Rashid Syed, Todd O. Yeates,
David Jacobson, and David J. Filman
Department of Molecular Biology, Research Institute of Scripps Clinic,
La Jolla, California 92037

I. Introduction

The polioviruses are the causative agents of poliomyelitis in man. Although there are a large number of strains of the virus, all known isolates can be grouped into one of three serotypes based on their reactivity with reference panels of neutralizing antisera. In 1985 we presented the three-dimensional structure of the Mahoney strain of type 1 poliovirus (P1/ Mahoney) and discussed the implications of the structure for the architecture, evolution, assembly, and immune recognition of picornaviruses (Hogle *et al.*, 1985). Since that time we have solved the structure of the Sabin strain of type 3 poliovirus (P3/Sabin) (Filman *et al.*, 1989), and more recently of an intertypic chimera called V510 (Martin *et al.*, 1988) in which a loop that constitutes a major antigenic site of poliovirus has been replaced (in P1/Mahoney) by the corresponding loop from the mouse-adapted Lansing strain of type 2 poliovirus (Yeates *et al.*, 1989).

P1/Mahoney is a commonly used neurovirulent laboratory strain of type 1 poliovirus and is the parent of the attenuated Sabin vaccine strain P1/Sabin. P3/Sabin is the attenuated vaccine strain of type 3 poliovirus which was derived from the neurovirulent P3/Leon strain. Genetic studies have shown that P3/Sabin and P3/Leon differ by only ten nucleotides in the 7500-nucleotide genome and that only two of these sequence changes, specifically a change at position 472 in the noncoding region at the 5' end of the genome and the substitution of Phe (P3/Sabin) for Ser (P3/Leon) at amino acid 91 of VP3, contribute significantly to the attenuation of P3/ Sabin (Westrop *et al.*, 1986; Minor *et al.*, 1989). P3/Sabin is also temperature sensitive, growing well at 37°C but not at 40°C. The temperature sensitivity of P3/Sabin has been shown to be due to the substitution of Phe for Ser at amino acid 91 of VP3 (Westrop *et al.*, 1986, 1989) and a number of second-site mutations which restore a nontemperature-sensitive phenotype to P3/Sabin have been characterized (Minor *et al.*, 1989). The V510

Use of X-Ray Crystallography in the Design of Antiviral Agents

chimera was constructed by Annette Martin, Czeslaw Wychowski, and Marc Girard at the Pasteur Institute (Martin *et al.*, 1988) and is one of several viable chimeric viruses in which heterologous sequences have been used to replace antigenic site 1 of type 1 polioviruses (Burke *et al.*, 1988; Murray *et al.*, 1988a,b; Martin *et al.*, 1988). The V510 chimera is of particular interest because the replacement of the loop confers mouse adaptation on the normally primate-specific Mahoney strain (Martin *et al.*, 1988; Murray *et al.*, 1988a).

Comparison of the three structures has provided insights into the structural determinants for serotype specificity in poliovirus. The high degree of sequence conservation between the three strains has allowed the serotype-specific structural differences to be correlated with specific sequence differences, providing a starting point for attempts to model loop structures computationally. The availability of three independently derived structures also has allowed the identification of several previously uninterpreted features which are common to all three electron density maps. These features include two sites of lipid substitution (a myristate molecule which is covalently bound to the amino terminus of VP4, and an extended hydrocarbon in the hydrophobic interior of the core of VP1), several residues from the otherwise disordered amino terminus of VP1, and two partially ordered nucleotides on the inner surface of the protein shell of the virus. Analysis of the locations of residues which regulate temperature sensitivity of P3/Sabin suggests that the ts phenotype of P3/Sabin is correlated with thermolability of the virion at some stage of its life cycle and provides important clues concerning the nature of conformational transitions in poliovirus. Finally, although our primary interest has been to correlate virus structure with biological properties, the structural studies also have provided valuable information relevant to vaccine development and to the design of antiviral agents.

II. Poliovirus Structure

Poliovirus is a member of the picornavirus family, which also includes the closely-related coxsackieviruses and rhinoviruses, the cardioviruses (EMC, Mengo, and Theiler's virus), the apthoviruses (foot-and-mouth disease virus), and hepatitis A virus. Like the other members of the family, the polioviruses are nonenveloped icosahedral viruses approximately 300 Å in diameter. Each virion contains one copy of a single-stranded, messenger-sense RNA genome of approximately 7500 nucleotides, enclosed in a roughly spherical capsid. The capsid is composed of 60 copies of each of the four coat protein subunits VP1 (33 kDa), VP2 (30 kDa), VP3

(26 kDa), and VP4 (7.4 kDa), arranged on a $T = 1$ icosahedral surface. In infected cells, the poliovirus genome is translated as a single large open reading frame. Early in translation, possibly cotranslationally, the resulting polyprotein is processed by virally encoded proteases, releasing the capsid precursor P1 (94 kDa), which contains the sequences of all four capsid proteins in the order VP4–VP2–VP3–VP1. Early in assembly, P1 is cleaved to yield VP1, VP3, and VP0 (a precursor in which VP4 and VP2 are covalently linked). This cleavage permits the amino-terminal extensions of the major capsid proteins to rearrange to form a network that appears to direct and stabilize the association of protomers in a pentameric assembly intermediate. Late in assembly, apparently after the encapsidation of the viral RNA, VP0 is cleaved to yield mature virus. The cleavage of VP0 to yield VP4 and VP2 is presumed to be autocatalytic in the intact virion, insofar as the cleavage site is located on the inner surface of the capsid, where it is inaccessible to exogenous proteases. This final cleavage appears to contribute substantially to the stability of the mature virion and may function to make particle assembly irreversible, though the mechanism by which the particle is stabilized is not yet well understood (see review in Rueckert, 1985).

Previously, we have described the three major capsid proteins (VP1, VP2, and VP3) of P1/Mahoney as having similar conserved cores (eight-stranded antiparallel β-barrels shaped roughly like triangular wedges), dissimilar extensions at their amino and carboxyl termini, and dissimilar loops connecting the regular secondary structural elements of the cores (Fig. 1) (Hogle *et al.*, 1985). The wedge-shaped eight-stranded β-barrel is apparently an ubiquitous structural motif in simple RNA viruses. A very similar structural motif also has been observed in related picornaviruses including rhinovirus 14 (Rossmann *et al.*, 1985), Mengo virus (Luo *et al.*, 1987), and foot-and-mouth disease virus (Acharya *et al.*, 1989); in several unrelated plant viruses including tomat bushy stunt virus (Harrison *et al.*, 1978), southern bean mosaic virus (Abad-Zapatero *et al.*, 1980), satellite tobacco necrosis virus (Liljas *et al.*, 1982), turnip crinkle virus (Hogle *et al.*, 1986), and cowpea mosaic virus (Stauffacher *et al.*, 1987); and in an insect virus: black beetle virus (Hosur *et al.*, 1988).

In the virions the cores of the capsid proteins form the continuous closed shell of the particle. The cores pack together so that the narrow edges of the VP1 β-barrels are clustered around the fivefold axes of the particle, and the narrow edges of the VP2 and VP3 β-barrels alternate around the particle threefold axes. The tilting of the cores outward along these axes produce significant radial protrusions at the fivefold and threefold axes, giving the particle the shape of the geometric solid shown in Fig. 2. The amino-terminal extensions and VP4 form a network on the inner

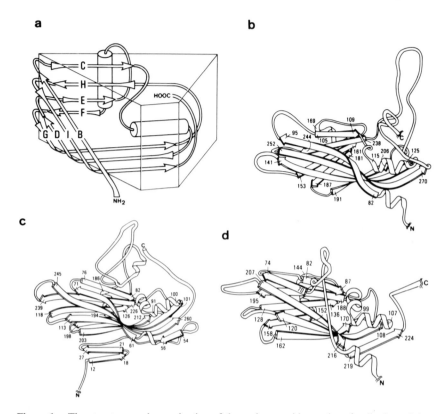

Figure 1. The structure and organization of the major capsid proteins of poliovirus. Schematic representation of the conserved wedge-shaped eight-stranded antiparallel β-barrel core motif shared by VP1, VP2, and VP3. Individual β-strands are shown as arrows and are labeled alphabetically from the amino terminus. Flanking helices are indicated by cylinders. (b–d) Ribbon diagrams of (b) VP1, (c) VP2, (d) VP3. Residue numbers have been included as landmarks. Extensions at the amino and carboxyl termini of VP1 and VP3 have been truncated for clarity. Adapted from Filman *et al.* (1989). The ribbon diagrams were drawn by E. D. Getzoff (Research Institute of Scripps Clinic).

surface of the protein shell which appears to direct and stabilize the assembly of the virus. The amino-terminal extensions are separated from the carboxyl termini by a closed shell of protein, and the free amino termini from different protomers associate into "assembly-dependent" structures. Both of these observations indicate that formation of the internal network is dependent on the proteolytic processing of the capsid protein precursors and provide an explanation for the role of proteolytic processing in early events in virus assembly. The connecting loops and carboxyl termini contribute to the major features exposed on the outer surface of the

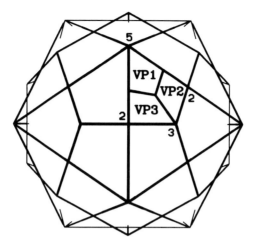

Figure 2. A geometric representation of the outer surface of the poliovirion is generated by superimposing an icosahedron and a dodecahedron. The symmetry axes of the particle and the positions of VP1, VP2, and VP3 in one protomer are indicated. Like the virion, the geometric figure has large radial projections at the fivefold axes and somewhat smaller projections at the threefold axes. Adapted from Filman *et al.* (1989).

virion and are the major contributors to the antigenic sites of the virus (Hogle *et al.*, 1985; Page *et al.*, 1988).

III. Comparison of the Structures of P1/Mahoney and P3/Sabin

Consistent with the high degree of sequence homology between the two strains (83%), the structures of P3/Sabin and P1/Mahoney are strikingly similar. The root mean square (rms) difference in α-carbon position is 0.78 Å for 846 sequentially equivalent residues, and the rms difference decreases to only 0.34 Å when 19 pairwise discrepancies greater than 2.0 Å are omitted. In the core structural motif, the packing of side chains within and between the β-barrels exhibits a precise atomic complementarity that tends to limit the viability of virus containing mutations in this region. Thus, it is not unexpected that the core structures of P3/Sabin and P1/Mahoney are nearly identical. A similar conservation of main chain structure is also seen in the amino-terminal extensions and in many of the connecting loops, despite significant local sequence differences (see Fig. 3). These sequence differences usually are accommodated either through localized adjustments of side chain conformation or by compensating

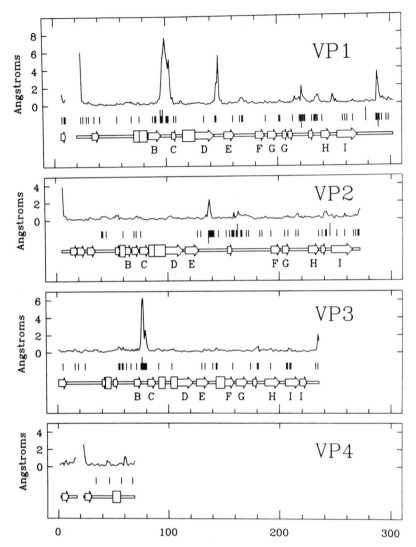

Figure 3. Comparison between the capsid protein structures of the P3/Sabin and P1/ Mahoney strains of poliovirus. At the top of each panel, the difference in position of structurally equivalent α carbons is plotted as a function of residue number. The vertical marks below each plot indicate differences in amino acid sequence. The larger marks extending upward indicate proline replacement, and those extending downward indicate insertions and deletions. At the bottom of each panel, the locations of α-helices and β-strands are indicated by wide rectangles and by arrows, respectively. The eight strands of the antiparallel β-barrel core motif of VP1, VP2, and VP3 are labeled B through I. Adapted from Filman *et al.* (1989).

changes in neighboring side chains. Because these loops and terminal extensions are not constrained by obvious packing considerations, their structural conservation serves as an indication that these residues are involved in the dynamic processes of the virus, such as assembly, receptor recognition, and uncoating.

Significant conformational differences in the main chain occur only in the exposed loops and chain termini of the virus (see Fig. 3). These structural differences fall into three general categories: (1) differences in loops due to insertions in one strain relative to the other, (2) loops which contain several sequence differences including the replacement of a proline residue, and (3) differences observed at points of transition between ordered and disordered structure. To these three general categories we can add one instance of an apparently flexible loop, the loop that connects the D and E strands of the VP1 β-barrel (residues 142–150, hereinafter designated the D–E loop), a relatively conserved sequence whose conformation is strongly influenced by its interactions with neighboring portions of the structure.

All of the insertions occur in exposed loops, or in the disordered amino terminus of VP1. The observable insertions occur at residue 222 of VP1 in P3/Sabin, at residue 289 of VP1 in P1/Mahoney, and at residue 138 of VP2 in P1/Mahoney. These insertions cause limited, highly localized structural perturbations which have little effect further than one or two residues on either side of the insertion (as shown in Fig. 3).

The largest structural differences between P3/Sabin and P1/Mahoney involve the replacement of proline residues in exposed loops with significant local sequence variability. Specifically, these are the B–C loop of VP1 (residues 95–105), the B–C loop of VP3 (residues 75–81), and the H–I loop of VP2 (residues 235–245). The importance of proline replacement is suggested by the E–F loop of VP2 (residues 128–187), which is nearly identical in conformation in both poliovirus structures. This loop is the site of numerous sequence differences between the two strains, but none of these differences involve the replacement of prolines.

Significant conformational differences between P3/Sabin and P1/Mahoney occur in all three of the major antigenic sites of the virus (as defined in Page *et al.*, 1988). In particular, structural changes are seen in the B–C loop of VP1 (which constitutes a major portion of antigenic site 1), in the insertion in the G–H loop of VP1 (site 2), in the insertion at position 289 of VP1 (site 3a), in and the B–C loop of VP3 (site 3b). This suggests that three-dimensional structural differences, as well as simple sequence changes, might play an important role in determining serotypes.

The most significant conformational difference between the P3/Sabin and P1/Mahoney structures occurs in the 10-residue loop that connects the

B and C strands of the VP1 β-barrel where the difference between equivalent α-carbon positions is as large as 8 Å. In poliovirus, this loop is the major contributor to neutralizing antigenic site 1, a site which has been shown to dominate the immune response to types 2 and 3 poliovirus in mice (Minor *et al.*, 1986; Icenogle *et al.*, 1986). Consistent with the immunodominance of this site in type 3 (but not in type 1) poliovirus, the B–C loop of VP1 is considerably more exposed on the surface of the virion in P3/Sabin than it is in P1/Mahoney. The exposure of this site, however, is not the sole determinant of immunodominance, since the degree of dominance of this site also has been shown to depend on the species and strain of animal immunized and perhaps on the route of immunization (Icenogle *et al.*, 1986).

IV. Serotype-Specific Conformations in Antigenic Site 1

Because of its importance as an antigenic site, the B–C loop of VP1 has been the focus of considerable attention. The sequence of this loop has been determined for a number of strains of all three serotypes of poliovirus (Minor *et al.*, 1987) (see also Table I). Recently, three research groups have reported the construction of viable intertypic chimeras in which the B–C loop of VP1 in type 1 poliovirus has been replaced the corresponding loop from a type 2 or type 3 strain (Burke *et al.*, 1988; Murray *et al.*, 1988a,b; Martin *et al.*, 1988). The resulting hybrids display the expected mosaic antigenicity and are able to induce neutralizing antibodies against both parental serotypes. In two of these hybrids, replacement of the B–C loop in the primate-specific P1/Mahoney strain with the corresponding loop from the mouse-adapted P2/Lansing strain produces a hybrid which is able to cause fatal paralysis in mice (Martin *et al.*, 1988; Murray *et al.*, 1988a). This demonstrates that the replacement of this 10-amino-acid loop (which includes only 6 sequence changes) is sufficient to confer mouse adaptation on the P1/Mahoney strain. In addition to these published reports, there are also a number of unpublished reports of viable constructs in which a variety of unrelated sequences have been inserted into this site. These studies raise the possibility that chimeric viruses ultimately may serve as novel vaccines for a variety of diseases.

We have solved the structure of the V510 chimera by molecular replacement methods (Yeates *et al.*, 1989). The 95–104 loop has a conformation significantly different from that observed in either the P1/Mahoney or the P3/Sabin structure. Somewhat surprisingly, the conformational changes in the heterologous B–C loop are accompanied by significant conformational changes in the D–E loop of VP1 (the third loop down

Table I. Consensus Sequences for the B–C Loop of VP1 in Poliovirus[a]

Consensus	91	92	93	94	95	96	97	98	99	100	101	102	103	104	105
Type 1 (11)	Thr	Val	Asp	Asn	Ser	Ala	Ser	Thr	Thr	Ser	Lys	Asp	Lys	Leu	Phe
Type 2 (2)	Glu	Val	Asp	Asn	Asp	Ala	Pro	Thr	Lys	Arg	Ala	Ser	Lys	Leu	Phe
Type 3 (13)	Glu	Val	Asp	Asn	Glu	Gln	Pro	Thr	Thr	Arg	Ala	Gln	Lys	Leu	Phe
Overall	—	Val	Asp	Asn	—	—	—	Thr	—	—	—	—	Lys	Leu	Phe

[a] The consensus sequences are based on sequence information from 11 strains of type 1 poliovirus, 2 strains of type 2 poliovirus and 13 strains of type 3 poliovirus. For type 1 and type 3 poliovirus the underlined residues indicate positions where there are two or more substitutions in the known strains. For type 2 the two sequences differ only at amino acid 103 (underlined), which is a lys in the Lansing strain and an arg in the Sabin strain. The sequence in the Mahoney strain of type 1 poliovirus is identical to the type 1 consensus sequence except at position 95, which is a lys in the Sabin strain. The sequence of type 1 poliovirus is identical to the type 1 consensus sequence except at position 95, which is a pro, and at position 100, which is an asn. The sequence of the Sabin strain of type 3 poliovirus is identical to the type 3 consensus sequence. Adapted from Minor *et al.* (1987).

at the narrow end of the wedge-shaped core), where the sequences of the chimera and the parental P1/Mahoney are identical. Interestingly, the structure observed in the D–E loop of the chimera is very similar to the corresponding loop in P3/Sabin, suggesting that there may be a coupling between conformational changes in the B–C and the D–E loops of VP1. Color Plate 15 compares the conformations of the loops of VP1 in all three structures (P1/Mahoney, P3/Sabin, and V510).

V. What Are the Determinants of Mouse Virulence?

Vincent Racaniello and his colleagues at Columbia University have demonstrated that the ability of the Lansing strain of type 2 poliovirus to infect mice maps to the capsid region of the viral genome (La Monica *et al.,* 1986), and that mutations in the B–C loop of VP1 attenuate mouse virulence (La Monica *et al.,* 1987). Mutations at Lys 99 and Arg 100 of VP1 are particularly effective in attenuating the mouse virulence of P2/Lansing (La Monica *et al.,* 1987). Our studies indicate that Lys 99 and Arg 100 are significantly more exposed in the chimera (compared with the equivalent residues in P1/Mahoney); and that Lys 99 (along with Asp 95, Pro 97, and Ser 102) makes specific interactions that appear to stabilize the conformation of the loop in the chimera. The observation of additional structural differences occurring outside of the heterologous B–C loop, however, raises the possibility that the B–C loop itself may not be the sole or dominant determinant of mouse adaptation. The increased exposure of neighboring residues in the H–I loop of VP1 or the conformational changes observed in the D–E loop might equally well be critical factors in regulating mouse virulence.

The factors which enable the chimera and the parental P2/Lansing strain to infect mice are as yet poorly understood. Indeed, the system has proven particularly difficult to analyze because neither strain is able to infect mice by any route other than intracerebral or intraspinal inoculation, and neither strain can infect common mouse cell lines in culture. Given the demonstrated importance of the highly exposed B–C loop (regardless of whether its contribution is direct or indirect) the simplest (but as yet unsubstantiated) model is that mouse virulence is determined at the level of the ability of the virion to interact with the viral receptor (or associated proteins) in the mouse central nervous system. Rossmann and his colleagues have presented a widely publicized hypothesis that the receptor-binding sites of polio and rhinoviruses are protected from their hosts' immune systems by being buried in the deep depression (or "canyon") which surrounds the fivefold axis (Rossmann *et al.,* 1985). However, the residues implicated in the control of mouse virulence, particularly Lys 99, Arg 100, and the D–E loop of VP1, are located very near the apex of the large protrusion at the fivefold axis. The observation that many of these residues are highly exposed and form parts of antigenic sites (rather than being shielded from the immune system) suggests that the premise underlying the canyon hypothesis may not be generally applicable. Indeed, in the structure of foot and mouth disease virus the canyon is not sufficiently deep to exclude antibodies, and there is evidence to suggest that at least part of the receptor binding site is located on a highly exposed and disordered loop on the surface of the virus (Acharya *et al.*, 1989).

VI. Molecular Modeling of Loop Structures in Poliovirus

The biological significance, the availability of sequence information from a large number of strains, and the large structural differences between strains make the B–C loop of VP1 a particularly attractive candidate for computational modeling studies designed to predict the effects of sequence variation on the conformation of the loop. These studies are analogous to previous calculations in which the structures of the complementarity-determining loops of antibodies have been predicted from sequence information by assuming structural conservation in the β-barrel cores of the Fv regions of Fabs (Chothia *et al.*, 1986; Fine *et al.*, 1986; Snow and Amzel, 1986; Bruccoleri and Karplus, 1987). In some ways, the loop-modeling calculations for poliovirus are more ambitious, because many of the loops (including the B–C loop of VP1) are longer than the loops which have been modeled in Fabs. In other respects, the calculations may prove to be more reliable, considering that the structures of the β-barrel cores of the polioviruses are better conserved than the β-barrels of the Fv regions of Fabs.

In anticipation of more rigorous modeling calculations, we have begun to analyze the B–C loop of VP1 in different strains of poliovirus in an effort to define specific features in the sequences (signatures) which are responsible for strain- and serotype-specific differences in the structure of the loop. As shown in Table I, the B–C loop begins with the sequence Val-Asp-Asn at its amino end and ends with the sequence Lys-Leu-Phe at its carboxyl end. Both sequences are highly conserved in all known naturally occurring strains of poliovirus. Between these highly conserved amino acids there are eight residues which are highly variable. In each of the three known structures, the two ends of the loop are connected by a main chain–main chain hydrogen bond. In P1/Mahoney (shown in Color Plate 16a) this hydrogen bond is donated by the amide nitrogen of Asp 93 and accepted by the carbonyl oxygen of Leu 104. In both P3/Sabin and V510 the analogous hydrogen bond is accepted by the carbonyl oxygen of Lys 103 (Plate 16b,c), reflecting the substantial conformational differences which occur in the loop structures.

The conserved sequence Val-Asp-Asn constitutes the last three residues of the B strand of the β-barrel of VP1. It is, therefore, not surprising that the conformation of these residues is very similar in all three strains. Although the regular pattern of main chain–main chain hydrogen bonding that is typical of antiparallel β-sheet does not extend beyond amino acid 94, residue 95 and the amide nitrogen of residue 96 are in nearly extended conformations in the P3/Sabin and V510 structures. In P1/Mahoney, however, the proline at position 95 is unable to adopt the main chain torsions required for an extended structure; and correspondingly, significant differences in the main chain conformation begin at this residue. This difference may prove not to be typical of most type 1 polioviruses, however, because the proline at position 95 is not conserved among the known type 1 sequences. The conserved sequence Lys-Leu-Phe is located at the carboxyl end of the B–C loop. Phe 105 is the first residue of the C strand of the β-barrel of VP1. In all three structures the aromatic side chain is buried in a hydrophobic pocket formed by the C strand and by the loop that connects the E and F strands of VP1 (Plate 16). Thus Phe 105 provides a conserved structural anchor for the loop. Unlike Phe 105, however, the sequentially conserved Lys 103 and Leu 104 are found in very different environments in the three structures. In P1/Mahoney, the side chain of Lys 103 extends from the lower surface of the B–C loop, forming hydrogen bonds with the side chain of Asn 94 and with the (main chain) carbonyl oxygen of Asn 246, while the side chain of Leu 104 extends from the upper surface of the loop (Plate 16a). In contrast, in P3/Sabin and in V510, the side chain of Lys 103 participates in an extensive network of charged side chains on the upper surface of the loop, while the side chain of Leu 104 lies on the lower surface, interacting with the hydrophobic side chain of Pro 97 (Plate 16b,c). The magnitude of the structural differences in the B–C loop and the dissimilarity in the conformations of highly conserved residues indicate that structural predictions based on simple interactive model building (which does not make explicit use of energy minimization and molecular dynamics) are likely to be seriously in error.

In each of the structures a number of specific interactions are observed that appear to be important for the stability of the loop conformation. The most significant stabilization of the B–C loop in P3/Sabin appears to be due to the side chain of Gln 102 (shown in Plate 16b), which extends across the loop to hydrogen bond to the side chain of Glu 95, to the main chain carbonyl oxygen of Gln 96, and to the amide nitrogen of Thr 98. Note that the positions of the latter two atoms are highly constrained by the proline residue (Pro 97) between them. The further observation that the side chains of Glu 95 and Gln 102 both are in extended conformations indicates that shorter side chains would be unable to make a similar interaction without significant alteration of the main chain. Additional stabilizing interactions are provided by hydrophobic contacts between the side chains of Pro 97 and Leu 104, as mentioned previously. The side chains which are involved in these stabilizing interactions (Glu 95, Pro 97, Gln 102, and Leu 104) are highly conserved among type 3 polioviruses (Table I), and thus may be diagnostic for the type 3 loop conformation.

Interactions made by the side chain of amino acid 102 also appear to play a central role in stabilizing the conformation of the B–C loop in V510 (Plate 16c). In V510, the side chain of Ser 102 participates in hydrogen bonds with the carboxylate side chain of Asp 95 and with Lys 252 from the nearby G–H loop. Other stabilizing interactions include one hydrogen bond between the side chains of Asp 95 and Lys 99, two hydrogen bonds made by the side chain of Asp 93 with the side chain of Thr 94 and with the amide nitrogen of Lys 103, and the main chain–main chain hydrogen bond between the carbonyl oxygen of Ala 96 and the amide nitrogen of Lys 99, which defines a β-turn. Although the residues which participate in these stabilizing interactions are conserved in the two known type 2 sequences, it will require the determination of additional type 2 sequences to assess the generality of these interactions in defining type 2 specific loop structures.

The effort to identify a conformational "signature" for the B–C loop in type 1 poliovirus is complicated by the fact that its principal stabilizing interactions involve residues that are not type 1-specific. As shown in Plate 16a, hydrogen bonds are formed between the amide nitrogen of Asp 93 and the carbonyl oxygen of Leu 104 and between the side chain carboxylate of Asp 93 and the amide nitrogen of Leu 104. The side chains of Lys 103 and Asn 94 are hydrogen bonded to one another and also to the main chain carbonyl oxygens of residues 246 and 247 in the H–I loop of VP1. These residues are highly conserved in all three serotypes of poliovirus. Other stabilizing influences shown in Plate 16a are the β-turn between residues 96 and 99 (which are relatively sequence-independent), hydrogen bonding to the main chain of residue 101 by the side chain of Thr 99 (which is not conserved), and the conformational constraints imposed by the presence of Pro 95 (which is a Ser in most type 1 sequences). Thus, the signature of the type 1 sequence might simply be the absence of a proline at position 97, together with the inability of residues 95 and 102 to form a hydrogen bond across the B–C loop.

VII. Newly Recognized Components of the Virion Structure

During the construction and refinement of the three poliovirus models, several interesting electron density features were identified which were not interpretable at the time of our original report of the P1/Mahoney structure (Hogle *et al.,* 1985). Although several of these features are relatively weak, and with one exception have yet to be corroborated by independent chemical evidence, the presence of nearly identical features in the refined electron density maps of three independently derived structures strongly suggests that they are significant and considerably increases the likelihood that the interpretation of these features is correct.

Two of the newly interpreted features were found in the assembly-dependent complex which is located on the inner surface of the capsid at each fivefold axis (Color Plate 17a). In this complex, five copies of the amino terminus of VP3 intertwine to form a twisted tube of parallel β-structure. This tube is flanked by five copies of the short antiparallel β-sheet formed by the amino termini of VP4. In all three electron density maps (P1/Mahoney, P3/Sabin, and V510) there is a roughly crescent-shaped electron density feature extending from the amino-terminal glycine of each copy of VP4. Chemical identification of this feature was established by radiolabeling experiments showing that the amino termini of VP4 and of its precursors (VP0 and P1) are myristoylated in all picornaviruses (Chow *et al.,* 1987). The myristate substituent, model-built with low-energy torsion angles, fits the electron density quite well. The myristate clearly plays an integral role in the capsid structure, mediating the interaction between the amino termini of VP3 and VP4 and shielding the hydrophobic side chains of Leu 2 and Pro 3 of VP3. Considering that myristate in other proteins often serves to permit the transient association of myristoylated proteins with membranes, we have also proposed that the myristate might play an additional role in the assembly of the virus or in its entry into the cell (Chow *et al.,* 1987). Myristoylation of viral proteins appears to be a fairly common covalent modification, which has been observed in a variety of virus families. In every virus where myristoylation has been observed (including poliovirus), mutations which prevent the modification are lethal. This suggests the possibility that appropriate inhibitors of myristolyation may represent a novel approach to broad-spectrum antiviral therapy.

All three electron density maps also indicate clearly the presence of a third β-strand, five residues long, extending the two-stranded β-sheet of VP4 towards the interior of the virus. Although the main chain unmistakably is making parallel β interactions, none of the electron density maps is as yet sufficiently well resolved in this area to allow for an unambiguous identification of the side chains in this additional beta strand. However, among the residues unaccounted for in the atomic models (residues 1–19 of VP1, 1–4 of VP2, 236–238 of VP3, and 17–22 of VP4) only the first 16 amino acids of VP1 could possibly be located in this vicinity.

A direct interaction between the amino termini of VP1 and VP4 has interesting implications. Upon attachment to susceptible cells, poliovirus and related picornaviruses undergo a conformational rearrangement in which VP4 is lost from the particle (see review in Rueckert, 1985). We have shown recently that this rearrangement also causes the amino terminus of VP1 to be extruded and that the exposed amino terminus of VP1 enables the altered particles to attach to liposomes (Fricks, 1988; Fricks and Hogle, 1989). This suggests that the exposed amino terminus might participate directly in membrane attachment during cell entry. The interaction between the amino termini of VP1 and VP4 seen in the native poliovirus structures makes it likely that the externalization of VP1 and the release of VP4 occur in a correlated fashion, at the same site in the particle.

The radiolabeling experiments which succeeded in identifying myristate linked to the amino terminus of VP4 actually were begun in an effort to identify an entirely different lipid substituent observed in the protein coat of the virus. In all three electron density maps, there is an elongated density feature, bent in the middle, which extends through the hydrophobic interior of the VP1 β-barrel. This feature can be model-built convincingly as a 16-carbon saturated chain with minimum-energy torsion angles. The 15 most deeply buried atoms in the chain make contact exclusively with hydrophobic side chains of the protein. Beyond the sixteenth carbon, the electron density for the substituent is discontinuous, and the environment is polar, consistent with a partially ordered polar head group. This end of the substituent is located near the outer surface of the virion, at the base of the G–H loop of VP1. In both the P1/Mahoney and P3/Sabin maps the shape of the electron density and the hydrogen bonding requirements of the polar end are consistent with the provisional identification of the lipid substituent as sphingosine. However, in the V510 maps, the density for the polar head group is smaller in size, located in a different portion of the pocket, and consistent with a provisional identification of the substituent as palmitate. At present, the identities of these substituents have not yet been confirmed by chemical means; we cannot rule out the presence of a mixture of lipids in any of the three structures; and the factors which control the nature of the bound species are under investigation.

The hydrocarbon binding site in poliovirus is nearly identical to the site which binds a class of antiviral drugs in rhinovirus 14 (Smith *et al.*, 1986; Badger *et al.*, 1988). The binding of these drugs is known to stabilize the virus against a variety of conformational rearrangements. In most picorna-viruses, these drugs do not interfere with the attachment of the virus to susceptible cells or to internalization but are known to prevent uncoating, apparently by inhibiting a conformational change associated with the penetration of virus across vesicular membranes or with the release of RNA. This poses the interesting possibility (discussed below) that the lipid molecule bound in the native poliovirus structures either mediates viral stability or plays an integral role in cell entry.

On the inner surface of the capsid, there are several electron density features which may correspond to portions of the crystallographically disordered RNA genome of the virus. The two most convincing of these features are distinctly planar and appear to be stacked with the aromatic side chains of Trp 38 and Tyr 41 of VP2 (Trp 38 and Phe 41 in P3/Sabin). A similar electron density feature in the vicinity of Trp 38 of VP2 in rhinovirus 14 has also been identified tentatively as a nucleotide (Arnold and Rossmann, 1988). Because all of the aromatic residues of the capsid proteins have been accounted for, we tentatively have identified the planar density features as nucleotide bases. Although many of the "fixed solvent" positions in this vicinity are separated from one another by fairly typical phosphate–phosphate distances, we have been unable to model convincingly the intervening density as the ribose–phosphate backbone of RNA. The identification of specific RNA-binding sites that are associated with the amino-terminal residues of VP2 may be relevant to understanding the mechanisms by which the encapsidation of RNA permits VP0 to be cleaved and by which this cleavage renders the assembly of pentamers irreversible.

After the encapsidation of a full complement of viral RNA, the autocatalytic cleavage of VP0 at a site on the inner surface of the capsid confers stability on the mature virion. Considering that the VP4–VP2 cleavage site, as it is seen in the native virion, is inaccessible to RNA-sized probes, we suspect that the maturation cleavage might first require an RNA-dependent rearrangement of VP0 so as to produce the putative nucleotide binding interactions that are observed in the native virion and that cleavage subsequently would permit some additional stabilizing rearrangements.

In the native virus structures there is a striking assembly-dependent structural feature located close to the putative nucleotide sites which could play a significant role in the stabilization of mature virions. This is a continuous seven-stranded β sheet (see Plate 17b), which is formed by residues from two different threefold-related protomers and appears to stabilize the association of pentamers in the native virion. One protomer contributes the outermost four strands and the innermost strand (respectively, the C, H, E, and F strands of the VP3 β-barrel and the extended chain comprising residues 36–38 in the amino-terminal extension of VP1.) The other protomer contributes the fifth and sixth strands of the extended sheet (specifically, the two-stranded antiparallel β-sheet formed by residues 14–25 in the amino-terminal extension of VP2.) Considering that the conformation of the innermost three strands is not sensible in the context of an isolated pentamer, it is likely that these strands are rearranged during assembly. As shown in Plate 17b, His 37 of VP1 in the seventh strand is located less than 5 Å from one of the presumed nucleotides. Several aspects of this interaction suggest that it may be relevant to the mechanism of assembly. The extended β-sheet is located close to the cleavage site of VP0; it contributes significantly to the stability of pentamer association; and it appears to be involved in the binding of nucleotides. These observations begin to provide a structural basis for resolving the previously unexplained correlation between the encapsidation of RNA, the cleavage of VP0, and the stability of mature virions. We anticipate that our current crystallographic studies of native-antigenic empty capsids will yield more definitive information about the nature of the rearrangements that take place in viral assembly.

VIII. Temperature Sensitivity in the P3/Sabin Strain

The temperature-sensitive phenotype of the P3/Sabin strain has been shown to be due to the substitution of phenylalanine (in P3/Sabin) for serine at amino acid 91 of VP3 during the derivation of the P3/Sabin strain from its neurovirulent and non-temperature-sensitive parent P3/Leon (Westrop *et al.*, 1986, 1989). In the P3/Sabin, P1/Mahoney, and V510 structures, residue 91 of VP3 is located in a turn of the helix at the carboxyl end of the C strand of the β-barrel. The loop that connects the G and H strands of the β-barrel folds up against this turn of helix, trapping a chain of three buried solvent molecules in a pocket. In P1/Mahoney and V510 (which, like P3/Leon, have serine residues at position 91 of VP3) the serine side chain points downward into this pocket, making a hydrogen bond with one of the trapped water molecules.

In P3/Sabin, however, the phenylalanine side chain points upward and is fully exposed to solvent at the base of the deep depression, or canyon, which surrounds the fivefold axes of the particle (Color Plate 18). Indeed, the P3/Sabin electron density maps contain clear indications for several ordered solvent molecules adjacent to the edge and to one of the faces of the aromatic group. Considering that solvation of this side chain is energetically unfavorable, we postulate that the substitution of Phe for Ser causes an increased temperature sensitivity in P3/Sabin by affecting the thermostability of the virus particle at some stage of its life cycle. Phe 91 thus might be expected to stabilize conformational states in which the aromatic side chain is buried and destabilize states in which it is exposed. It is not yet clear whether the effect on thermostability is expressed at a level of the intact virion or whether it affects the dynamic processes of virus assembly and cell entry.

Phil Minor and his colleagues at the National Institute of Biological Standards and Control, London in collaboration with Jeff Almond and his colleagues at Reading University have characterized a number of second site non-ts revertants of P3/Sabin (Minor *et al.*, 1989). We have located the mutated residues in the three-dimensional structure of the capsid (Plate 18) and found that the mutations tend to occur in the interfaces between protomers. These residues might restore thermostability by stabilizing the interfaces against conformational rearrangements. Note that a particularly striking cluster of mutations occurs in the interface between fivefold-related protomers close to the position of Phe 91 (Plate 18). It may be relevant that the corresponding interface in the structurally analogous $T = 3$ icosahedral plant viruses (for example, tomato bushy stunt virus) is disrupted during the expansion of the particle, which is induced by the depletion of divalent cations at basic pH (Robinson and Harrison, 1982) and that many of the loops containing second-site mutations in poliovirus correspond to loops in the plant viruses which contribute to the binding of divalent cations (Hogle *et al.*, 1983). The analogy suggests that this interface may be involved in the control of structural transitions in a variety of viruses. An additional mutation observed in one of the non-ts revertants occurs at Leu 18 of VP2, which is located in the assembly-dependent seven-stranded β sheet. Mutations in this residue may thus modulate the stability of pentamer association in the capsid.

A notable exception to the general observation that the second-site suppressors occur in the interfaces between protomers is the replacement of Phe 134 of VP1 by leucine. This mutation has been observed in three of four non-ts revertants of virus produced from a construct in which the Ser at position 91 of VP3 of P3/Leon was replaced by a Phe. Phe 134 is located in the hydrocarbon-binding pocket of VP1 and interacts directly with the bound lipid molecule. Simple model-building experiments indicate that a leucine side chain in the same position would have a more extensive van der Waals interaction with the ligand and presumably would influence the stability of the particle by increasing the binding energy of the ligand.

The mechanism by which the binding of lipid stabilizes the conformation of the virus is suggested by the crystallographic results of Rossmann and his colleagues (Smith *et al.*, 1986; Badger *et al.*, 1988) showing that the binding of the antiviral compound WIN 51711 and related compounds to rhinovirus causes (as its primary structural effect) the movement of some seven or eight residues at the carboxyl end of the G–H loop. In rhinovirus (without the antiviral compound bound) these residues fold into the binding pocket, in a conformation unlike that seen in poliovirus. When the binding pocket is occupied, however, steric interference with the ligand precludes these residues from assuming their native conformation, and they are forced outward into the interface between protomers, in a conformation more similar to that seen in poliovirus with lipid bound. The corresponding residues in poliovirus are located very close to several of the stabilizing second-site mutations, and in the vicinity of amino acid 91 of VP3. The contacts which these residues make with the neighboring protomer (principally with the G–H loop of VP3) presumably stabilize this interface. The analogy suggests that the occupant of the hydrophobic pocket regulates structural transitions of the virus simply by inducing a localized conformational change that causes additional stabilizing interactions to be formed in the interface between protomers. It suggests further that the antiviral drugs may exploit a site normally used by the virus to regulate its stablity and its conformational transitions.

Acknowledgments

This work was supported by NIH grants AI20566 and GM3879 to J.M.H. The work on the V510 chimera has also been support in part by a postdoctoral training grant (T32 NS07078) to T.O.Y. We thank T. Critchlow for excellent technical assistance and M. Graber for help in preparing the manuscript. We would particularly like to acknowledge the contributions of M. Chow (Massachusetts Institute of Technology) in the studies of P1/Mahoney and P3/Sabin, P. Minor and A. Macadm (National Institute of Biological Standards and Control, London) in the studies of structural markers for thermostability in P3/Sabin, and A. Martin, C. Wychowski, and M. Girard (Pasteur Institute, Paris) in the studies of the V510 chimera. This is publication no. 5854-MB of the Research Institute of Scripps Clinic.

References

Abad-Zapatero, A., Abdel-Meguid, S. S., Johnson, J. E., Leslie, A. G. W., Rayment, I., Rossmann, M. G., Suck, D., and Tsuikihara, T. (1980). Structure of southern bean mosaic virus at 2.8 Å resolution. *Nature (London)* **286,** 33–39.

Acharya, R., Fry, E., Stuart, D., Fox, G., Rowlands, D., and Brown, F. (1989). The three-dimensional structure of foot-and-mouth disease virus at 2.9 Å resolution. *Nature (London)* **337,** 709–716.

Arnold, E., and Rossmann, M. G. (1988). The use of molecular-replacement phases for the refinment of the human rhinovirus 14 structure. *Acta Crystallogr., Sect. A* **A44,** 270–282.

Badger, J., Minor, I., Kremer, M. J., Olivera, M. A., Smith, T. J., Griffith, J. P., Guerin, D. M. A., Krishnaswamy, S., Luo, M., Rossmann, M. G., McKinlay, M. A., Diana, G. D.,

Dutko, F. J., Fancher, M., Rueckert, R. R., and Heinz, B. A. (1988). Structural analysis of a series of antivial agents complexed with human rhinovirus 14. *Proc. Natl. Acad. Sci. USA* **85**, 3304–3308.

Bruccoleri, R. E., and Karplus, M. (1987). Prediction of the folding of short polypeptide segments by uniform conformational sampling. *Biopolymers* **26**, 137–168.

Burke, K. L., Dunn, G., Ferguson, M., Minor, P. D., and Almond, J. W. (1988). Antigen chimaeras of poliovirus: potential novel vaccines against picornavirus infections. *Nature (London)* **332**, 81–82.

Clothia, C., Lesk, A. M., Levitt, M., Amit, A. G., Mariuzza, R. A., Phillips, S. E. V., and Poljak, R. J. (1986). The predicted structure of immunoglobulin D1.3 and its comparison with the crystal structure. *Science* **233**, 755–758.

Chow, M., Newman, J. F. E., Filman, D. J., Hogle, J. M., Rowlands, D., and Brown, F. (1987). Capsid protein VP4 of picornavirus particles is covalently modified at its N-terminus by myristic acid. *Nature (London)* **327**, 482–486.

Filman, D. J., Syed, R., Chow, M., Macadm, A. J., Minor, P. D., and Hogle, J. M. (1989). Structural factors that control conformational transitions and serotype specificity in type 3 poliovirus. *EMBO J.* **8**, 1567–1579.

Fine, R. M., Wang, H., Shenkin, P. S., Yarmush, D. L., and Levinthal, C. (1986). Predicting antibody hypervariable loop conformations II, minimization and molecular dynamics studies of MCPC603 from many randomly generated loop conformations. *Proteins* **1**, 342–362.

Fricks, C. E. (1988). Studies of the conformational changes of poliovirus during virus neutralization, cell entry and infection. Ph.D. Thesis, Univ. of California, San Diego.

Fricks, C. E., and Hogle, J. M. (1989). The cell-induced conformational change of poliovirus: Externalization of the amino terminus of VP1 is responsible for liposome binding. Submitted, *J. Virol.*

Harrison, S. C., Olson, A. J., Schutt, C. E., Winkler, F. K., and Bricogne, G. (1978). Tomato bushy stunt virus at 2.9 Å resolution. *Nature (London)* **317**, 368–373.

Hogle, J. M., Kirchhausen, T., and Harris, S. C. (1983). The divalent cation binding sites of tomato bushy stunt virus: difference maps at 2.9 Å resolution. *J. Mol. Biol.* **171**, 95–100.

Hogle, J. M., Chow, M., and Filman, D. J. (1985). The three-dimensional structure of poliovirus at 2.9 Å resolution. *Science* **229**, 1358–1365.

Hogle, J. M., Maeda, A., and Harrison, S. C. (1986). The structure of turnip crinkle virus at 3.2 Å resolution. *J. Mol. Biol.* **191**, 625–638.

Hosur, M. V., Schmidt, T., Tucker, R. C., Johnson, J. E., Gallagher, T. M., Selling, B. H., and Rueckert, R. R. (1988). Structure of an insect virus at 3.0 Å resolution. *Proteins* **2**, 167–176.

Icenogle, J. P., Minor, P. D., Ferguson, M., and Hogle, J. M. (1986). Modulation of the humoral response to a 12 amino acid site on the poliovirus virion. *J. Virol.* **60**, 297–301.

La Monica, N., Meriam, C., and Racaniello, V. R. (1986). Mapping of sequences required for mouse neurovirulence of poliovirus type 2 Lansing. *J. Virol.* **57**, 515–525.

La Monica, N., Kupsky, W. J., and Racaniello, V. R. (1987). Reduced mouse neuvorirulence of poliovirus type 2 Lansing antigenic variants selected with monoclonal antibodies. *Virology* **161**, 429–437.

Liljas, L., Unge, T., Jones, T. A., Fridborg, K., Lovgren, S., Skoglund, U., and Strandberg, B. (1982). Structure of satellite tobacco necrosis virus at 3.0 Å resolution. *J. Mol. Biol.* **159**, 93–108.

Luo, M., Vriend, G., Kamer, G., Minor, I., Arnold, E., Rossmann, M. G., Boege, U., Scraba, D. G., Duke, G. M., and Palmenberg, A. C. (1987). The atomic structure of Mengo virus at 3.0 Å resolution. *Science* **235**, 182–191.

Martin, A., Wychowski, C., Couderc, T., Crainic, R., Hogle, J. M., and Girard, M. (1988).

Engineering a poliovirus type 2 antigenic site on a type 1 capsid results in a chimaeric virus which is neurovirulent for mice. *EMBO J.* **7**, 2839–2847.

Minor, P. D., Ferguson, M., Evans, D. M. A., and Icenogle, J. P. (1986). Antigenic structure of polioviruses of serotypes 1, 2, and 3. *J. Virol.* **67**, 1283–1291.

Minor, P. D., Ferguson, M., Phillips, A., McGrath, D. I., Huovilainen, A., and Hovi, T. (1987). Conservation *in vivo* of protease cleavage sites in antigenic sites of poliovirus. *J. Gen. Virol.* **68**, 1857–1865.

Minor, P. D., Dunn, G., Evans, D. M. A., Magrath, D. I., John, A., Howlett, J., Phillips, A., Westrop, G., Wareham, K., Almond, J. W., and Hogle, J. M. (1989). The temperature sensitivity of the Sabin type 3 vaccine strain of poliovirus: molecular and structural effects of a mutation in the capsid protein VP3. *J. Gen. Virol.* **70**, 1117–1123.

Murray, M. G., Bradley, J., Yang, X.-F., Wimmer, E., Moss, E. G., and Racaniello, V. R. (1988a). Poliovirus host range is determined by a short amino acid sequence in neutralization antigenic site I. *Science* **241**, 213–215.

Murray, M. G., Kuhn, R. J., Arita, M., Kawamura, N., Nomoto, A., and Wimmer, E. (1988b). Poliovirus type 1/type 3 antigenic hybrid virus constructed *in vitro* elicits type 1 and type 3 neutralizing antibodies in rabbits and monkeys. *Proc. Natl. Acad. Sci. USA* **85**, 3203–3207.

Page, G. S., Mosser, A. G., Hogle, J. M., Filman, D. J., Rueckert, R. R., and Chow, M. (1988). Three-dimensional structure of the poliovirus serotype 1 neutralizing determinants. *J. Virol.* **62**, 1781–1794.

Robinson, I. K., and Harrison, S. C. (1982). Structure of the expanded state of tomato bushy stunt virus. *Nature (London)* **297**, 563–568.

Rossmann, M. G., Arnold, E., Erickson, J. W., Frankenberger, E. A., Griffith, J. P., Hecht, H.-J., Johnson, J. E., Kamer, G., Luo, M., Mosser, A. G., Rueckert, R. R., Sherry, B., and Vriend, G. (1985). Structure of a human common cold virus and functional relationship to other picornaviruses. *Nature (London)* **317**, 145–153.

Rueckert, R. R. (1985). Picornaviruses and their replication. *In* "Virology" (B. Fields, ed.), Raven, New York. pp. 705–738.

Smith, T. J., Kremer, M. J., Luo, M., Vriend, G., Arnold, E., Kamer, G., Rossmann, M. G., McKinlay, M. A., Diana, G. D., and Otto, M. J. (1986). The site of attachment in human rhinovirus 14 for antiviral agents that inhibit uncoating. *Science* **233**, 1286–1293.

Snow, M. E., and Amzel, L. M. (1986). Calculating three-dimensional changes in protein structure due to amino-acid substitutions: the variable regions of immunoglublins. *Proteins* **1**, 267–279.

Stauffacher, C., Usha, R., Harrington, M., Schmidt, T., Hosur, M. V., and Johnson, J. E. (1987). The structure of copwpea mosaic virus at 3.5 Å resolution. *In* "Crystallography in Molecular Biology" (D. Moras, J. Drenth, B. Strandberg, D. Suck, and K. Wilson, eds.), pp. 293–308. Plenum, New York.

Westrop, G. D., Evans, D. M. A., Minor, P. D., Magrath, D., Schild, G. C., and Almond, J. W. (1986). Investigation of the molecular basis of attenuation in the Sabin type 3 vaccine using novel recombinant polioviruses constructed frominfectious cDNA. *In* "Molecular Biology of the Positive Strand RNA Viruses" (D. J. Rowlands, B. W. J. Mahy, and M. Mayo, eds.), pp. 53–60. Academic Press, London.

Westrop, G. D., Wareham, K. A., Evans, D. M. A., Dunn, G., Minor, P. D., Magrath, D. I., Taffs, F., Marsden, S., Skinner, M. A., Schild, G. C., and Almond, J. W. (1989). Genetic basis of attenuation of the Sabin type 3 oral poliovirus vaccine. *J. Virol.* **63**, 1338–1344.

Yeates, T. O., Syed, R., Filman, D. J., Martin, A., Wychowski, C., Girard, M., and Hogle, J. M.. (1989). Structure of a mouse virulent intertypic chimera of type 2 and type 1 poliovirus. In preparation.

12 Structure of Foot-and-Mouth Disease Virus

Ravi Acharya, Elizabeth Fry, Derek Logan, and David I. Stuart
Laboratory of Molecular Biophysics, Oxford OX1 3QU United Kingdom

Graham Fox, Dave Rowlands, and Fred Brown
Wellcome Biotech, Beckenham, Kent, BR3 3BS United Kingdom

I. Introduction

We have determined the three-dimensional structure of foot-and-mouth disease virus (FMDV) at close to atomic resolution (Acharya *et al.*, 1989a). The virus structure is of interest for a variety of reasons. First, the disease remains a major problem in large areas of the world; we therefore hope that structural information will lead to the development of effective antiviral agents targeted against the virus. Second, we would hope for a structural explanation for the peculiar success of synthetic peptide vaccines against FMDV which might then have broad implications for the design of synthetic peptide vaccines. The structure that has been revealed by our studies has a number of interesting features of a more academic nature which have already been described (Acharya *et al.*, 1989a) and our focus here will be on the structural information as a guide in the design of therapeutic agents.

A. The Disease

Foot-and-mouth disease, one of the most important viral diseases of farm animals, was first documented some four centuries ago. It is highly contagious and infection results in the appearance of lesions in the mouth and on the feet, fever, anorexia, depression, and a fall in meat and milk production. Mortality is low, but of greater consequence to farmers is the loss in productivity and indirect losses caused by the interruption of trading in meat and dairy products. It is thus not surprising that the causative agent of such an economically devastating disease has been the subject of considerable study, but disappointing that it still remains a major problem.

In countries where the disease does not normally occur, the usual means of control is by the slaughter of infected animals. Where the disease is endemic (every continent except North America and Australia) chemically inactivated virus vaccines are used. Since protection requires high levels of antibody, two or three vaccinations per year are required. Vaccination is further complicated by the existence of seven serotypes that afford no cross-protection, and antigenic variation within serotypes, which can result in incomplete protection. There are also problems associated with the large quantities of live virus required, occasional incomplete chemical inactivation of the virus, and the maintenance of the integrity and thus immunogenicity of the vaccine by storage at refrigeration temperatures, which is often difficult under field conditions. Synthetic vaccines would overcome most of these problems, but clearly a better understanding of the antigenic determinants is required if such products are to approach the potency of current vaccines and perhaps improve on them by having a broader antigenic spectrum.

B. The Virus

FMDV is a member of the picornavirus family of single-stranded, positive-sense, small RNA viruses (Cooper *et al.*, 1978). The capsid is composed of four proteins, VP1–VP4, 60 copies of each being present in the intact virion. The proteins VP1–VP3 are partly exposed on the capsid surface and comprise (across the family) 200–300 residues. VP4 is internal, appearing to mediate between the RNA and the other proteins; it is much smaller, 69–85 residues in length, and has an amino-terminal myristic acid (Chow *et al.*, 1987; Paul *et al.*, 1987). The picornaviruses fall into four genera (Cooper *et al.*, 1978; Newman *et al.*, 1973): the enteroviruses, for example, poliovirus; the rhinoviruses, for example, human rhinovirus (HRV); the cardioviruses, for example, Mengo virus; and the aphthoviruses (FMDV). The rhino and entero genera are the most closely related, while the cardioviruses seem to lie approximately midway between these and the aphthoviruses (Palmenberg, 1989). The picornaviruses are well studied crystallographically and structures are now known for representatives of all four genera (Rossmann *et al.*, 1985; Hogle *et al.*, 1985; Luo *et al.*, 1987; Acharya *et al.*, 1989a). We are therefore in an excellent position to understand the family and look for common features and variations not only directly in their structure but also in how this structure relates to function.

It has been proposed that the site of cell attachment lies in an antibody-inaccessible depression on the surface of the picornaviruses (the canyon hypothesis) (Rossmann *et al.*, 1985). There is evidence (Colonno *et al.*, 1988) that this is the case for the human rhinoviruses, and 'homologous' depressions have been found in polio and Mengo virus. We do not find a corresponding depression on FMDV. All antigenic sites mapped in the picornaviruses are exposed on the surface. The major antigenic site of

FMDV, the so-called FMDV loop (approximately residues 140–160 of VP1) has additional unique properties. As a synthetic peptide it can induce high levels of neutralizing and protective antibodies (Strohmaier *et al.*, 1982; Bittle *et al.*, 1982; Pfaff *et al.*, 1982). Furthermore, part of the loop has been implicated in cell attachment (Fox *et al.*, 1989). We find that this loop is disordered in the crystal structure, and based on this observation we have proposed an explanation of its apparently contradictory properties (Acharya *et al.*, 1989a,b).

II. Determination of the Three-Dimensional Structure

The structure of crystals of virus of serotype O_1 strain BFS 1860 has been determined at 2.9 Å resolution. The virus crystallizes in space group I23 ($a = 345$ Å) with five icosahedral subunits in the crystallographic unit (Fox *et al.*, 1987). The structure determination was made more difficult by the small size of the crystals ($\sim 0.15 \times 0.15 \times 0.08$ mm^3) and the short crystal lifetime. However, using the high quality X-ray beam at the Daresbury synchrotron we were able to obtain a set of data that was almost complete ($\sim 95\%$) by examining nearly 500 crystals. We found that we could obtain improved data at the slightly shorter than usual wavelength of 0.9 Å. Initial estimates of the phase angles to a resolution of 8 Å were obtained by placing suitably oriented models of both HRV14 and Mengo virus at the origin of the I23 cell, summing the electron density, and then inverting the resultant map. Standard procedures (Bricogne, 1976) then produced a good set of low-resolution phases. Successful phase extension was achieved despite some initial difficulties and yielded an electron density map of outstanding quality. The model that was built into this map has now been refined (Brunger *et al.*, 1987) to give an R-value of 17% against all data (including weak and negative intensities) in the range 5–2.9 Å. No water molecules have yet been added to the model. In summary, we believe that we now have a model for the protein coat of the virus that for the most part is accurate to within about 0.2 Å.

III. An Overview of the Structure of the Capsid

The FMDV capsid forms a thin spherical shell, with the larger proteins VP1–3 forming the surface 'crystal' of the virus and VP4 mediating between these and the RNA in the interior. In common with the other picornaviruses we have not observed any ordered RNA. The overall structure of the capsid and the arrangement of proteins within it are similar in gross terms to that described for the other picornaviruses (Rossmann *et al.*, 1985; Hogle *et al.*, 1985; Luo *et al.*, 1987); however, there are also important differences. FMDV has a rather thinner capsid with a somewhat smoother surface than either HRV14 or Mengo virus. The differences arise in large part from a radical truncation of many of the surface loops (the individual proteins are smaller in FMDV; see below). Detailed comparison of FMDV, Mengo, and HRV14 reveals, as expected, that Mengo is intermediate in structure between FMDV and HRV14. Color Plate 19 shows in detail the folds of the polypeptide chains of VP1–4.

1. VP1 (FMDV 213 residues, HRV14 279 residues): The truncation of the protein is largely accommodated by 'trimming' the loops at the fivefold end of the β-barrel. The effect of this is to produce a 'hole' at the fivefold axis of the virion. Residues 1–132 and 159–209 are well ordered; some density is seen for the 4 C-terminal residues and residues 133, 134, 135, 136, 157, and 158; however, there is nothing other than low-level, diffuse density for residues 137–156, which correspond to the major antigenic site. Note also that the C-terminal residues are quite antigenic (Bittle *et al.*, 1982) and rather disordered. Furthermore, the C-terminal residues form a long arm which reaches across from one VP1 to its fivefold-related neighbor and which ends nestled close to the FMDV loop of this adjacent subunit.

2. VP2 (FMDV 218 residues, HRV14 262 residues): The first five residues cannot be located and the electron density for residues 130–132 is rather weak; otherwise the chain is well defined.

3. VP3 (FMDV 220 residues, HRV14 238 residues): This protein is the most highly conserved between the picornaviruses and all of the residues are well defined in the electron density map.

4. VP4 (FMDV 85 residues, HRV14 69 residues): This protein varies considerably between the picornaviruses and is often not as well ordered as VP1–3. In FMDV only about half the residues are clearly visible (residues 15–39 and 65–85).

IV. Antigenic Surface of Foot-and-Mouth Disease Virus

Many of the antigenic properties of proteins are likely to be explicable in relatively simple structural terms if we understand the nature of the antigen, the antibody, and the complex they form. In spite of the availability of reliable structural information for all three of these (although as yet we do not have a complete matched set of all three for one antigen–antibody system), there is no general consensus as to the source of antigenicity. One possibility is that the antigenicity of a piece of structure is related to its flexibility (Tainer *et al.*, 1986; Westhof *et al.*, 1984). More specifically it has been proposed that the crystallographic B-factor may be used to predict the antigenicity of any residue. The competing hypothesis simply states that antigenicity is dependent on whether a given part of the structure is available on the surface to interact with an antibody (the static accessibility model) (Novotný *et al.*, 1986).

By far the best known antigenic region of FMDV is the FMDV loop, and since this shows unmeasurably high B-factors and also lies at an extremely exposed position, it adds little to the argument. However, the majority of antibodies raised against FMDV do not in fact bind to the 140–160 loop; although many of these are not neutralizing, some are, and information on the location of their epitopes can be gleaned from the positions of substitutions in monoclonal resistant mutants. We present this information and a detailed discussion elsewhere; however, to display the information we have devised a simple method of representing the surface features of the virus, which may be of broader interest. Essentially, the protomeric unit of the icosahedron is shown viewed in a direction normal to the surface. Each residue is then represented by a circle whose area corresponds to the surface accessible to a spherical probe. By choosing a probe radius of 10 Å we may crudely approximate an antibody combining site, which normally interacts with a 'footprint' of 20–30 Å in any direction. The results are shown in Fig. 1 for FMDV and HRV14, where we have marked the canyon as defined by Rossmann's group on HRV14 (note that there is no obvious canyon or pit on the surface of FMDV). We find the correlation of these accessibilities with the mutant-mapping results convincing. The correlation with B-factors is less convincing.

V. Peptide Vaccines

The success of a given peptide in eliciting a neutralizing antibody response against an intact antigen is likely to be governed by principles different from those considered above, the key requirement being that the peptide can mimic a sufficiently large area on the surface of the virus to form a good antibody combining site. Experience with antigen–antibody complexes suggests that the total area of interaction will be of the order of 700 $Å^2$ (Amit *et al.*, 1985); it has been concluded by other workers that virtually no such area of a protein surface will be made from a single continuous stretch of polypeptide chain (Barlow *et al.*, 1986). The crystallographic result for FMDV seems to provide a satisfying explanation for the activity of the region 140–160 in VP1. Either the loop is mobile, in which case its appear-

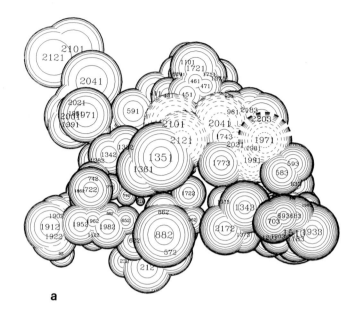

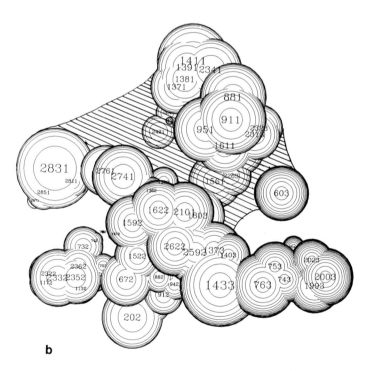

Figure 1. Views of the surface accessibility to a 10 Å probe of (a) FMDV, (b) HRV14. Residues lying behind the center of gravity of the protomer are not displayed. In (b), the cross-hatching indicates the extent of the canyon identified by Rossmann *et al.* (1985).

ance on the virus will be similar to that of a free peptide in solution, or the loop may have a well-defined conformation that arises from interactions within the loop rather than primarily with the rest of the capsid: Again, the free peptide should appear antigenically similar to the peptide as part of the virus (Acharya *et al.*, 1989b). In summary, FMDV presents an unusual structure, a small continuous portion of VP1 appearing to behave largely independently of the rest of the virus and, since it is exposed to an extreme extent, acting immunodominantly. If this result is generally true (and we would like to confirm it by reference to the structure of another serotype), it has important implications for the development of peptide vaccines and suggests that success stories such as FMDV are likely to remain in a small minority unless careful attention is paid to the conformational properties of the peptide.

VI. Cell Attachment Site

Not only does VP1 possess the major immunogenic site but there is a growing body of evidence that it is heavily involved in attachment of the virus to the cell:

1. All antibodies which are known to bind to the FMDV loop region prevent attachment.
2. Proteolytic cleavage of this loop or the C-terminal region of VP1 abolishes cell attachment (Wild *et al.*, 1969; Cavanagh *et al.*, 1977).
3. Short peptides including the highly conserved sequence Arg-Gly-Asp at residues 145–147 inhibit virus attachment to susceptible cells (Fox *et al.*, 1989).
4. The Arg-Gly-Asp sequence has been shown to have cell adhesion properties in a range of other systems, and indeed there is a family of molecules, the integrins, many of which interact with this sequence (Ruoslahti and Piersbacher, 1987).

We expect that receptor binding residues will be relatively conserved, and fortunately there are a large number of sequences available for VP1s of FMDV. Table I and Color Plate 20 present some of this information. Plate 20 shows the remarkable segregation in the degree of conservation of residues: Those on the exterior of the capsid are almost all extremely variable whereas the core residues are well conserved, as are the bulk of the internal residues. Ironically, the variability of the surface residues gives us a clear background of variation upon which two conserved surface regions stand out. The first of these, in the FMDV loop, cannot be shown on the structure in Plate 20, but is represented in Table I, in which the

Table I. FMDV Amino Acid Sequences in Region of VP1 Residues 141–152

V P N L	R G D	L Q V L A	141-152 FMDV VP1 (O_1BFS)
V P N V	R G D	L Q V L A	141-152 FMDV VP1 (O_1camp)
V P N V	R G D	T Q V L D	141-152 FMDV VP1 (O_6V_1)
M S N V	R G D	L Q V L T	141-152 FMDV VP1 (O_{HK})
T A S T	R G D	L A H L T	141-152 FMDV VP1 (C_3obb)
A S A R	R G D	L A H L A	141-152 FMDV VP1 (C_3 ind)
R E N I	R G D	L A T L A	141-152 FMDV VP1 (SAT 1)
V T A I	R G D	R E V L A	141-152 FMDV VP1 (SAT 2)
V T P R	R G D	M A V L A	141-152 FMDV VP1 (SAT 3)
E P T M	R G D	R A V L A	141-152 FMDV VP1 (ASIA 1)
G G P R	R G D	M G S A A	141-152 FMDV VP1 (A_5)
G S G V	R G D	F G S L A	141-152 FMDV VP1 ($A_{12}P1$)
G S G R	R G D	M G S L A	141-152 FMDV VP1 (A_{24}BW)
G S G R	R G D	M G T L A	141-152 FMDV VP1 ($A_{24}P1$)
A S D S	R S G D	L G S I A[a]	141-153 FMDV VP1 (A_{10}61)

[a] Has RSGD in place of RGD.

significance of the Arg-Gly-Asp conservation in the region 145–147 is immediately apparent. The second region is the C-terminus of VP1; although the conservation is not as rigid as for residues 145–147 it is a striking feature of Plate 20. We feel that these structural features, taken together with the evidence already presented, make the case for involvement of the residues 145–147 extremely compelling and also imply the likely involvement of some of the C-terminal residues of VP1. In the light of this it is very interesting to note that there appears to be some similarity between the sequence of the C-terminal residues of VP1 and the C-terminal residues of the γ chain of fibrinogen, a member of the integrin binding family.

VII. Uncoating, Assembly, and Disassembly

Our X-ray structure is not in itself informative about these aspects of the viral life cycle; however, work by Michael Rossmann's group (Smith *et al.*, 1986; Badger *et al.*, 1988) has led to insights into these processes in human rhinoviruses (HRV) and to proposals that similar mechanisms may apply far more widely, perhaps even to HIV (Rossmann, 1988). Essentially, by examining crystallographically the interaction of compounds that prevent uncoating of rhinovirus, Rossmann's group have concluded that a pocket which exists within the hydrophobic interior of the β-barrel of VP1 plays a crucial role in the disassembly of the capsid. They proposed that the capsid is normally locked together rather like a jigsaw; compression of the barrel allows a key piece to unlock, triggering the dissolution of the capsid. The success of certain anti-HRV compounds is then explained by their action in filling the pocket and thus preventing the unlocking of VP1. A key problem is to determine whether the compounds (or related compounds) bind to FMDV. Although we do not yet have an answer to this question, the hypothesis would predict that even if binding was prevented by, for instance, the entrance to the binding pocket being occluded, then the pocket itself should still exist. The drug binding pocket is illustrated in Color Plate 21 for HRV14, Mengo virus, and FMDV. It is striking that in both FMDV and Mengo virus three aromatic residues invade the volume occupied by the WIN compounds, and indeed the rings of the WIN compounds as bound to HRV14 seem almost to be mimicking the side chains. This "mimicking" may occur *in vivo* since it has been observed that the pocket is occupied in the native poliovirus crystals, and it has been proposed (J. M. Hogle, personal communication) that *in vivo* the binding site is occupied by cellular material. In FMDV the volume of the pocket is reduced by approximately 40% compared to HRV14, suggesting that simple compression of the β-barrel of VP1 may not be a general mechanism for the uncoating of the picornaviruses.

We intend to attempt to exploit the particular physical and chemical properties of FMDV, in particular its acid lability, in designing anti-FMDV compounds. FMDV is the most acid-labile of the picornaviruses; below pH 7 the virus disrupts into pentameric arrays of VP1–3, releasing VP4 and the RNA. Since this acid lability is conserved for all FMDVs we suspect that it is functionally important. It is interesting that along the seams where the pentameric assemblies fit together in the virus there appears to be an unusually high concentration of histidine residues which would, by nature of their ionizability in the required pH range, be ideal as pH-dependent "switches."

VIII. Conclusions

The structure of FMDV has provided a satisfying understanding of many of its properties although there remain a number of outstanding questions. Regarding the use of the structure as a target for drug design by the method of receptor fit, we are only beginning to plan our first steps along a path which may or may not lead to a rainbow of therapeutic agents.

References

Acharya, R., Fry, E., Stuart, D. I., Fox, G., Rowlands, D., and Brown, F. (1989a). The three-dimensional structure of foot and mouth disease virus at 2.9 Å. *Nature (London)* **337**, 709–716.

Acharya, R., Fry, E., Stuart, D., Fox, G., Rowlands, D., and Brown, F. (1989b). Implications of the three-dimensional structure of foot and mouth disease virus for its antigenicity and cell attachment. (R. A. Lerner, H. Ginsberg, R. M. Chanock, and F. Brown eds.) pp. 1– 7. *Vaccines '89, Cold Spring Harbor Lab.*

Amit, A. G., Maruizza, R. A., Phillips, S. E. V., and Polijak, R. J. (1985). Three-dimensional structure of an antigen–antibody complex at 6 Å. *Nature (London)* **313**, 156–158.

Badger, J., Minor, I., Kremer, M. J., Oliveria, M. A., Smith, T. J., Griffith, J. P., Guerin, D. M. A., Krishnaswamy, S., Luo, M., Rossmann, M. G., McKinlay, M. A., Diana, G. D., Dutko, F. J., Fancher, M., Rueckert, R. D., and Heinz, B. A. (1988). Structural analysis of a series of antiviral agents complexed with human rhinovirus 14. *Proc. Natl. Acad. Sci. USA* **85**, 3304–3308.

Barlow, D. J., Edwards, M. S., and Thornton, J. M. (1986). Continuous and discontinuous protein antigenic determinants. *Nature (London)* **322**, 747–750.

Bittle, J. L, Houghten, R. A., Alexander, H., Shinnick, T. M., Sutcliffe, J. G., Lerner, R. A., Rowlands, D. J., and Brown, F. (1982). Protection against foot-and-mouth disease by immunization with a chemically synthesized peptide predicted from the viral nucleotide sequence. *Nature (London)* **298**, 30–33.

Bricogne, G. (1976). Methods and programs for direct-space exploitation of geometric redundancies. *Acta Crystallogr., Sect. A* **A32**, 832–847.

Brunger, A. T., Kuriyan, J., and Karplus, M. (1987). Crystallographic R factor refinement by molecular dynamics. *Science* **235**, 458–460.

Cavanagh, D., Sangar, D. V., Rowlands, D. J., and Brown, F. (1977). Immunogenic and cell attachment sites in FMDV: Further evidence for their location in a single capsid polypeptide. *J. Gen. Virol.* **35**, 149–158.

Chow, M., Newman, J. F. E., Filman, D., Hogle, J. M., Rowlands, D. J., and Brown, F. (1987). Myristylation of picornavirus capsid protein VP4 and its structural significance. *Nature (London)* **327**, 482–486.

Colonno, R. J., Condra, J. H., Mizutani, S., Callahan, P. L., Davies, M., and Murcko, M. A. (1988). Evidence for the direct involvement of the rhinovirus canyon in receptor binding. *Proc. Natl. Acad. Sci. USA* **85**, 5449–5453.

Cooper, P. D., Agol, V. I., Bacharach, H. L., Brown, F., Ghendon, Y., Gibbs, A. J., Gillespie, J. H., Lonberg-Holm, K., Mandel, B., Melnick, J. L., Mohanty, S. B., Povey, R. C., Rueckert, R. R., Schaffer, F. L., and Tyrrell, D. A. (1978). Picornaviridae: Second report. *Intervirology* **10**, 165–180.

Fox, G., Stuart, D., Acharya, K. R., Fry, E., Rolands, D., and Brown, F. (1987). Crystallization and preliminary X-ray diffraction analysis of foot and mouth disease virus. *J. Mol. Biol.* **196**, 591–597.

Fox, G., Parry, N. V., Barnett, P. V., McGinn, B., Rowlands, D. J., and Brown, F. (1989). The cell attachment site on foot and mouth disease virus includes the sequence RGD. *J. Gen. Virol.* **70**, 625–637.

Hogle, J. M., Chow, M., and Filman, D. J. (1985). Three-dimensional structure of poliovirus at 2.9 Å resolution. *Science* **229**, 1358–1365.

Luo, M., Vriend, G., Kamer, G., Minor, I., Arnold, E., Rossmann, M. G., Boege, U., Scraba, D. G., Duke, G. M., and Palmenberg, A. C. (1987). The atomic structure of Mengovirus at 3.0 Å resolution. *Science* **235**, 182–191.

Newman, J. F. E., Rowlands, D. J., and Brown, F. (1973). A physico-chemical sub-grouping of the mammalian picornaviruses. *J. Gen. Virol.* **18**, 171–180.

Novotný, J., Handschumacher, M., Haber, E., Bruccoleri, R. E., Carlson, W. B., Fanning, D. W., Smith, J. A., and Rose, G. D. (1986). Antigenic determinants in proteins coincide with surface regions accessible to large probes (antibody domains). *Proc. Natl. Acad. Sci. USA* **83**, 226 230.

Palmenberg, A. C. (1989). Sequence Alignment of Picornavirus Capsid Proteins in Molecular Aspects of Picornavirus Infection and Detection (Bert L. Semter and Ellie Ehrenseld, eds), pp. 211–241. American Society for Microbiology, Washington D.C.

Paul, A. V., Shultz, A., Pincus, S. E., Oroszlan, S., and Wimmer, E. (1987). Capsid protein VP4 of poliovirus is N-myristoylated. *Proc. Natl. Acad. Sci. USA* **84**, 7827–7831.

Pfaff, E., Mussgay, M., Bohm, H. O., Schulz, G. E., and Schaller, H. (1982). Antibodies against a preselected peptide recognize and neutralize foot-and-mouth disease virus. *Embo J.* **1**, 869–874.

Rossmann, M. G. (1988). Antiviral agents targeted to interact with viral capsid proteins and a possible application to HIV. *Proc. Natl. Acad. Sci. USA* **85**, 4625–4627.

Rossmann, M. G., Arnold, E., Erickson, J. W., Frankenberger, E. A., Griffith, J. P., Johnson, J. E., Kamer, G., Luo, M., Mosser, A. C., Rueckert, R. R., Sherry, B., and Vriend, G. (1985). Structure of a human common cold virus and functional relationship to other picornaviruses. *Nature (London)* **317**, 145–153.

Ruoslahti, E., and Piersbacher, M. D. (1987). New perspectives in cell adhesion: RGD and integrins. *Science* **238**, 491–497.

Smith, T. J., Kremer, M. J., Luo, M., Vriend, G., Arnold, E., Kamer, G., Rossmann, M. G., McKinlay, M. A., Diana, G. D., and Otto, M. J. (1986). The site of attachment in human rhinovirus 14 for antiviral agents that inhibit uncoating. *Science* **233**, 1286–1293.

Strohmaier, K., Franze, R., and Adam, K. H. (1982). Location and characterization of the antigenic portion of the FMDV immunizing protein. *J. Gen. Virol.* **59**, 295–306.

Tainer, J. A., Getzoff, E. D., Alexander, H., Houghten, R. A., Olson, A. J., Lerner, R. A., and Hendrickson, W. A. (1986). The reactivity of anti-peptide antibodies is a function of the atomic mobility of sites in a protein. *Nature (London)* **312**, 127–134.

Westhof, E., Altschuh, D., Moras, D., Bloomer, A. C., Mondragon, A., Klug, A., and Van Regenmortel, M. H. V. (1984). Correlation between segmental mobility and the location of antigenic determinants in proteins. *Nature (London)* **311**, 123–126.

Wild, T. F., Burroughs, J. N., and Brown, F. (196). Surface structure of foot-and-mouth disease virus. *J. Gen. Virol.* **4**, 313–320.

13 Escape Mutant Analysis of a Drug-Binding Site Can Be Used to Map Functions in the Rhinovirus Capsid

Beverly A. Heinz, Deborah A. Shepard, and
Roland R. Rueckert
Institute for Molecular Virology, University of Wisconsin, Madison,
Wisconsin 53706

I. Introduction

WIN compounds are a second generation of neutralizing antivirals derived from arildone (McSharry *et al.*, 1979); they neutralize infectivity of picornaviruses by binding reversibly to the virus capsid. These drugs either block uncoating, as reported for poliovirus types 1 and 2 and for human rhinovirus (HRV) type 2 (Fox *et al.*, 1986; Zeichhardt *et al.*, 1987), or they prevent attachment to the cell receptor, as observed for HRV14 (Pevear *et al.*, 1989). We selected the compound WIN 52084 (Fig. 1) for study because it is highly active against our target virus, HRV14, whose protein structure is known in atomic detail (Rossmann *et al.*, 1985). WIN 52035, though less active against HRV14, was used to study how drug structure influences virus resistance.

The protein shell of HRV14 (Fig. 2) is composed of 60 protomers, each containing three external polypeptides (VP1, VP2, VP3) and one internal polypeptide (VP4, not shown). The WIN compounds insert into a hydrophobic pocket within the β-barrel of VP1; this pocket lies just beneath the floor of the "canyon" (Smith *et al.*, 1986), the region postulated to contain the cell receptor site (Colonno *et al.*, 1988; Rossmann *et al.*, 1985; Rossmann and Palmenberg, 1988; Rueckert *et al.*, 1986). Each of the 60 protomers in the shell is capable of binding a drug molecule (Smith *et al.*, 1986). In HRV14, drug binding induces conformational changes in three regions of VP1 that comprise both the roof of the drug-binding pocket and the canyon floor (Badger *et al.*, 1988). WIN 52084 enters the pocket with the isoxazole ring pointing toward the viral fivefold axis. Many related WIN compounds, including WIN 52035, bind in the reverse orientation; nevertheless, all of the active compounds examined so far induce similar conformational changes in wild-type virus (Badger *et al.*, 1988).

Our goals are to investigate the mode of action of these drugs and to use resistant mutants as tools to map virus functions such as attachment and

Use of X-Ray Crystallography in the Design of Antiviral Agents
173

oxazoline   phenoxy   aliphatic chain   isoxazole

WIN 52084

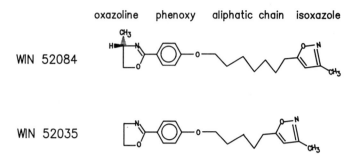

WIN 52035

Figure 1. Structures of WIN 52084, (−)-5-[7-[4-(4,5-dihydro-4-methyl-2-oxazolyl)phe-noxy]heptyl]-3-methylisoxazole, and WIN 52035, 5-[5-[4-(4,5-dihydro-2-oxazolyl)phenoxy]-pentyl]-3-methylisoxazole, the compounds used to select and characterize HRV14 escape mutants.

uncoating on the rhinovirus capsid. We have previously identified the residues on the virus structure that can interfere with drug function (Heinz *et al.*, 1989). Here we provide evidence for two different mechanisms of resistance in HRV14 and show how characterizing the phenotypes of WIN-resistant mutants can be used to identify regions of the capsid involved in different stages of infection.

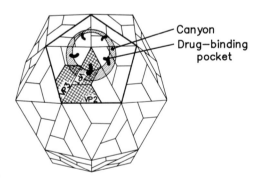

Canyon
Drug–binding
pocket

Figure 2. Schematic representation of the HRV14 capsid and location of drug-binding sites in the floor of the canyon circling each viral fivefold axis (outlined in boldface). The protomer, composed of VP1, VP2, VP3, and VP4 (not shown), is hatched. Insertion of WIN compound into the binding pocket pushes up the floor of the canyon about 4 Å, roughly one-sixth the canyon depth (Smith *et al.*, 1986). Correlated with this upheaval is a block in attachment of virus in a cell-free receptor system (Pevear *et al.*, 1989) and to HeLa cell suspensions (see Fig. 7).

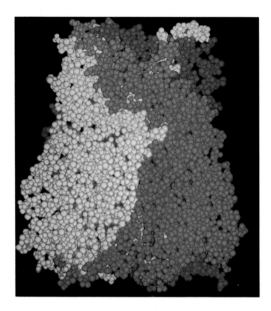

Color Plate 1. The hexon trimer, with each subunit separately colored to illustrate the intersubunit binding topology in hexon. The viewing orientation is perpendicular to the threefold symmetry axis. Each subunit "clamps" its partner on its left from below with its N-terminal loop and from above with its ℓ_2 and ℓ_1 loops. The ℓ_2 loop rests upon its left neighbor's ℓ_1 domain, and its ℓ_1 loop rests upon the same neighbor's ℓ_4 loop.

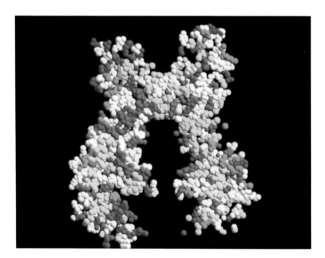

Color Plate 2. A vertical cross section, 10 Å thick, cut through the center of hexon. The atoms, represented by shaded spheres, have been colored in the Eisenberg "consensus" hydrophobicity scale (Eisenberg, 1984). Hydrophobic amino acids are shown in shades of yellow-orange and hydrophilic amino acids in shades of blue. The bottom of the section cuts through the P1 and P2 domains from different subunits, and the top cuts the interiors of two of the three tower domains, T. Each tower includes parts of the ℓ_2, ℓ_1, and ℓ_4 loops in descending order from its top.

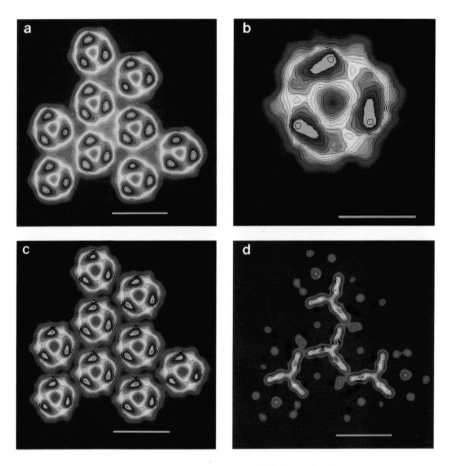

Color Plate 3. Contour plots of the images used in the STEM investigation. (a) The average of 57 STEM images of GONs. Note the equivalent trimeric appearance of the three independent hexon images, and the four large and three small cavities between the hexons at the local threefold axes. (b) The hexon image as derived from the X-ray model. (c) The artificial GON constructed by placing nine copies of the X-ray hexon shown in (b) at the positions observed in the average GON image (a). Note the similarity of this image with (a), except within the four large cavities between hexons. (d) The difference image formed by subtracting (c), the scaled artificial GON image of hexons alone, from (a), that of the STEM image of natural GONs containing both hexons and polypeptide IX. Polypeptide IX is revealed as four clusters of three molecules, located in the large cavities between hexons and extending along the hexon–hexon interfaces. Scale bars equal 100 Å.

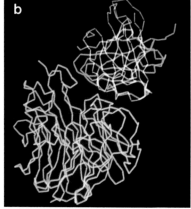
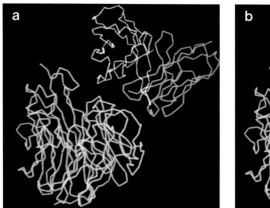

Color Plate 4. C_{α} chain tracings of the complexes between (a) whale NA and antibody NC10 Fab and (b) tern NA and antibody NC41 Fab. Only the V module is shown; in the case of the NC10 complex the C module is not visible in the electron density map (Colman *et al.*, 1989).

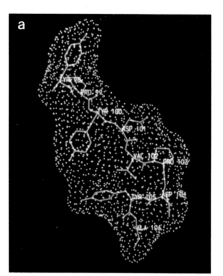

Color Plate 5. Diagrammatic illustration of the problem raised by the dual recognition of both peptide and protein by antipeptide antibodies. Although the peptide is flexible in solution and completely accessible for interacting with the antibody, only a restricted portion of the peptide in a defined conformation is available for interaction when embedded in the intact protein. (a) Conformation of sequence 98–106 as found in the crystal structure of the influenza virus. The dot surface represents the solvent-accessible surface. This conformation differs from that found in solution, which is notable for the presence of a β-turn involving residues 98–101 (Dyson *et al.*, 1985, 1988a). (b) Solvent-accessible surface around sequence 98–106 in the hemagglutinin monomer structure. Within the sequence 98–106 only Tyr 100, Asp 101, Pro 103, Asp 104, Tyr 105, and Ala 106 are accessible (gray shading). These residues correspond to those which have been assigned to the determinant based on antibody binding data and represent a surface area of only 16 × 15 × 7 Å (From Wilson *et al.*, 1984.)

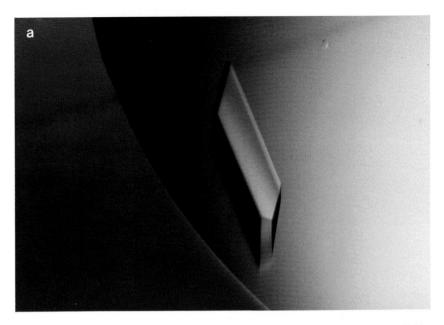

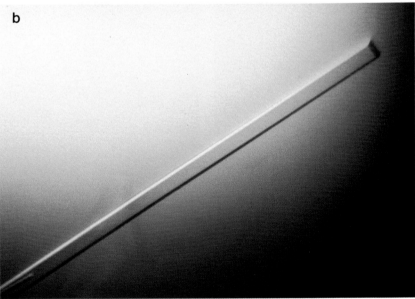

Color Plate 6. (a) Photomicrograph of a native Fab 17/9 crystal. These high quality prismatic crystals were grown from PEG 600 imidazole–malate buffer, pH 5.6–6.5. These crystals diffract to at least 1.9 Å resolution. The crystals grow to dimensions of 0.6 × 0.2 × 0.2 mm. (b) Photomicrograph of an Fab 17/9–peptide complex crystal. These crystals grow as long, thin platelike rods in sodium acetate buffer, pH 5.5, containing PEG 600. The crystals diffract to 2.1 Å resolution. The crystals grow to dimensions of 1.0 × 0.02 × 0.01 mm.

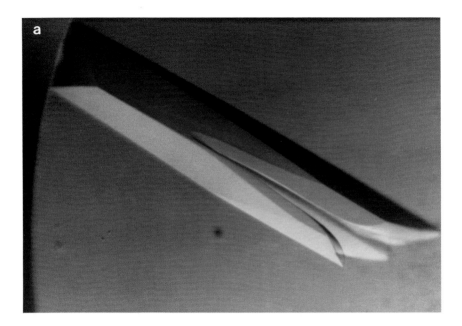

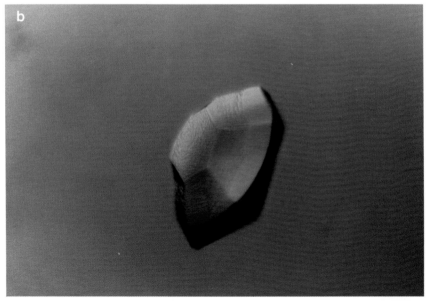

Color Plate 7. (a) Photomicrograph of a native Fab I2 crystal. The crystals are grown from sodium citrate solutions, pH 5.0–7.0, containing small amounts of MPD and diffract to 2.6 Å resolution. The crystals grow as rectangular rods with dimensions of up to 1.0 × 0.2 × 0.2 mm. (b) Photomicrograph of an Fab I2–peptide complex crystal. These crystals are grown from sodium phosphate solutions, pH 5.75, and diffract to 2.6 Å resolution. The crystals grow as hexagonal plates with dimensions of up to 1.0 × 0.2 × 0.2mm.

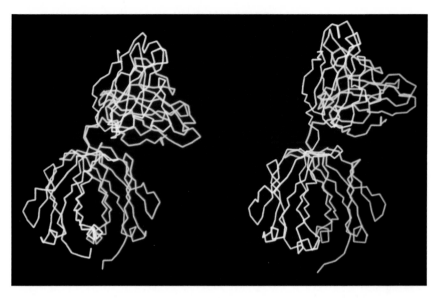

Color Plate 8. Comparison of the C$_\alpha$ tracings of the Mcg Bence-Jones dimer crystallized in ammonium sulfate (left) and in deionized water (right). Monomer 1, the heavy chain analog, is colored orange and monomer 2 is green. The "constant" (C) domains of monomer 1 (lower right domain in each model) are placed in the same orientations.

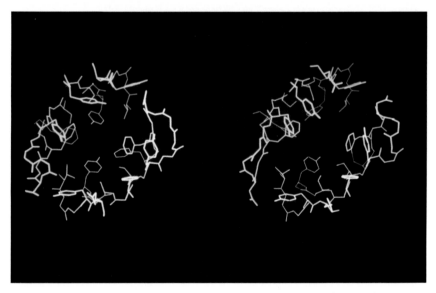

Color Plate 9. End-on views into the main binding cavities of the two forms of the Mcg Bence-Jones dimer. The trigonal form (ammonium sulfate) is on the left and the orthorhombic form (deionized water) is on the right. As in Color Plate 8, monomer 1 constituents are colored orange and monomer 2 components are in green. Note the differences in the shapes and sizes of the cavities and in the orientations of the side chains.

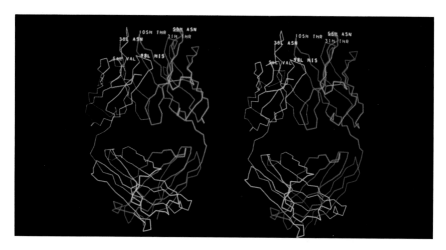

Color Plate 10. Stereo diagram of the C_α skeletal models of the light (blue) and heavy (red) chain components of the BV04-01 Fab, which binds single-stranded DNA. Interchain disulfide bonds within individual domains are represented by gold bars between appropriate strands of the β-pleated sheets. The interchain disulfide bond at the bottom of the tracing connects the carboxyl terminal residue (219) of the light chain with residue 136 of the heavy chain. The three complementarity-determining regions (CDR) (Kabat *et al.*, 1977) are identified by the labeling of residues within these loops (e.g., Asn 35, Val 56, and His 96 for CDR 1, 2, and 3 of the light chain; Thr 31, Asn 56 and Thr 105 for the heavy chain.

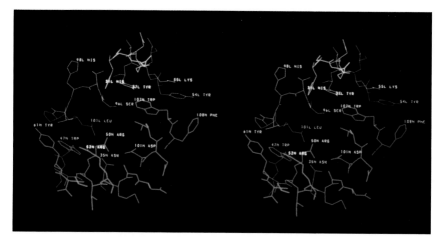

Color Plate 11. Stereo diagram of skeletal models of selected residues in the putative binding site for single-stranded DNA in the BV04-01 Fab. Note the set of aromatic residues from left to right at the top of the diagram: histidines 98 and 31 and tyrosines 37 and 54 of the light chain and tryptophan 107 and phenylalanine 108 of the heavy chain. In addition to the histidines, which would be protonated at lower pH values, there are three positively charged side chains available for possible interactions with phosphate groups of DNA: lysine 55 of the light chain and arginines 50 and 52 of the heavy chain. Access to a potential binding cavity in the $V_L - V_H$ interface is blocked by a cluster of side chains; serine 96 and leucine 101 of the light chain and the ion pair formed by arginine 50 and aspartic acid 101 of the heavy chain.

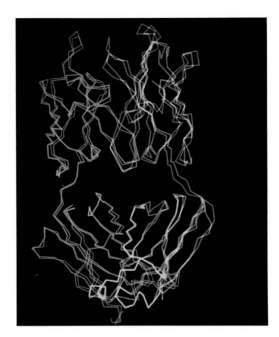

Color Plate 12. Superposition of the C_α tracings of the unliganded BV04-01 and liganded 4-4-20 Fabs to illustrate the strong similarities in overall structures. The BV04-01 Fab was crystallized in ammonium sulfate and the fluorescein-Fab (4-4-20) complex was crystallized in 2-methyl-2,4-pentanediol.

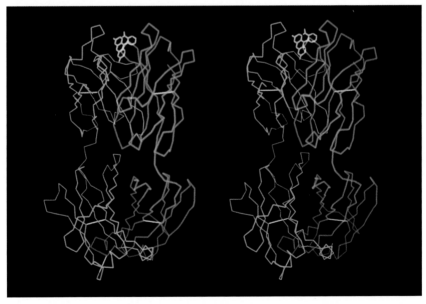

Color Plate 13. Stereo diagram of the C_α model of the 4-4-20 Fab, cocrystallized with fluorescein. The light chain is colored blue, the heavy chain is red, and the hapten is green.

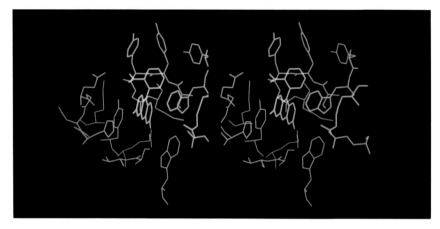

Color Plate 14. Stereo diagram of the hapten binding site of the 4-4-20 Fab. Light (L) chain constituents are blue, heavy (H) chain components are red, and fluorescein is green. The xanthonyl group (three ring) of fluorescein is involved in a stacking interaction with tyrosine L37 on the left and is flanked by tryptophan H33 on the right. Tryptophan L101 forms the bottom of the slot. Enolic oxygen atoms on opposite corners of the xanthonyl ring interact with arginine L39 (ion pair) and histidine L31 (hydrogen bond). The phenyl carboxyl group of fluorescein (single ring) is located below tyrosines H103 (left) and H102 (right), which form the top of the binding slot.

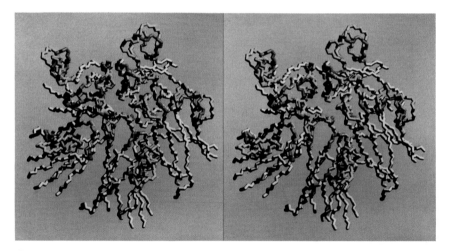

Color Plate 15. Structural differences in three strains of poliovirus in the vicinity of the particle fivefold axis. This stereo representation shows main chain atoms from five symmetry-related copies of the narrow end of the β-barrel of capsid protein VP1, viewed from the outside of the virion. Note that only a small portion of VP1 is included. Structurally conserved atoms (principally in the eight strands of the antiparallel β-barrel) are shown in white. Conformational differences (which are observed in three of the loops that connect the β-strands) are indicated in cyan (P1/Mahoney), yellow (P3/Sabin), and magenta (type 2/type 1 chimera). Structural differences as large as 8 Å are seen in the B–C loops (the colored loops furthest from the center of view). Smaller, but significant, structural differences are also seen in the D–E loops (which are the colored loops closest to the center of view).

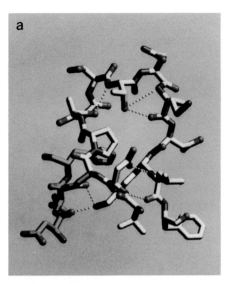

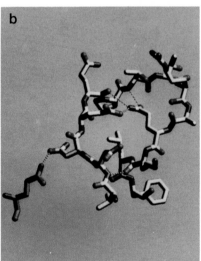

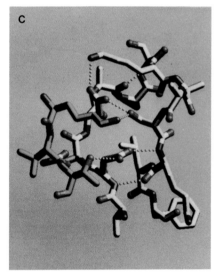

Color Plate 16. The atomic structure of the B–C loop in P1/Mahoney (a), P3/Sabin (b), and the V510 chimera (c). Oxygen atoms are red and nitrogens are dark blue. The carbon atoms shown in white belong to the B–C loop, while those shown in light blue belong to adjacent portions of the protein which make structurally important interactions with the loop. The yellow dotted lines represent hydrogen bonds which stabilize the B–C loop and which appear to play significant roles in determining its conformation.

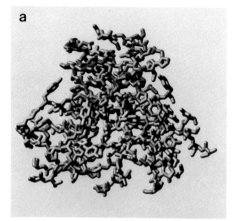
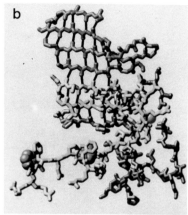

Color Plate 17. Assembly-dependent structures which contribute substantially to the stability of the poliovirion. (a) The complex formed by the interaction of the amino termini of VP3, VP4, and VP1 around each fivefold axis appears to be important for directing the assembly of protomers to form pentamers. The inner surface of the capsid (which is located roughly 90 Å from the center of the particle) is shown at the bottom. The top of the complex is located at a radius of about 130 Å (as compared with the outer surface of the virion, which extends to a radius of 165 Å near the fivefold axis). At the top of the view, five intertwined amino termini of VP3 (residues 1–11, shown in red) form a twisted tube of parallel β-structure. Five copies of the amino terminus of VP4 (residues 2–8 and 23–33, shown in green with yellow sidechains) form short segments of two-stranded antiparallel β-sheet on the inner surface of the capsid. The amino-terminal glycine of each copy of VP4 is linked via an amide bond to a myristic acid moiety (shown in magenta). Each of the five-residue segments of polypeptide chain (shown in blue) is believed to be a portion of the amino terminus of VP1. Although the electron density for this chain is consistent with a parallel β-main chain hydrogen bonding pattern, the electron density for the side chains are not sufficiently well resolved to correlate with the capsid sequence of the virus. (b) The seven-stranded extended β-sheet which is located near the threefold axis of the virion stabilizes the association of neighboring pentamers. The outer surface of the capsid is at the top, and the inner surface of the capsid is at the bottom. Black lines indicate the main chain hydrogen bonds of the extended β-sheet. At the top, the C, H, E, and F strands of the VP3 β-barrel (shown in red) form the outermost four strands of the extended sheet. The seventh strand (shown in blue) is formed by residues 36–38 of VP1 from the same protomer. In the foreground, residues from a threefold-related protomer are shown. The amino-terminal extension of VP2 (yellow) forms the fifth and sixth strands of the β-sheet, interacts with the carboxyl terminus of VP4 (green), and binds two planar electron density features (shown in magenta) which tentatively have been identified as nucleotides of the viral RNA. The large orange sphere represents the side chain hydroxyl group of Ser 10 of VP2. This group is hydrogen bonded to the cleaved carboxyl terminus of VP4 in the mature virion and may participate in the mechanism of VP0 cleavage. In the background, the carboxyl terminus of VP4 from the first protomer is shown in green.

Color Plate 18. Mutations observed in nontemperature sensitive revertants of type 3 poliovirus, showing the portion of the P3/Sabin structure which is within 40 Å of Phe 91 of VP3. The virion is viewed from the outside and is oriented so that a fivefold axis is at the top, a threefold axis is at the bottom right, and a twofold axis is at the bottom center. Main chain atoms are depicted as thin tubes. Large spheres indicate the atoms of side chains that have been found to mutate in one or more of the revertants. Phe 91 is shown in green. VP1 is blue; VP2 and VP4 are yellow; VP3 is red. The extended hydrocarbon which occupies the center of the VP1 β-barrel is magenta. The intact model has been expanded to separate the protomers and illustrate the tendency of the mutations to cluster in the interfaces between protomers, or in the hydrocarbon binding site.

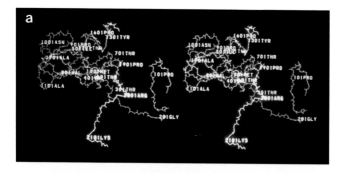

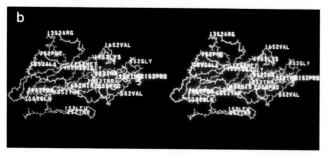

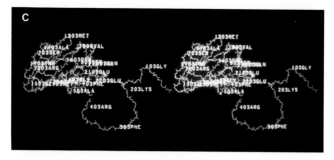

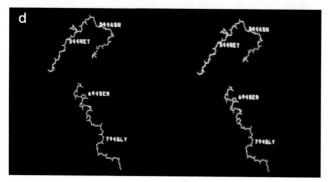

Color Plate 19. Stereo views of main chain atom traces of the individual capsid proteins of FMDV; (a) VP1, (b) VP2, (c) VP3, (d) VP4.

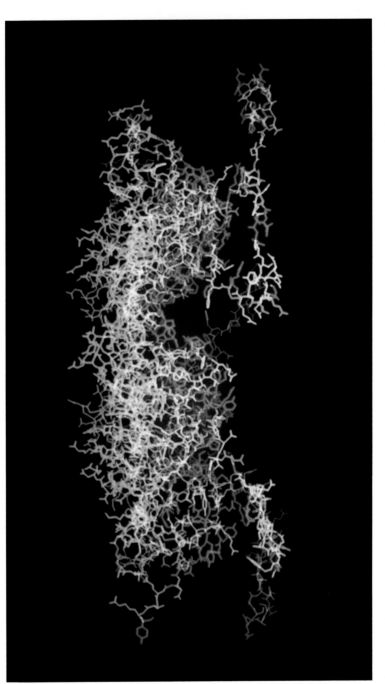

Color Plate 20. A side view of a pentamer of VP1s of FMDV illustrating the segregation in the degree of conservation of residues. The orientation is such that the capsid surface is horizontal with the external surface uppermost. Residues conserved across all the serotypes are colored blue, those which are highly variable are colored red, and those intermediate are colored purple.

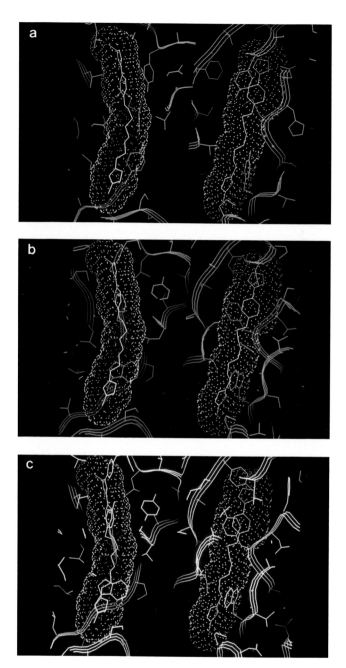

Color Plate 21. The WIN compound binding pocket in (a) HRV14, (b) Mengovirus and (c) FMDV. The coordinates and van der Waals surface displayed are as determined by Badger *et al.* (1988). Part of the pocket is occupied by three aromatic residues on FMDV and Mengo.

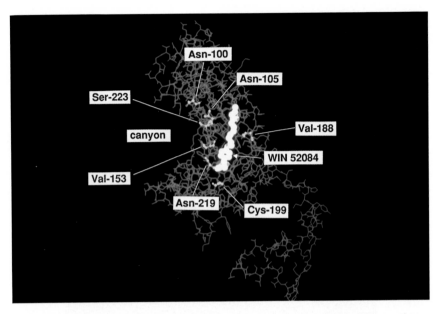

Color Plate 22. Molecular graphics diagram of VP1 showing the locations of HR and LR mutations relative to WIN 52084 and the canyon floor.

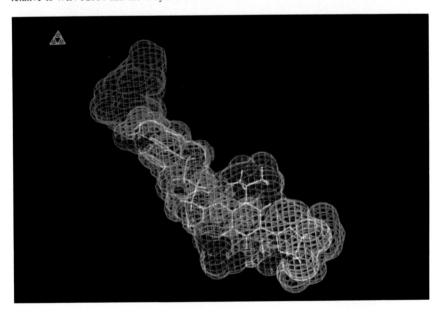

Color Plate 23. Overlay of active compounds (blue) and inactive compounds (red). The structures of seven compounds active against HRV14 were overlaid by the program SYBYL (Tripos Associates, Inc., St. Louis, MO) according to their X-ray coordinates when bound to HRV14. Volume maps were generated from van der Waals surfaces and compared to volume maps generated from seven inactive compounds.

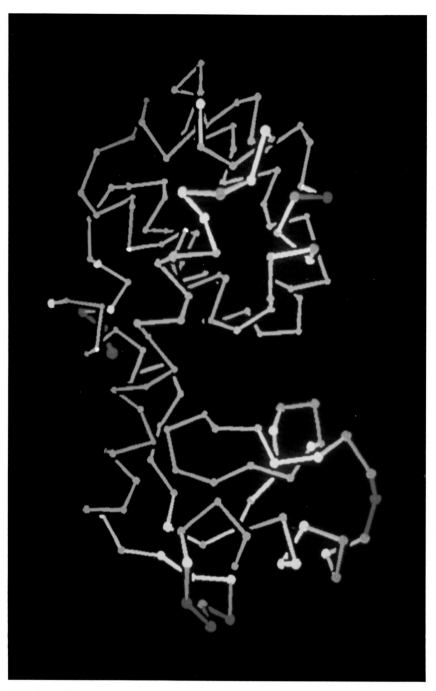

Color Plate 24. Backbone of phage T4 lysozyme colored according to the mobility displayed in the crystal structure. Red regions have greatest mobility, blue regions have least.

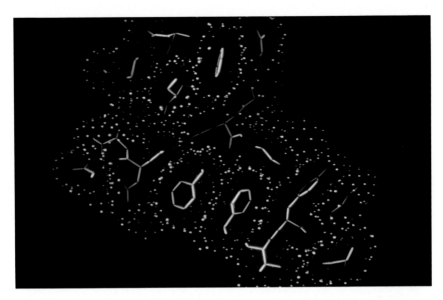

Color Plate 25. A section through the interior of bovine pancreatic trypsin inhibitor. The internal van der Waals dot surfaces mesh together with minimal interpenetration, illustrating the tight packing of side chains. Amino acids are color coded as follows: Hydrophobics are green, acidics red, basics blue, and alcohols orange. Note the "herringbone" packing of the two phenylalanines at the top.

Color Plate 26. Illustration of the method for calculating shared surface area. Each dot from the surface of the phenylalanine is assigned (indicated by a white line) to the closest atom of a nonbonded neighbor. If no atom is within 3.75 Å the dot is assigned a water, which is tested for overlap. The various assignments are then tabulated and converted to area. The color coding is by atom type as follows: Carbons are green, nitrogens blue, and oxygens red.

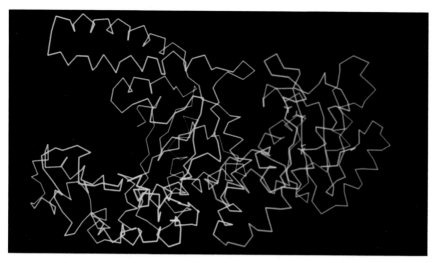

Color Plate 27. Backbone drawing of the structure of Klenow fragment determined by X-ray crystallography (Ollis *et al.*, 1985). The polymerase domain is colored yellow and the 3'-5'-exonuclease domain is magenta.

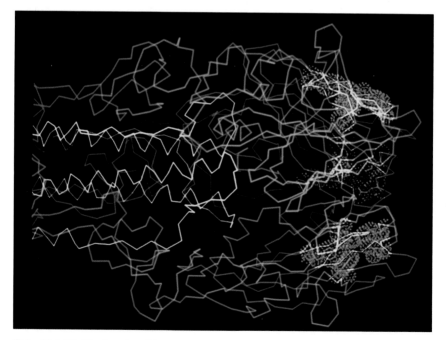

Color Plate 28. Distal portion of the influenza hemagglutinin trimer (Wilson *et al.*, 1981) showing the location of the receptor recognition site. Recognition of sialic acid on the receptor occurs in a recessed pocket within the site, shielded from neutralizing antibodies. The HA$_1$ chains are colored orange, the HA$_2$ chains yellow, and the receptor attachment site residues and van der Waals surface green.

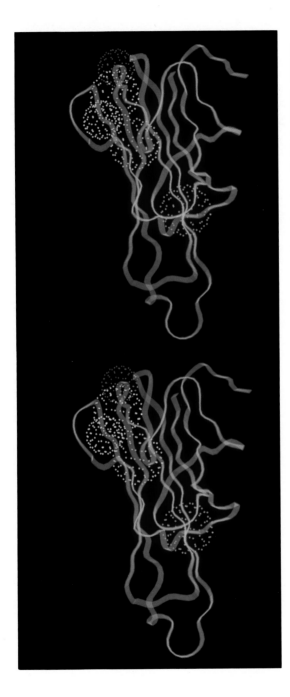

Color Plate 29. The locations of proposed receptor binding sites in HA, HRV14 and FMDV mapped onto the polypeptide fold for TNF. The FMDV loop is marked by a red sphere. Areas which form the canyon floor in VP1 of HRV14 are denoted by yellow dots, and the approximate position of the sialic acid binding site of HA is indicated by the light blue sphere.

II. Results

A. Isolation of Mutants and Identification of High and Low Resistance

Our first goal was to determine whether drug-resistant mutants occurred at a high enough frequency that they might be due to single mutational events. To this end, we hoped to determine the fraction of drug-resistant mutants that exist in a population of HRV14 by measuring the proportion of virus that could form plaques in the presence of different concentrations of drug (Fig. 3). As the drug concentration was increased from 0.05 to 1.0 μg/ml, the curve descended rapidly until it reached a plateau corresponding to about 4×10^{-5} survivors. This frequency was in the same range as that previously observed for single-step mutations with antibody escape mutants of HRV14 (Rueckert *et al.*, 1986).

The distinctive shape of the dose–response curve (Fig. 3) suggested that there were two classes of drug resistance: a high-resistance (HR) class, able to form plaques in the presence of 2 μg/ml of WIN 52084; and a low-resistance (LR) class only able to form plaques in the presence of $\leq$0.4 μg/ml. The HR mutants occurred with a frequency of about

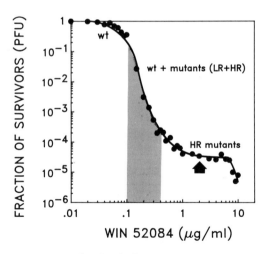

Figure 3. Dose–response curve showing the frequency of plaque formers resistant to different concentrations of WIN 52084. A stock of HRV14 (produced using a low multiplicity of infection) was pretreated and plaqued in the presence of drug, 0.4% bovine serum albumin, and 0.1% dimethylsulfoxide. Surviving plaques were counted after 48 hr incubation at 35°C. High-resistance (HR) mutants were selected in 2 μg/ml drug (arrow); low-resistance (LR) mutants were isolated in 0.1–0.4 μg/ml drug (shaded region); wt, wild-type virus.

4×10^{-5}, while the LR mutants occurred with a 10- to 30-fold higher frequency. In order to identify the locations of the mutations, we isolated over 80 spontaneous drug-resistant mutants of HRV14. Each mutant was selected from an individually amplified wild-type plaque, thus ensuring that each isolate represented a separate mutational event.

B. High-Resistance Mutations Map to Only 2 of the 16 Amino Acids Lining the Drug Binding Pocket

The RNA genomes of 64 HR mutants and 17 LR mutants were sequenced by primer extension analysis in regions encoding the walls of the drug-binding site. Nearly all of the mutations encoded single amino acid substitutions and many were identical (Table I). In fact, all of the HR mutations mapped to just 2 of the 16 amino acid residues that form the walls of the drug-binding pocket (Fig. 4, Color Plate 22). The side chains of these two residues, V188 and C199 of VP1, were invariably replaced by bulkier groups that pointed into the drug-binding pocket. These findings suggested

Table I. HRV14 Mutants Resistant to WIN 52084

Phenotype	Mutation	No. Mutants 52084[a]	52035[b]
HR	Cys 199 → Trp	37	0
	Arg	7	0
	Tyr	3	1
	Phe	2	0
	Val 188 → Met	5	1
	Leu	2	0
LR	Ser 223 → Gly	9	5
	Asn 100 → Ser	0	1
	Asn 105 → Ser	2	1
	Asn 219 → Ser	1	0
	Val 153 → Ile	0	3
	Val 176 → Ala &	1	0
	Asn 198 → Ser[c]		

[a] HR mutants selected in 2 µg/ml WIN 52084 except for C199→F and V188→L; these two HR mutants and all LR mutants selected in 0.1–0.4 µg/ml WIN 52084.

[b] Selected in 7 µg/ml WIN 52035.

[c] Double mutant.

that mutations at these two sites confer high resistance by hindering entry or seating of the drug within the binding pocket.

C. Low-Resistance Mutations Map to the Drug-Deformable Region of the Canyon Floor

Although only 17 in number, single mutations that conferred low resistance occurred at five different positions of VP1: S223, N100, N105, N219, and V153 (Table I); a double mutant at N198 and V176 was also observed. The much higher proportion of altered sites in LR mutants (5/17 = 29%), relative to that of HR mutants (2/64 = 3%), suggests there are substantially more mutational pathways to low resistance than to high. This hypothesis would account for the higher incidence of LR mutants noted above and suggests that a substantial number of low-resistance mutations remain yet to be discovered. In fact, substitutions in 10 additional LR mutants have not yet been located, indicating that some resistance mutations lie outside of the region of viral RNA sequenced (i.e., those regions encoding amino acids within 5 Å of the drug) (Heinz *et al.*, 1989). Most significantly, however, all of the LR mutations mapped to drug-deformable regions of the polypeptide chain near the canyon floor (Fig. 4, Plate 22). The side chains of some of these residues point away from the drug (Fig. 4b) and were often replaced with smaller side chains (Table I). These results suggested that some LR mutants might have altered attachment phenotypes and thus would be useful for mapping capsid residues involved in virus attachment. The role of S223 in HRV14 attachment has already been reported by Colonno *et al.* (1988), who demonstrated that substituting alanine for serine decreases virus binding affinity for HeLa cells.

D. Thermostabilization by Bound WIN Compounds

The foregoing studies suggested that drug resistance might be acquired in two ways: by decreasing binding affinity of the drug, perhaps by excluding it altogether, and by compensating for the effect of bound drug. A direct test of this hypothesis required a method of measuring drug binding to virus. Because arildone had been shown to protect poliovirus against heat inactivation (Caliguiri *et al.*, 1980), we explored the thermostabilizing effect of WIN compounds on HRV14. In this simple assay (see Fig. 5), virus samples are complexed with drug and heated; the drug is then dissociated from virus by dilution and the titer of surviving virus is determined. In the absence of drug, wild-type virus was inactivated to 0.2% survivors within 2 min at 52°C (Fig. 5); in the presence of WIN 52084, virus

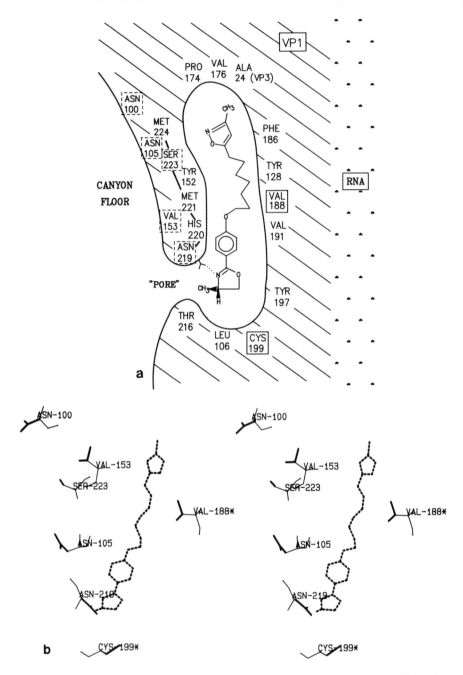

Figure 4. (a) Orientation of WIN 52084 and identity of the amino acid residues lining the wall of the drug-binding pocket in VP1 (Badger *et al.*, 1989). The isoxazole ring (above) points toward the viral fivefold axis. Squares with solid borders identify amino acid residues which

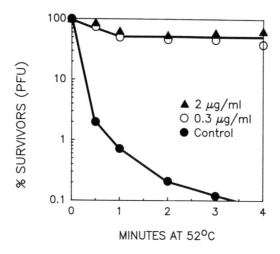

MINUTES AT 52°C

Figure 5. Thermostabilization of HRV14 by WIN 52084. Virus was neutralized by diluting into growth medium containing the indicated concentration of drug (plus 0.4% bovine serum albumin and 0.1% dimethylsulfoxide); samples were incubated 1 hr at 20°C, then overnight at 4°C. Aliquots (0.15 ml) were heated at 52°C for different lengths of time, chilled, diluted 100-fold, and allowed to stand at least 1 hr to permit drug release from virus before plating. Plaques were developed after 48 hr under nutrient agar. PFU, plaque-forming units.

infectivity was almost fully protected. We have found that the drug also stabilizes strongly against acid inactivation (not shown), but because that method requires extensive virus dilution, the thermoprotection assay is more sensitive and versatile.

E. Evidence for Two Mechanisms of Drug Resistance in HRV14

To determine whether drug resistance can occur by excluding drug from the binding pocket, we studied two of the most highly resistant mutants, C199→Y and C199→W. The C199→Y mutant is able to form plaques in the presence of high concentrations of both WIN 52084 and 52035 (Heinz *et al.*, 1989). Crystallographic studies on this mutant (Badger *et al.*, 1989) showed that the tyrosine side chain points into the drug-binding pocket (Fig. 6) and partially blocks the proposed entryway used by the drug (the

were altered in high-resistance (HR) mutants; squares with dashed lines mark low-resistance (LR) mutants (from Heinz *et al.*, 1989). (b) Stereo diagram showing the orientation of WIN 52084 and the seven residues that mutate to drug resistance; side chains are shown in boldface. The two HR mutants are labeled with an asterisk (*). Note that the side chain of C199 does not point into the pocket, but the side chain of Y199 *does* (see Fig. 6).

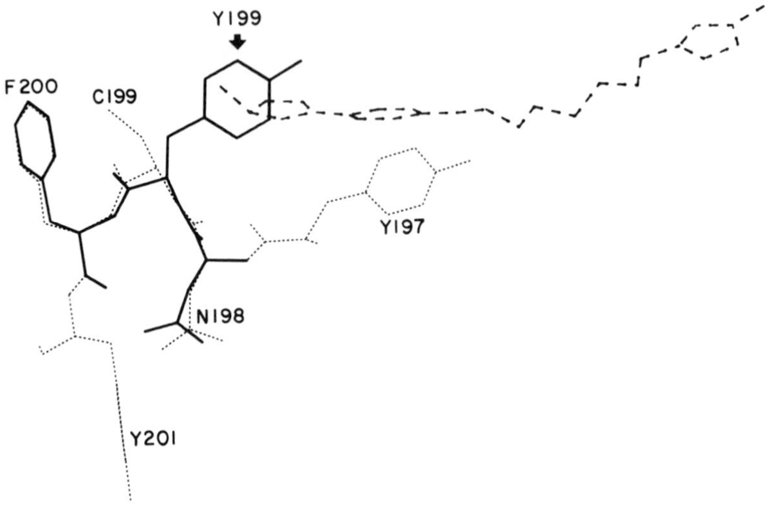

Figure 6.   Molecular model of WIN 52084 (dashed) superimposed in the drug-binding pocket of wild-type HRV14 (dotted) and the C199→Y mutant (solid). The viral fivefold axis is to the right.

"pore") (Fig. 4A). Modeling studies also indicated that the tyrosine residue makes the pocket too short to accommodate either compound, WIN 52084 or 52035 (Badger *et al.*, 1989) (see spatial overlap between drug and the tyrosine residue in Fig. 6). It is instructive to note, therefore, that plaque formation of the C199→W mutant is inhibited by WIN 52035 (Heinz *et al.*, 1989), indicating the drug can enter the pocket despite the bulkier tryptophan side chain. Although crystal stability problems have so far blocked efforts to solve the atomic structure of the C199→W mutant, it is evident that the tryptophan side chain does not block entry of the shorter drug.

Results of thermostabilization assays of these two mutants verified these conclusions (Table II). As expected for a mutation that prevents drug entry, both drugs failed to protect the C199→Y mutant against heat inactivation. The C199→W mutant, however, which is inhibited by WIN 52035 but not by 52084, was thermoprotected by the former drug but not by the latter. Thus, we conclude that decreasing the binding affinity of the pocket for the drug *is* an actual mechanism of drug resistance in HRV14. These results also reassured us that thermostabilization requires drug bound *within* the binding pocket and is not a nonspecific effect of the WIN compounds.

We next tested the hypothesis that some mutants could infect even

Table II.   Use of the Thermostability Assay to Measure Drug
Binding by Wild-Type and Mutant HRV14

Drug	Wild-type	C199→Y	C199→W	V153→I	N100→S
	Surviving PFU[a] (%)				
None	0.4	0.7	0.1	0.01	0.1
52084	69	1	1	56[b]	58[b]
52035	70	2	35	41[c]	51[c]

[a] After heating 2 min at 52°C. Drug concentration was 28 μM unless
otherwise indicated.
[b] 2 μg/ml (6 μM); blocks plaque formation.
[c] 7 μg/ml (22 μM); does not block plaque formation.

when drug was bound. To this end, we examined the thermostabilizing
effect of drugs on two mutants (V153→I and N100→S) that have low
resistance (by plaque assay) to WIN 52084 but are highly resistant to WIN
52035 (unpublished observations). We wondered whether these viruses
were highly resistant to 52035 because they fail to bind this drug or because
their mutations compensate for the attachment-inhibitory effects of bound
drug. As seen in Table II, both mutants were strongly protected against
thermal inactivation by either drug. Thus, it appears that these mutants are
able to infect even when drug is bound. This conclusion was verified using
single-cycle growth assays of one of these two mutants, V153→I (Section
II,F).

F. Single-Cycle Growth Curves

Single-cycle growth curves in the presence and absence of WIN compounds identify the effects of mutations and drugs on different stages of the
infectious cycle. In these experiments, virus complexed with drug was
attached to HeLa cell suspensions by agitating gently at room temperature
for 30 min; unattached virus was removed by sedimenting and rinsing the
cells before resuspending them in culture medium at 35°C. Infected cul-
tures were sampled at 30- to 60-min intervals and assayed for plaque titer.
   Growth curves conducted with wild-type and mutant HRV14 (Fig. 7)
illustrate the effect of high concentrations of these WIN compounds on the
virus cycle. The controls in all four experiments show a typical decline of
10-fold or more, presumably corresponding to uncoating of attached virus.
The flat eclipse phase was then followed by a rise period, indicating that
assembly of progeny was complete within 7–8 hr. With wild-type HRV14

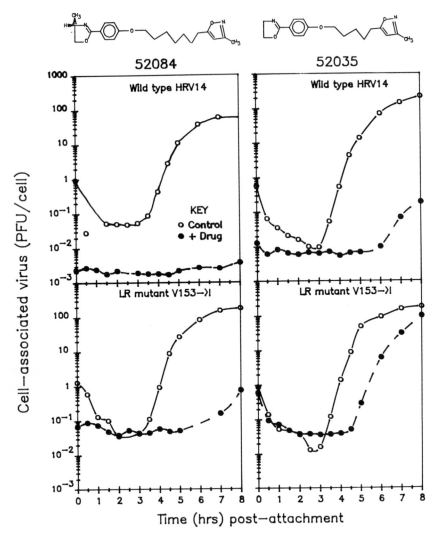

Figure 7. Single-step growth curves of virus in HeLa cell suspensions in the presence (filled circles) and absence (open circles) of WIN compounds. (Top) Wild-type virus. (Bottom) A low-resistance mutant selected for ability to grow in the presence of WIN 52035 at 7 μg/ml. Virus was complexed with drug by incubating 1 hr at 20°C then overnight at 4°C (2 μg/ml of WIN 52084 or 7 μg/ml of WIN 52035 in 0.1% dimethylsulfoxide and 0.4% bovine serum albumin). After an attachment (10 PFU/cell) period of 30 min at room temperature in medium containing drug, cells were sedimented to remove free virus, resuspended in fresh medium containing drug, and incubated at 35°C. In the absence of drug, about 10% of the wild-type virus input attached under these conditions. Aliquots were removed at intervals and stored frozen at −70°C; cell-associated virus was released by three cycles of freezing and thawing and infectivity was titered by plaque assay.

(top panels), WIN 52084 reduced attachment about 1000-fold, and there was little evidence of multiplication. The effect of WIN 52035 (a less active drug against HRV14) was similar except that attachment was inhibited only about 100-fold and there was some evidence of multiplication after a prolonged delay period.

The effects of high concentrations of WIN 52084 and 52035 on the resistant mutant V153→I are shown in the lower panels. This mutation only partially relieved the attachment-inhibiting effect of WIN 52084 (attachment was still inhibited about 10-fold), and some progeny were produced after a long delay. On the other hand, this mutation completely relieved the attachment-inhibiting effect of WIN 52035. Nevertheless, the rise period was delayed 1–1.5 hr. Whether this delay is due to an effect upon uncoating, inhibition of virus assembly, or drug cytotoxicity can now be determined, using biochemical assays specific for each stage of infection.

III. Discussion

A. Use of Escape Mutations to Map Viral Functions

We have shown that LR mutations appear in the deformable part of the canyon, a region that seems to be intimately involved with the attachment step (Pevear *et al.*, 1989) (Fig. 7, top panels). Some of our drug-resistant mutants, both HR (not shown) and LR, partially or completely relieve the attachment-inhibitory effect of bound drug. Other mutants, particularly some HR mutants, attach poorly to cells in our growth curve assays even in the absence of drug (not shown). Thus, drug-resistant mutants are useful for mapping regions of the capsid involved in attachment.

Our success in this regard encourages us to believe that escape mutant analysis can also be used to map other viral functions. For example, the compound WIN 51711 is reported to block uncoating, rather than attachment, of HRV2 (Fox *et al.*, 1986). Thus, capsid regions involved in uncoating might be identified by analysis of HRV2 mutants resistant to this drug. We are also keenly interested in the possibility of using HRV14 to map capsid regions involved in uncoating, either by designing uncoating inhibitors for wild-type HRV14 or by identifying drugs that block uncoating of HRV14 drug-resistant mutants. Growth curves of several of our mutants show a prolonged eclipse period which may reflect a delay in uncoating or assembly. Since, as already noted above, bound drug also enhances acid stability of HRV14, it is possible that the delayed rise period reflects depression by bound WIN drug of the endosomal pH required to trigger uncoating. Alternately, the delay may reflect the time required for release of drug before uncoating can take place.

The ubiquitous presence of hydrophobic pockets within the β-barrels of icosahedral RNA plant and animal viruses (Rossmann, 1988) suggests that they play a role in virus replication. It has been proposed (Rossmann, 1988; Smith *et al.*, 1986), that loose packing of polypeptide chains comprising an empty drug-binding pocket is needed to accommodate conformational changes during disassembly. On the other hand, the drug-binding pockets in poliovirus type 1, Mahoney, appear to be naturally occupied by a molecule resembling a long-chain hydrocarbon (Hogle *et al.*, 1987). Whether this molecule plays a role during poliovirus replication, and whether an analogous molecule ever occupies the drug-binding pocket in HRV14, remains unclear.

B. Further Analysis of Drug Resistance by Site-Directed Mutagenesis

Although our collection of spontaneous mutations is unlikely to be exhaustive, it appears that each of the single mutations reported here is sufficient to confer drug resistance because most appeared repeatedly in independent isolates (Table I). Moreover, crystallographic analysis of two of the mutants, C199→Y and V188→L (Badger *et al.*, 1989), detected only the predicted mutation in the crystallographically visible regions of the coat protein; the only invisible regions of the protomer are 16 residues at the amino end of VP1 and 25 residues at the amino end of VP4 (Rossmann *et al.*, 1985). The conclusion that resistance results from single mutations can now also be verified by engineering site-directed mutations into cDNA clones of wild-type virus.

Our results focus attention upon various regions of the capsid that might be fruitfully studied by site-directed mutagenesis. For instance, why were HR mutations observed at only two positions? The drug-binding pocket consists of 16 amino acids with side chains lying within 4 Å of the bound drug (Badger *et al.*, 1989). Nine of these residues are potential sites for HR mutations in that their side chains point into the pocket and could be replaced by a larger side chain through a single nucleotide change (Badger *et al.*, 1989). There are three possible explanations for absence of mutations at any of the other amino acids lining the pocket: Mutations may be lethal; they may not confer drug resistance; or the rates of spontaneous mutation at these other codons may be much lower than that of the C199 and V188 codons. These hypotheses can also be tested by genetic engineering. In addition, this approach could be used to assess the importance of the proposed hydrogen bond interaction (Smith *et al.*, 1986) (Fig. 4A) between N219 and the oxazoline nitrogen of WIN 52084, to investigate how residues not directly lining the canyon (such as C199) might play a role in virus attachment, and to determine the viability of viruses that contain multiple mutations within the drug-binding pocket.

C. Applications of the Thermostabilization Assay

We have used the thermostabilization assay to determine whether our drug-resistant mutants retain their ability to bind drug. This approach has a number of advantages over the use of radioactively labeled drug for measuring drug uptake. First, each labeled drug must be custom-synthesized and purified. Second, near milligram quantities of purified virus are required even if tritium is used as the label. Third, virus-bound drug must be separated from free drug before it can be quantitated, and an undetermined amount of bound drug could be lost during this process. Fourth, virus infectivity can be detected with much higher sensitivity than can radioactivity. Use of radiolabeled drugs will, of course, ultimately be crucial for determining the number of drug molecules per virus particle required to thermostabilize virus or block attachment or uncoating.

Thermostabilization experiments have identified two mechanisms by which mutations can confer resistance: (1) by inserting bulky side chains that reduce available space within the pocket and prevent binding; (2) by alleviating the attachment-inhibiting effect of bound drug. This approach will also be useful for studying drug-binding kinetics, the permeability of cell membranes to WIN compounds, and as a rapid screen for sorting drug-binding mutants from drug-excluding mutants.

Acknowledgments

This study was supported by NIH Postdoctoral Training Grant in Viral Oncology CA09075 (BAH) and NIH Grant AI24939 (RRR).

References

Badger, J., Minor, I., Kremer, M. J., Oliveira, M. A., Smith, T. J., Griffith, J. P., Guerin, D. M. A., Krishnaswamy, S., Luo, M., Rossmann, M. G., McKinlay, M. A., Diana, G. D., Dutko, F. J., Fancher, M., Rueckert, R. R., and Heinz, B. A. (1988). Structural analysis of a series of antiviral agents complexed with human rhinovirus 14. *Proc. Natl. Acad. Sci. USA* **85,** 3304–3308.

Badger, J., Krishnaswamy, S., Kremer, M. J., Oliveira, M. A., Rossmann, M. G., Heinz, B. A., Rueckert, R. R., Dutko, F. J., and McKinlay, M. A. (1989). The three-dimensional structures of drug-resistant mutants of human rhinovirus 14. *J. Mol. Biol.* **207,** 163–174.

Caliguiri, L. A., McSharry, J. J., and Lawrence, G. W. (1980). Effect of arildone on modifications of poliovirus *in vitro. Virology* **105,** 86–93.

Colonno, R. J., Condra, J. H., Mizutani, S., Callahan, P. L., Davies, M. E., and Murko, M. A. (1988). Evidence for the direct involvement of the rhinovirus canyon in receptor binding. *Proc. Natl. Acad. Sci. USA* **85**, 5449–5453.

Fox, M. P., Otto, M. J., and McKinlay, M. A. (1986). The prevention of rhinovirus and poliovirus uncoating by WIN 51711: A new antiviral drug. *Antimicrob. Agents Chemother.* **30**, 110–116.

Heinz, B. A., Rueckert, R. R., Shepard, D. A., Dutko, F. J., McKinlay, M. A., Fancher, M., Rossmann, M. G., Badger, J., and Smith, T. J. (1989). Genetic and molecular analysis of spontaneous mutants of human rhinovirus 14 resistant to an antiviral compound. *J. Virol.* **63**, 2476–2485.

Hogle, J. M., Chow, M., Fricks, C. E., Minor, P. D., and Filman, D. J. (1987). The three-dimensional structure of poliovirus: its biological implications. *In* "Protein Structure, Folding and Design 2" (D. L. Oxender, ed.), pp. 505–519. Alan R. Liss, New York.

McSharry, J. J., Caliguiri, L. A., and Eggers, H. J. (1979). Inhibition of uncoating of poliovirus by arildone, a new antiviral drug. *Virology* **97**, 307–315.

Pevear, D. C., Fancher, M. J., Felock, P. J., Rossmann, M. G., Miller, M. S., Diana, G., Treasurywala, A. M., McKinlay, M. A., and Dutko, F. J. (1989). Conformational change in the floor of the human rhinovirus canyon blocks adsorption to HeLa cell receptors. *J. Virol.* **63**, 2002–2007.

Rossmann, M. G. (1988). Antiviral agents targeted to interact with viral capsid proteins and a possible application to human immunodeficiency virus. *Proc. Natl. Acad. Sci. USA* **85**, 4625–4627.

Rossmann, M. G., and Palmenberg, A. C. (1988). Conservation of the putative receptor attachment site in picornaviruses. *Virology* **164**, 373–382.

Rossmann, M. G., Arnold, E., Erickson, J. W., Frankenberger, E. A., Griffith, J. P., Hecht, H.-J., Johnson, J. E., Kamer, G., Luo, M., Mosser, A. G., Rueckert, R. R., Sherry, B., and Vriend, G. (1985). Structure of a human common cold virus and functional relationship to other picornaviruses. *Nature (London)* **317**, 145–153.

Rueckert, R. R., Sherry, B., Mosser, A., Colonno, R., and Rossmann, M. (1986). Location of four neutralization antigens on the three-dimensional surface of a common-cold picornavirus, human rhinovirus 14. *In* "Virus Attachment and Entry into Cells" (R. L. Crowell and K. Lonberg-Holm, eds.), pp. 21–27. ASM Press, Washington, D.C.

Smith, T. J., Kremer, M. J., Luo, M., Vriend, G., Arnold, E., Kamer, G., Rossmann, M. G., McKinlay, G. D., Diana, G. D., and Otto, M. J. (1986). The site of attachment in human rhinovirus 14 for antiviral agents that inhibit uncoating. *Science* **233**, 1286–1293.

Zeichhardt, H., Otto, M. J., McKinlay, M. A., Willingmann, P., and Habermehl, K.-O. (1987). Inhibition of poliovirus uncoating by disoxaril. *Virology* **160**, 281–285.

14 Quantitative Structure–Activity Relationships and Biological Consequences of Picornavirus Capsid-Binding Compounds

Frank J. Dutko, Guy D. Diana, Daniel C. Pevear,
M. Patricia Fox, and Mark A. McKinlay
Department of Virology and Oncopharmacology, Sterling Research Group,
Rensselaer, New York 12144

I. Introduction

The Picornaviridae family includes several important human pathogens such as human rhinoviruses (HRV) and enteroviruses. As the leading cause of the common cold, the approximately 100 serotypes of HRV remain a significant cause of morbidity in people and represent an especially difficult therapeutic challenge (Sperber and Hayden, 1988).

The picornaviruses are among the smallest viruses, with a diameter of approximately 30 nm. As small viruses with limited genetic coding capacity, rhinoviruses have solved the problem of how to produce their virion capsid by using multiple copies of a protomeric subunit. As described for HRV (Rossmann et al., 1985), poliovirus (Hogle et al., 1985), and Mengo virus (Luo et al., 1987), the picornaviral capsid consists of 60 identical protomers. Each protomeric unit contains one copy of four viral proteins (VP1, VP2, VP3, and VP4).

A new class of antiviral agents with activity against a broad spectrum of human rhinoviruses and enteroviruses has been developed by the Sterling Research Group (McKinlay, 1985). These compounds (oxazolinylphenyl isoxazoles) inhibit virus replication as a consequence of binding to the viral capsid protein. The structure of compounds (Fig. 1) bound to HRV14 has been solved by X-ray crystallography (Badger et al., 1988). The results of these studies have allowed us to examine interactions of these compounds with virions on a molecular level, and also to develop hypotheses concerning the mechanism of action(s) of these compounds.

II. Enantiomeric Effects

One of the initial compounds examined by X-ray crystallography was Compound I, which inserts into the β-barrel of VP1 with the isoxazole end

Use of X-Ray Crystallography in the Design of Antiviral Agents

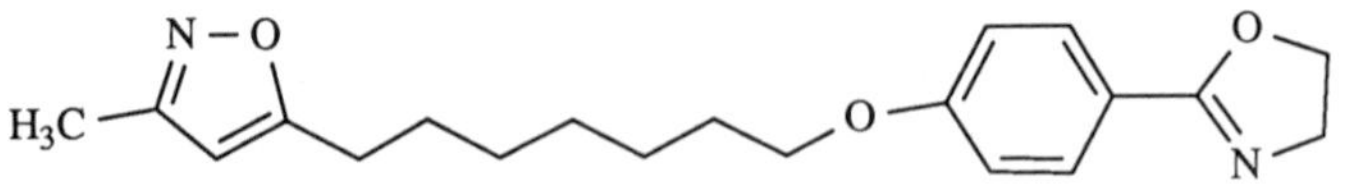

DISOXARIL (WIN 51711)

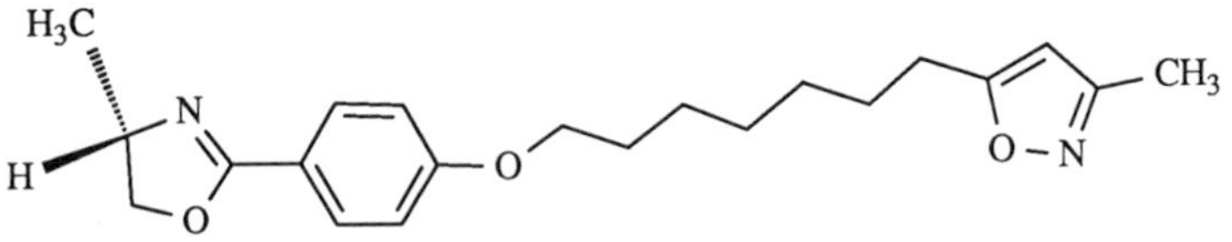

COMPOUND I

Figure 1. The chemical structures of disoxaril (WIN 51711; 5-[7-[4-(4,5-dihydro-2-oxa-zolyl)phenoxy]heptyl]-3-methylisoxazole) and Compound I, S-(−)-5-[7[(4,5-dihydro-4-methyl-2-oxazolyl)phenoxy]heptyl]-3-methylisoxazole, are shown in their orientation in the hydrophobic pocket in VP1 of HRV14 (From Badger *et al.,* 1988; see also Chapter 10 by Rossmann in this volume.)

at the deep end of the pocket. The methyloxazolinyl moiety is located below the canyon floor with a presumed hydrogen bond between the nitrogen of the oxazoline and asparagine 1219. Compound I has an asymmetric center at the 4-position of the oxazoline ring. The (*S*) isomer was approximately 10 times more active in antiviral (plaque reduction) tests than the (*R*). This was consistent with the result in which preferential binding of the (*S*) isomer to the β-barrel in VP1 occurred when HRV14 crystals were soaked in a racemic mixture of both (*S*) and (*R*) isomers (Diana *et al.,* 1988). Additional homologs (Table I) were prepared and tested against HRV14 in order to determine the extent of this enantiomeric effect. In every case, the (*S*) isomer was more active than the (*R*). Optimum activity was achieved with the ethyl and propyl homologs.

The interaction of the (*R*) and (*S*) conformers in the binding site were analyzed by performing an energy profiling study using the X-ray structure of Compound I in the HRV14 binding site as a starting point and using residues within 8 Å of any atom of Compound I. The purpose of this study was to determine the location of energy minima as the oxazolinyl ring was rotated 360 degrees. The results of this study showed a narrow valley at a twist angle between the phenyl and oxazoline rings of 10–30 degrees for each of the (*S*) conformers, suggesting a rigidly confined stable conformation. This was consistent with the results of the X-ray data, which

Table I. Comparative Evaluation of
Enantiomers against HRV14

	MIC (μM)[a]	
X	(S)[b]	(R)[b]
CH_3	0.05	0.56
C_2H_5	0.03	0.16
n - C_3H_7	0.03	0.18
i - C_3H_7	0.08	1.57
n - C_4H_9	0.15	1.31

[a] The minimal inhibitory concentration (MIC) was defined as the concentration of compound which inhibited the number of plaque-forming units of HRV14 by 50%.
[b] Absolute conformation.

revealed a torsion angle of 10–12 degrees between the phenyl and oxazo-line rings. The (R) enantiomers, however, displayed a wide energy minimum between 30–120 degrees. In performing these energy profiling studies, hydrophobic interactions were taken into account and considered to be the major contributor to the binding energy.

III. Structure–Activity Studies Based on Chain Length

X-ray crystallography studies have been performed on several compounds in the isoxazole series with both five and seven carbon chains connecting both ends of the molecule. A series of homologs in the disoxaril series and analogs containing a chlorine on the phenyl ring were evaluated against HRV14 in an effort to establish a structure–activity relationship between chain length and activity (Table II). In both series, a chain length of seven carbon atoms was required for optimum activity for HRV14. It is interesting to note that in both series, the six-carbon homolog was less active than both the five- and seven-carbon homologs. These results suggest a spatial requirement for optimum activity. The significance of these findings will become more obvious further on in the discussion.

We have also examined the effect of substituents on the phenyl ring on activity against HRV14. Twelve compounds shown in Table III were screened. In order to evaluate the significance of the data, a regression

Table II. Relationship between Chain Length
and Activity against HRV14

	MIC (μM)[a]	
n	*X* = H	*X* = Cl
4	NA	9.2
5	0.7	2.4
6	2.9	3.9
7	0.4	1.1
8	3.9	14.3

[a] The minimal inhibitory concentration (MIC) was defined as the concentration of compound which inhibited the number of HRV14 plaque-forming units by 50%.

analysis was performed to establish the correlation between activity and certain physicochemical parameters. Log P, an indicator of lipophilicity, molecular weight (MW) of the compound, which was chosen as an indicator of bulk, and σ_m, representing an inductive effect, were used. Using a stepwise regression analysis, the results indicated a poor correlation between log[1/MIC], where MIC is the mean of the minimum inhibitory concentration for five HRV serotypes in plaque reduction assays, and log P ($r = 0.58$) and no correlation with MW ($r = 0.29$) or σ_m ($r = 0.14$). However, a combination of log P and MW resulted in a correlation of $r = 0.80$ as described in the following equation:

$$\mathrm{Log}_{10}[1/\mathrm{MIC}] = 0.476[\log \mathrm{P}] - 4.09[\mathrm{MW}] + 1.49$$

In this equation, MW makes a negative contribution to activity; that is, the larger the substituent, the less active the compound. This observation is consistent with the hydrophobic nature of the drug-binding pocket in VP1 and its limitations on the size of ligand which may bind. Binding affinity under these circumstances can only be a function of the size, hydrophobicity, and possibly the "flexibility" of the molecule.

IV. A Model for HRV14 Activity

The elucidation of the three-dimensional structure of HRV14, the identification of the compound binding site on the capsid protein, and the X-ray

Table III. Effect of Phenyl Ring Substituents on Activity against HRV14

X	MIC[a] (μM)
CH₃	0.6
H	0.6
C₂H₅	1.1
CH₃O	1.7
F	1.7
NO₂	1.8
Br	2.0
CHO	2.3
Cl	2.4
CF₃	3.0
NH₂	3.3
COCH₃	7.3
CH₂OH	15.7

[a] The MIC was defined as the concentration of compound which inhibited the number of plaques by 50%.

structures of several compounds bound to HRV14 have allowed for the development of a model representing the requirements for activity for this series of compounds. The model was developed by examining seven active compounds and seven inactive compounds. A composite of active and inactive structures was generated using the program SYBYL and then compared using volume maps comprising van der Waals surfaces (Color Plate 23). The volume maps were examined and differences between the active and inactive structures noted. The results of this comparison revealed two major differences. First, inactive compounds displayed additional bulk around the phenyl ring when compared to active compounds. Exceeding a critical bulk requirement results in inactive compounds, possible due to spatial constraints within the compound binding site. Second, the active compounds occupy space in the binding site beyond that of the inactive compounds. The most active compounds extend well into the pore area of the binding site, suggesting that filling this space increases the strength of the hydrophobic interactions and thereby increasing binding affinity.

The conclusions drawn from this model substantiate the results of the SAR studies on the chain length previously described. A chain length of seven carbon atoms satisfies the space-filling requirements of the model. The negative effect of molecular weight on activity resulting from the QSAR emphasizes the bulk limitations. It should be noted that this analysis defines the structural features which are required for optimal activity against HRV14. Unfortunately, superior activity against HRV14 does not correlate with excellent activity against the majority of HRV serotypes.

V. Binding of Disoxaril

A. Binding to HRV14

Previous work with disoxaril and poliovirus had indicated that disoxaril inhibited the replication of poliovirus by a direct interaction with the virions. We have quantitated the binding of radioactive disoxaril to purified HRV14 in order to determine the relationship of binding to antiviral activity. The Scatchard analysis in Fig. 2 shows that the K_D (the affinity

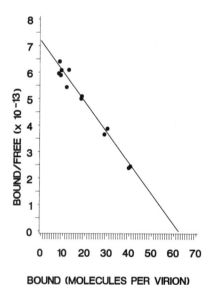

Figure 2. Scatchard analysis of the binding of ^{14}C-labeled disoxaril to purified HRV14. HRV14 was purified by gradient centrifugation and 10 μg of HRV14 was incubated for 4 hr with ^{14}C-labeled disoxaril. Virions with bound compound were separated from free compound by chromatography on Sephadex G-50.

binding constant) is approximately $4.5 \times 10^{-7} M$ (0.15 μg/ml). This value is virtually identical to the MIC as determined in plaque reduction tests, which is 0.14 μg/ml. Data are needed for more compounds and viruses to determine fully the relationship of K_D to the antiviral activity (MIC).

B. Binding to Drug-Resistant HRV14

The relationship of the binding of compound to antiviral activity can also be investigated by using drug-resistant HRV14 (see chapter by Rueckert in this volume). The binding of radioactive disoxaril has been determined for two drug-resistant viruses: Leu1188, in which valine at residue number 188 in VP1 of HRV14 has been replaced by leucine, and Trp1199, in which cysteine at residue number 199 in VP1 of HRV14 has been replaced by tryptophan. No detectable binding of ^{14}C-labeled disoxaril was found with Leu1188. This finding is consistent with the MIC of disoxaril in plaque reduction tests with Leu1188, which is greater than 6.2 μg/ml (i.e., not active at the maximum testable level of compound). With Trp1199 and disoxaril, the K_D was 1.4 μg/ml and the MIC was 3.4 μg/ml. These results suggest, as expected, that the binding of compound to HRV14 is an important determinant of the antiviral activity. Furthermore, the K_D of compound and HRV14 may be directly proportional to the antiviral activity as quantitated by the MIC in plaque reduction assays.

C. Rates of Association and Dissociation

The rates of association (K_{ON}) and dissociation (K_{OFF}) were separately quantitated for disoxaril and HRV14 in order to calculate a K_D K_D(CALC)] and then compare K_D(CALC) to the K_D determined by saturation binding experiments and Scatchard analysis [K_D(SAT)]. The K_{OFF} for disoxaril and HRV14 was 0.023 min^{-1} (standard deviation = 0.002 min^{-1}) and the K_{ON} was 167,000 M^{-1} min^{-1} (standard deviation = 6000 M^{-1} min^{-1}). The K_D calculated from the K_{OFF} and K_{ON} was $1.4 \times 10^{-7} M$ and was within threefold of the K_D(SAT). Thus, the quantitation of K_D(SAT) in the saturation binding experiments has been validated.

An examination of the K_{ON} (167,000 M^{-1} min^{-1}) shows that the association of disoxaril and HRV14 is a relatively slow process. This rate of association is what is expected for a hydrophobic compound to desolvate prior to displacing the water in the hydrophobic binding pocket in VP1 and HRV14.

VI. Mechanism of Action

An appreciation of the mechanism of action of WIN compounds against the picornarviruses requires an understanding of the early events in the virion replication cycle. In the case of poliovirus, the virus adsorbs to a specific receptor on the surface of permissive cells and enters the host via receptor-mediated endocytosis into clathrin-coated pits (Dales, 1973; Lonberg-Holm, 1975; Zeichhardt *et al.*, 1987). Once inside the cell, the virus is passed through endosomes and lysosomes at progressively lower pH. At an as yet undetermined point in this progression, conformational alterations in the virion capsid occur which lead to the release or uncoating of the virion RNA into the cell cytoplasm. The picornavirus RNA is of positive polarity and thus directly serves as message for the initial round of translation of virion proteins.

The early events in the human rhinovirus replication cycle have not been studied as thoroughly as those for polioviruses. However, differences between rhinoviruses and polioviruses are likely to be subtle. Only two unique receptor groups have been identified for the over 100 known HRV serotypes (Lonberg-Holm and Korant, 1972; Colonno *et al.*, 1986). The major HRV receptor, which is used by 78 of 88 HRV serotypes and some of the group A coxsackieviruses to initiate infection (Colonno *et al.*, 1986) is a 90-kDa protein found on cells of human and chimpanzee origin (Tomassini and Colonno, 1986). Ten other rhinovirus serotypes bind to the minor HRV receptor (HRV1A, 1B, 2, 29–31, 44, 47, 49, and 62), which is present on both human and mouse (NIH 3T3) cells (Colonno *et al.*, 1986). It is likely that the HRVs also infect cells by a pH-mediated uncoating event, but the hypothesis has not been as rigorously tested as for poliovirus.

Mechanism of action studies for the WIN compounds have focused on adsorption and uncoating, the early events in the replication cycle. One of these compounds, disoxaril, has previously been shown to block the uncoating of neutral red-encapsidated PV2 and HRV2 (a minor binding group virus), probably by stabilizing the virion capsid against the pH-mediated conformational change necessary for release of the viral RNA from the virion core (Fox *et al.*, 1986; Zeichhardt *et al.*, 1987). HRV2 and PV2 adsorption and penetration are unaffected at concentrations of compound which completely inhibit virus replication in HeLa cells.

X-ray crystallographic analysis of HRV14 (a major binding group virus) complexed with disoxaril has determined that this, and a number of homologs, bind within a hydrophobic pocket in the VP1 β-barrel, directly beneath the putative virion receptor binding site, or canyon (Smith *et al.*, 1986; Rossmann *et al.*, 1985). Upon binding, the compounds induce dramatic conformational changes in the neighboring polypeptide chain of as much as 4.5 Å in C_α positions from the native crystalline structure (Badger *et al.*, 1988). These conformational changes translate to the floor of the canyon, resulting in a more shallow depression in the virion surface. As a test of the role of the canyon in receptor binding, we reexamined the mechanism of action of various WIN compounds in HRV14 to determine what affect they had on virion attachment.

In order to examine the capacity of WIN compounds to block HRV14 attachment to receptors, a modification of a published HeLa cell membrane binding assay was used (Colonno *et al.*, 1986). Initially, saturation binding experiments were done in order to determine the kinetics of binding of [^{35}S]methionine-labeled HRV14, and a high resistance mutant (Leu-1188) to the HeLa cell receptors. A K_D could then be calculated from the Scatchard equation and a virus concentration of the K_D was used in all experiments. The kinetics of binding of wild-type HRV14 and the Leu1188 mutant were indistinguishable (data not shown). Neither virus bound to identically prepared mouse L cell membranes or to bovine serum albumin controls.

Pretreatment of HRV14 with any of the compounds which had been found to induce conformational changes in the canyon floor resulted in marked inhibition of virion attachment to HeLa cell receptors. As shown in Fig. 3, HRV14 binding to HeLa cell receptors was inhibited in a dose-

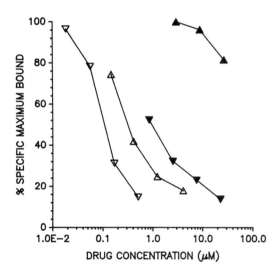

Figure 3. Dose–response curves of the inhibition of the adsorption of HRV14 (open symbols) and the Leu1188 resistant virus (filled symbols) in the presence of disoxaril (triangles) and Compound I (inverted triangles).

dependent manner by several compounds. The IC_{50} (inhibitory concentraton 50%: the concentration of compound required to block 50% attachment versus mock-treated controls) correlated strongly (Fig. 4) with the level of drug required to reduce plaque numbers by 50% (the minimal inhibitory concentration or MIC) indicating a cause and effect relationship between these two measures of antiviral activity. For the Leu1188 mutant, which is highly resistant to the compounds as measured by plaque reduction assays, much higher drug levels were required to reach an IC_{50}. Because the compounds are lipophilic, the results with the Leu1188 mutant are consistent with a direct effect of the compounds on the virion capsid and not on the cellular receptor.

VII. Antiviral Activity in Virus-Infected Animals

This class of compounds is of interest because the biological consequences of the binding of compound to picornaviruses are not only the inhibition of virus replication in cell culture, but also the prevention of morbidity and mortality in mice induced by enterovirus infection. Because disoxaril was

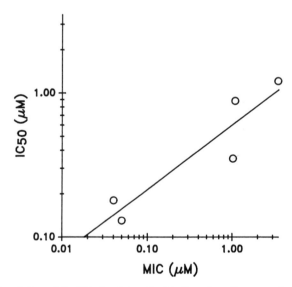

Figure 4. Correlation of the IC_{50} (as determined in the adsorption assay in Fig. 3) and the MIC (as determined in plaque reduction assays as the concentration of compound which inhibited the number of plaque forming units of HRV14 by 50%). Data for five compounds are plotted: disoxaril, Compound I, the (S)-C_2H_5 compound in Table I, and two compounds in Table III with $X = $ H and $X = $ Cl.

active in cell culture against poliovirus type 2 as well as coxsackieviruses and echoviruses (Otto *et al.*, 1985), the effect of oral medication with disoxaril was determined in mouse models of infection using these viruses. Disoxaril was found to be orally efficacious in mice infected intracerebrally with human poliovirus type 2 (McKinlay and Steinberg, 1986) or in suckling mice infected with echovirus type 9 (McKinlay *et al.*, 1986).

VIII. Conclusions

The oxazolinylphenyl isoxazoles are an interesting group of antiviral compounds that inhibit the replication of picornaviruses. These compounds bind to a specific site in VP1 of the viral capsid and cause conformational changes which results in inhibition of adsorption of HRV14. Both an analysis of compound binding using X-ray crystallography and the generation of structure–activity relationships using *in vitro* antiviral assays have contributed to the design of agents with improved antiviral activity. While this information is not by itself helpful in directing the synthesis of broad-spectrum antipicornavirus agents, it is the essential fist step in understanding the "ground rules" for designing hydrophobic capsid-binding compounds.

References

Badger, J., Minor, I., Kremer, M. J., Oliveira, M. A., Smith, T. J., Griffith, J. P., Guerin, D. M. A., Krishnaswamy, S., Luo, M., Rossmann, M. G., McKinlay, M. A., Diana, G. D., Dutko, F. J., Fancher, M., Rueckert, R. R., and Heinz, B. A. (1988). Structural analysis of a series of antiviral agents complexed with human rhinovirus 14. *Proc. Natl. Acad. Sci. USA* **85**, 3304–3308.
Colonno, R. J., Callahan, P. L., and Long, W. J. (1986). Isolation of a monoclonal antibody that blocks attachment of the major group of human rhinoviruses. *J. Virol.* **57**, 7–12.
Dales, S. (1973). Early events in cell–animal virus interactions. *Bacteriol. Rev.* **37**, 103–135.
Diana, G. D., Otto, M. J., Treasurywala, A. M., McKinlay, M. A., Oglesby, R. C., Maliski, E. G., Rossmann, M. G., and Smith, T. J. (1988). Enantiomeric effects of homologues of disoxaril on the inhibitory activity against human rhinovirus-14. *J. Med. Chem.* **31**, 540–544.
Fox, M. P., Otto, M. J., and McKinlay, M. A. (1986). Prevention of rhinovirus and poliovirus uncoating by WIN 51711, a new antiviral drug. *Antimicrob. Agents Chemother.* **30**, 110–116.
Hogle, J. M., Chow, M., and Filman, D. J. (1985). Three-dimensional structure of poliovirus at 2.9 Å resolution. *Science* **229**, 1358–1365.
Lonberg-Holm, K. (1975). The effect of concanavalin A on the early events of infection by rhinovirus type 2 and poliovirus type 2. *J. Gen. Virol.* **28**, 313–327.
Lonberg-Holm, K., and Korant, B. D. (1972). Early interaction of rhinoviruses with host cells. *J. Virol.* **9**, 29–40.

Luo, M., Vriend, G., Kamer, G., Minor, I., Arnold, E., Rossmann, M. G., Boege, U., Scraba, D. G., Duke, G. M., and Palmenberg, A. C. (1987). The structure of mengo virus at atomic resolution. *Science* **235,** 182–191.

McKinlay, M. A. (1985). Win 51711: A new systemically active broad-spectrum antipicornavirus agent. *J. Antimicrob. Chemother.* **16,** 284–286.

McKinlay, M. A., and Steinberg, B. A. (1986). Oral efficacy of WIN 51711 in mice infected with human poliovirus. *Antimicrob. Agents Chemother.* **29,** 30–32.

McKinlay, M. A., Frank, J. A., Benziger, D. P., and Steinberg, B. A. (1986). Use of WIN 51711 to prevent echovirus type 9-induced paralysis in suckling mice. *J. Infect. Dis.* **154,** 676–681.

Otto, M. J., Fox, M. P., Fancher, M. J., Kuhrt, M. F., Diana, G. D., and McKinlay, M. A. (1985). *In vitro* activity of WIN 51711, a new broad-spectrum antipicornavirus drug. *Antimicrob. Agents Chemother.* **27,** 883–886.

Pevear, D. C., Fancher, M. J., Felock, P. J., Rossmann, M. G., Miller, M. S., Diana, G. D., Treasurywala, A. M., McKinlay, M. A., and Dutko, F. J. (1989). Conformational change in the floor of the human rhinovirus canyon blocks adsorption to HeLa cell receptors. *J. Virol.* **63,** 2002–2007.

Rossmann, M. G., Arnold, E., Erickson, J. W., Frankenberger, E. A., Griffith, J. P., Hecht, H. J., Johnson, J. E., Kramer, C., Luo, M., Mosser, A. G., Rueckert, R. R., Shevey, B., and Vriend, G. (1985). Structure of a human common cold virus and functional relationship to other picornaviruses. *Nature (London)* **317,** 145–153.

Smith, T. J., Kremer, M. J., Luo, M., Vriend, G., Arnold, E., Kamer, G., Rossmann, M. G., McKinlay, M. A., Diana, G. D., and Otto, M. J. (1986). The site of attachment in human rhinovirus 14 for antiviral agents that inhibit uncoating. *Science* **233,** 1286–1293.

Sperber, S. J., and Hayden, F. G. (1988). Chemotherapy of rhinovirus colds. *Antimicrob. Agents Chemother.* **32,** 409–419.

Tomassini, J. E., and Colonno, R. J. (1986). Isolation of a receptor protein involved in attachment of human rhinoviruses. *J. Virol.* **58,** 290–295.

Zeichhardt, H., Otto, M. J., McKinlay, M. A., Willingmann, P., and Habermehl, K.-O. (1987). Inhibition of poliovirus uncoating by disoxaril (WIN 51711). *Virology* **160,** 281–285.

15 Comparative Structures of Two Lysozyme–Antilysozyme Complexes

David R. Davies, Eduardo A. Padlan, and Steven Sheriff
Laboratory of Molecular Biology, National Institute of Diabetes and Digestive
and Kidney Diseases, National Institutes of Health, Bethesda, Maryland
20892

I. Introduction

During the last three years there has been a spectacular advance in our understanding of the nature of antibody interactions with protein antigens through X-ray diffraction analysis. There have been three crystal structure determinations of lysozyme–antilysozyme complexes (Amit *et al.*, 1986; Sheriff *et al.*, 1987; Padlan *et al.*, 1989), and a determination of the structure of the influenza virus neuraminidase with an anti-neuraminidase Fab. We here describe our work on two of the lysozyme–Fab complexes (which has been extensively reviewed) and compare the results with the work on the third complex.

The nature of the forces involved in antibody–antigen interaction are basically no different from those observed in the general area of protein–protein interactions. They can be related to the work of many investigators on the factors determining specificity in other multi-subunit protein systems, enzyme–inhibitor complexes, etc. However, there are several aspects that apply particularly to the antibody system, to questions relating to the antigenicity of proteins, to the mapping of the antigenic surfaces, and to the use of peptides in the development of vaccines.

The first of these concerns the nature of the epitope. Does it consist entirely or mostly of a single segment of polypeptide chain or is it made up of a number of distant segments brought together by the folding of the protein? Some previous work (Atassi, 1978), in epitope determination by the use of competition assays with peptides had suggested that the single segments would be dominant. However, several theoretical analyses (Barlow *et al.*, 1986; Fanning *et al.*, 1986) showed that, given the expected size of the epitope, it was likely that it would consist of more than one segment.

Second, what is the role of the relative mobility of the protein surface in determining antigenicity? It has been suggested (Moore and Williams, 1980; Westhof *et al.*, 1984; Tainer *et al.*, 1984) that the most mobile parts of the protein would be the most antigenic. Other factors such as accessibility had also been implicated in antigenicity. On the basis of studies with monoclonal antibodies, Benjamin and co-workers (Benjamin et al., 1984) had proposed that any part of the protein surface that was accessible to the antibody would be potentially antigenic.

Third, the question of conformational change has been linked to the functional properties of antibodies (Huber et al., 1976). It was hypothesized that the different angles of the elbow bend could be used to transmit a signal to the effector (Fc) part of the molecule that the antibody was engaged in contact with the antigen. Colman et al. (1987) have reported that a realignment of V_H relative to V_L takes place upon binding to the neuraminidase, making a significant change in quaternary structure. Other smaller changes are also possible, such as small adjustments in backbone and side chains as normally seen in crystal packing. Alternatively, there might be no conformational change, with the antibody and antigen fitting together while retaining their preformed conformations.

In addition to these questions there were other questions of a more general nature having to do with the kind of complementarity that would be observed in the interaction of the surface and of the charge. These involve the role of water molecules in the interface, the importance of salt bridges, the role of hydrogen bonds, etc.

Much of this material has been reviewed (Colman, 1988; Davies et al., 1988; Alzari et al., 1988).

II. HyHEL-5 and HyHEL-10 Fab Complexes with Lysozyme

We have previously reported preliminary results on the crystal structures of complexes of lysozyme with two monoclonal Fabs to lysozyme. These were HyHEL-5 (Sheriff et al., 1987) and HyHEL-10 (Padlan et al., 1989). Each of these complexes is characterized by remarkable complementarity in shape between the interacting surfaces of the antibody and antigen, such that water is almost entirely excluded from the interface.

In both complexes the epitope is clearly discontinuous. In HyHEL-5 it is dominated by a segment of polypeptide chain 13 amino acid residues long (41–53), but it also includes two other regions from different parts of the sequence (residues 67–70 and 79–85). It should be noted, however, that a change of R68→K, as in bobwhite quail, can produce a dramatic decrease in the binding (Smith-Gill et al., 1984). In HyHEL-10, the epitope consists of a helix, 88–99, together with some surrounding residues. Here too, with five separate segments (residues 15–16, 20–21, 63 and 74–75), the epitope is discontinuous.

In both complexes, only small conformational changes are observed in the lysozyme as a result of binding to the antigen. There are movements of the backbone atoms of as much as 2 Å at the point of contact, but no gross movement of the structure is seen.

In HyHEL-5, as in McPC603 (Satow *et al.,* 1986) charge neutralization appears to play an important role in the interaction. Two arginines in lysozyme, R45 and R68, form a ridge on the surface of the molecule. This ridge fits into a groove on the antibody, at the bottom of which there are two glutamic acids, E35 and E50, both from the heavy chain. However, charge neutralization cannot by itself provide the necessary interaction, since when the R68→K change occurs there is a 1000-fold decrease in the association constant, showing that packing considerations and the formation of specific hydrogen bonds must also have a significant role. In HyHEL-10, in spite of the presence of two lysines in the middle of the epitope (K96,K97), charge neutralization does not seem to play as significant a role. There is only one weak salt bridge formed and this is at the edge of the interface, exposed to solvent.

The HyHEL-5 complex was determined in two different crystal forms. The structure of the Fab in these two forms differs by a change in the elbow bend, the angle between the V and C modules, of 7 degrees. This observation, together with the lack of correlation between elbow bend and whether or not antigen is bound (Davies *et al.,* 1988), lead to the conclusion that the variation in this angle is simply an indication of flexibility in this part of the Fab. A recent analysis of the elbow includes a proposal of a mechanism for the observed movement (Lesk and Chothia, 1988).

An additional complex of lysozyme with a monoclonal antibody Fab has been reported (Amit *et al.,* 1986). The properties of this complex are not significantly different from those observed in HyHEL-5 and HyHEL-10. Fortuitously, the epitopes from these three complexes are virtually non-overlapping, and together they cover over 40% of the lysozyme surface. These data would appear to support the conclusion (Benjamin *et al.,* 1984) that any part of a protein surface that is accessible to antigen is potentially antigenic.

References

Alzari, P. M., Lascomb, M. B., and Poljak, R. J. (1988). *Annu. Rev. Immunol.* **6,** 555.
Amit, A. G., Mariuzza, A. R., Phillips, S. E. V., and Poljak, R. J. (1986). *Science* **233,** 747.
Atassi, M. Z. (1978). *Immunochemistry* **12,** 423.
Barlow, D. J., Edwards, M. S., and Thornton, J. M. (1986). *Nature (London)* **322,** 747.

Benjamin, D. C., Berzofsky, J. A., East, I. J., Gurd, F. R. N., Hannum, C., Leach, S. J., Margoliash, E., Michael, J. G., Miller, A., Prager, E. M., Reichlin, M., Sercarz, E. E., Smith-Gill, S. J., Todd, P. E., and Wilson, A. C. (1984). *Annu. Rev. Immunol.* **2**, 67.

Colman, P. M. (1988). *Adv. Immunol.* **434**, 99.

Colman, P. M., Laver, W. G., Varghese, J. N., Baker, A. T., Tulloch, P. A., Air, G. M., and Webster, R. G. (1987). *Nature (London)* **326**, 358.

Davies, D. R., Sheriff, S., and Padlan, E. A. (1988). *J. Biol. Chem.* **263**, 10541.

Fanning, D. W., Smith, J. A., and Rose, G. D. (1986). *Biopolymers* **25**, 863.

Huber, R., Deisenhofer, J., Colman, P. M., Matsushima, M., and Palm, W. (1976). *Nature (London)* **264**, 415.

Lesk, A. M., and Chothia, C. (1988). *Nature (London)* **335**, 188.

Moore, G. R., and Williams, R. J. P. (1980). *Eur. J. Biochem.* **103**, 543.

Padlan, E. A., Silverton, E. W., Sheriff, S., Cohen, G. H., Smith-Gill, S. J., and Davies, D. R. (1989). *Proc. Natl. Acad. Sci. USA* **86**, 5938.

Satow, Y., Cohen, G. H., Padlan, E. A., and Davies, D. R. (1986). *J. Mol. Biol.* **190**, 593.

Sheriff, S., Silverton, E. W., Padlan, E. A., Cohen, G. H., Smith-Gill, S. J., Finzel, B., and Davies, D. R. (1987). *Proc. Natl. Acad. Sci. USA* **84**, 8075.

Smith-Gill, S. J., Lavoie, T. B., and Mainhart, C. R. (1984). *J. Immunol.* **133**, 384.

Tainer, J. A., Getzoff, E. D., Alexander, H., Houghton, R. A., Olson, A. J., Lerner, R. A., and Hendrickson, W. A. (1984). *Nature (London)* **312**, 127.

Westhof, E., Altsuhuh, D., Moras, D., Bloomer, A. C., Mondragon, A. A., Klug, A., and van Regenmortel, M. H. V. (1984). *Nature (London)* **311**, 123.

16 Structural Basis of Antigen– Antibody Recognition

R. A. Mariuzza, P. M. Alzari, A. G. Amit, G. Bentley,
G. Boulot, V. Chitarra, T. Fischmann, M.-M. Riottot,
F. A. Saul, H. Souchon, D. Tello, and R. J. Poljak
Département d'Immunologie, Institut Pasteur, 75724 Paris Cedex 15, France

I. Introduction

Antigen recognition by the immune system is mediated by two distinct, but related, classes of receptor molecules: antibodies and T cell receptors. Whereas antibodies recognize intact proteins (as well as other macromolecular antigens), T cell receptors recognize only processed antigen in the form of peptides in association with molecules of the major histocompatibility complex (MHC). Our current knowledge of the molecular basis of immunological recognition is based largely on studies of the three-dimensional structure of a MHC class I molecule (Bjorkman *et al.*, 1987) and of several antigen–antibody complexes, which are the subject of the present paper.

Antibody molecules are composed of two identical light (L) and two identical heavy (H) polypeptide chains, each having variable (V) and constant (C) portions. The amino-terminal regions of the H and L chains, V_H and V_L, each contain three hypervariable or complementarity-determining regions (CDR1, CDR2, and CDR3) responsible for antigen recognition. These are flanked by less variable framework regions (FR1, FR2, FR3, and FR4) (Kabat *et al.*, 1983) which support the antigen-contacting residues. X-ray crystallographic studies of several antigen binding fragments (Fab) of myeloma immunoglobulins (Amzel and Poljak, 1979) have shown that the Fab fragment can be divided into two structural domains, V and C, each composed of two compact, independent units of three-dimensional structure (V_L and V_H; C_L and C_H1) related by a pseudo-twofold axis of symmetry. Each of these so-called homology subunits consists of two stacked β-pleated sheets surrounding an internal volume tightly packed with hydrohobic side chains; the two β-sheets are linked by an intrachain disulfide bridge in a direction approximately perpendicular to the plane of the sheets. The three-dimensional structure of the β-sheet strands, which are composed mostly of FR residues, is highly conserved

Use of X-Ray Crystallography in the Design of Antiviral Agents

between the different homology subunits as well as between the V_L and V_H subunits of different antibodies. Most of the structural differences between the V_L and V_H domains of different antibodies occur in the segments connecting β-sheet strands. The six CDRs of the two V subunits form six connecting loops in close spatial proximity which determine the conformation of the combining site. The CDRs may also include the ends of β-sheet strands, suggesting that residues adjacent to the connecting segments influence their conformation (Chothia *et al.*, 1986).

The first detailed information on specific binding interactions at antibody combining sites came from X-ray crystallographic studies of hapten–antibody complexes. A complete characterization of antibody combining sites, however, could not be obtained from these studies because the ligands, vitamin K_1OH (Amzel *et al.*, 1974) and phosphorylcholine (Segal *et al.*, 1974), are relatively small and occupied only part of the potential binding area. Important questions also remained about the possibility of conformational changes following complex formation and about the nature of protein antigenic determinants (epitopes) recognized by antibodies. For example, are epitopes composed of continuous sequences of amino acids, or are they discontinuous, topographical features assembled by the three-dimensional folding of the antigen molecule? Such questions could only be answered by the study of antigen–antibody complexes.

In this paper, we summarize our results in determining the three-dimensional structure of an antigen–antibody complex in the context of two other recently determined complex structures (Sheriff *et al.*, 1987; Colman *et al.*, 1987) and discuss what they have taught us about the structure of epitopes and about conformational flexibility in antigen–antibody recognition. We also describe our current work on an idiotype–anti-idiotype complex and attempts to generate a water-soluble form of the T cell antigen receptor for eventual use in crystallographic studies.

II. Crystallographic Studies of Fab–Lysozyme Complexes

Hen egg-white lysozyme (HEL) was chosen as the protein antigen because its three-dimensional structure is known to high resolution (Blake *et al.*, 1965) and because it has served as a favorite model antigen for immunologists (Benjamin *et al.*, 1984). A systematic search of 27 murine anti-HEL antibodies yielded several crystalline Fab–lysozyme complexes suitable for high-resolution X-ray diffraction analysis. Antibodies D44.2 and F10.6.6 recognize an antigenic determinant of HEL that includes Arg 68 (Fischmann *et al.*, 1988), as does HyHEL-5, whose structure has been determined by Sheriff *et al.* (1987); it will therefore be of interest to compare the extent to which the epitopes recognized by these antibodies overlap. Another complex being studied is that between antibody D11.15 and pheasant egg lysozyme (Guillon *et al.*, 1987). This is an example of a heteroclitic antibody, or one whose affinity for a heterologous antigen is greater than that for the immunizing antigen (HEL). Its structure determination should provide a test for the prediction that increases in affinity may be achieved through more effective packing of the antigen–antibody interface (Amit *et al.*, 1986; Mariuzza *et al.*, 1987).

Our current understanding of the molecular basis of antigen–antibody recognition is based largely on the structures of three Fab–lysozyme complexes (Amit *et al.*, 1986; Sheriff *et al.*, 1987; Davies *et al.*, 1988) and on that between the NC41 Fab fragment and influenza virus neuraminidase (Colman *et al.*, 1987). The three antilysozyme antibodies, D1.3, HyHEL-5, and HyHEL-10, recognize three distinct epitopes which, taken together, cover about half the solvent-accessible surface of lysozyme. Here we present the results of our structure determination of the FabD1.3–lysozyme complex in relation to those obtained for the other three antigen–antibody complexes.

A. The Antigen–Antibody Interface

The epitope recognized by D1.3 is composed of two segments of the polypeptide chain of lysozyme (residues 18–27 and 116–129) which form a contiguous patch of about 20 × 30 Å on the antigen surface. The antibody combining site appears as an irregular, relatively flat surface with protuberances and depressions formed by the amino acid side chains of the H and L chain CDRs. There is, in addition, a small pocket between the CDR3s of V_H and V_L which corresponds to the classical binding site in hapten–antibody complexes (Amzel *et al.*, 1974; Segal *et al.*, 1974). It is occupied by the side chain of Gln 121 of HEL, which has been shown by immunochemical techniques to be a critical residue in complex formation. Altogether 17 antibody residues from each of the six CDRs make close contacts with 16 antigen residues. A comparable number of antigen residues contact antibody in the other three complexes, although they derive from three stretches of the antigen polypeptide chain in the case of HyHEL-5 and four stretches in the cases of HyHEL-10 and NC41. The involvement of all six CDRs in direct contacts with antigen has also been observed for the two other Fab–lysozyme complexes, but in the FabNC41–neuraminidse complex V_L CDR1 does not appaer to be implicated (Colman *et al.*, 1987). In addition to the CDRs, certain FR residues contribute directly to antigen binding in the D1.3 and HyHEL-5 complexes.

The interacting surfaces in the FabD1.3–HEL complex are remarkably complementary, with protruding side chains of one surface lying in depressions of the other, to the complete or nearly complete exclusion of water molecules. Many van der Waals interactions are interspersed with hydrogen bonds, as for example HEL residue Gln 121, which makes close contacts with three aromatic side chains (V_L Tyr 32, V_L Trp 92, and V_H Tyr 101) and whose buried amide nitrogen forms a strong hydrogen bond to the main chain oxygen of V_L Phe 91. The specificity of D1.3 is greatly enhanced by an additional 11 hydrogen bonds, which require precise juxtaposition of donor and acceptor groups at the antigen–antibody interface. Although the contacting surfaces are highly complementary, there are some areas of seemingly less-than-perfect fit, which might explain the occurrence of heteroclitic antibodies and of the observed ability of somatic mutation to increase antibody affinity. In addition, at least for the three Fab–lysozyme complexes for which the structures have been partially refined, not all potential hydrogen bonds are actually made; for example, in the FabD1.3–HEL complex, the polar parts of lysozyme residues Arg 21 and Thr 118 are unbonded in the interface.

Although all six CDRs of D1.3 contact antigen, not all of them make an equal contribution: V_H CDRs, and V_H CDR3 in particular, have more interactions with lysozyme than V_L CDR3s, as measured by number of hydrogen bonds and van der Waals contacts. All four V_H CDR3 residues of D1.3 in contact with antigen (residues 99–102) are encoded by the D segment, which illustrates its importance in the generation of functionally different combining sites. In the HyHEL-5 complex, on the other hand, V_L CDR3 and V_H CDR2 are responsible for most of the contacts (Sheriff *et al.*, 1987).

In summary, while the four structures are broadly similar in terms of size of the antigen–antibody interface and degree of complementarity between the interacting surfaces, they differ with respect to the relative contributions of different parts of the antibody combining site to antigen binding.

B. The Nature of Protein Antigenic Determinants

Immunochemists have traditionally divided antigenic determinants into two structural categories: (1) continuous determinants, in which the antibody contacting residues are contained primarily in a single segment of the amino acid sequence of the antigen and (2) discontinuous determinants, in which the contact residues derive from several distinct segments of the antigen polypeptide chain brought together by the folding of the protein in its native conformation (Sela *et al.*, 1967). The probability of occurrence of each of these types of determinants depends on what proportion of the surface of a globular protein is composed of linear arrays of residues. Barlow *et al.* (1986) computed this proportion as a function of increasing radius of the recognition site and found that even when the contact radius is 6 Å, only approximately 25% of the surface is continuous. When the contact radius is over 10 Å, as in all four antigen–antibody complexes studied to date, virtually none of the surface is continuous. It is therefore unlikely that there will be many antibodies that recognize purely continuous epitopes; this at least partially explains the difficulty in obtaining monoclonal anti-peptide antibodies which also react with the native protein, and vice versa.

It has been proposed that atomic mobility is a critical factor in determining antigenicity (Tainer *et al.*, 1985). An alternative view is that any part of the solvent-accessible surface of a protein is potentially antigenic (Benjamin *et al.*, 1984). Novotny *et al.* (1986a) have refined the notion of accessibility by using a large probe of diameter roughly equal to that of an antibody combining site (20 Å). A good correlation was observed between large-probe accessibility and the presence of an antigenic site, as identified by immunochemical studies. Indeed, the antigenic site of lysozyme recognized by D1.3 corresponds to two contact-area maxima, with peaks at position 21 and 114, neither of which lies in a region of above-average temperature factors. For the HyHEL-5 and HyHEL-10 complexes, both high- and low-mobility regions form parts of the antigenic determinant, while the correlation with high accessibility is generally maintained (Davies *et al.*, 1988). We conclude that atomic mobility is not the determining factor in antigen–antibody recognition or, alternatively, that the minimal observed atomic mobilities of residues located on the solvent-exposed areas of a protein are largely sufficient for antibody binding. Furthermore, as antibodies D1.3, HyHEL-5,and HyHEL-10 alone cover about half the solvent-accessible surface of lysozyme, it appears likely that the remainder of its surface will also prove antigenic, at least under the appropriate conditions of host genetic background, route of immunization, etc.

C. Conformational Flexibility in Antigen–Antibody Recognition

Antigen–antibody recognition may be explained by two alternative (but not mutually exclusive) mechanisms (Berzofsky, 1985): (1) the lock and key model, in which antigen and antibody essentially behave as rigid bodies without the need for significant conformational changes in either and (2) the induced fit model, in which antigen and antibody achieve greater complementarity through some degree conformational change. The latter mechanism would operate if the predominant conformation(s) of the antigen does not make an optimum fit with the antibody combining site. In this case, an induced fit movement would occur if the free energy gain resulting from improved complementarity is greater than the free energy cost of distorting a protein from its native or lowest-energy conformation(s).

The availability of several high-resolution structures of HEL in different crystal forms allows a direct evaluation of the effect of complex formation on antigen conformation. For the Fab D1.3–HEL complex, a least squares fit of C_α atoms of complexed lysozyme to those of free lysozyme in the tetragonal crystal form gives a root-mean-square (rms) deviation of 0.64 Å, which is comparable to the estimated error in atomic positions in the complex. Some differences in side chain conformation can be observed between bound and unbound lysozyme, but these are similar to those between different crystal structures of free HEL. More significant changes in antigen conformation have been reported for the HyHEL-5–HEL and NC41–neuraminidase complexes. In the former, the C_α of Pro 70 has been displaced by 1.7 Å, probably due to formation of a hydrogen bond with V_H Tyr 97. The overall rms difference in the atomic position of main chain atoms between bound and unbound lysozyme, however, remains small (0.48 Å). In the Fab NC41–neuraminidase complex, the C_α atoms of neuraminidase residues 368–371 have moved by 1 Å or more from their positions in the unbound antigen. It thus appears that antibody binding can produce a range of local effects on antigen conformation (or vice versa), including limited displacements of the polypeptide backbone.

With regard to conformational changes in the antibody, Colman et al. (1987) reported the relative disposition of V_L and V_H to be significantly different in the NC41–neuraminidase complex from that in other Fab structures and suggested V_L–V_H "sliding" as a further mechanism for increasing antigen–antibody complementarity. The inclusion in the comparison of V_L–V_H orientations of the α-carbon coordinates of four additional Fabs (HyHEL-5, J539, R19.9, and D1.3), however, reveals that the difference in relative disposition of V_H and V_L in the neuraminidase complex falls within the observed range for Fabs, independently of their liganded or unliganded states (Lascombe et al., 1989). Thus, it is unnecessary to invoke antigen binding to explain the particular V_H–V_L orientation of Fab NC41.

III. Idiotype–Anti-Idiotype Interactions

Antilysozyme antibody D1.3 has been used as an immunogen to obtain a number of syngeneic monoclonal anti-idiotypic antibodies. The complex between the Fab fragment of one of these (E225) and FabD1.3 has been crystallized in a form suitable for high-resolution X-ray analysis (Boulot *et al.*, 1987). To determine whether E225 is directed mainly against the CDRs of D1.3, or whether FR residues also contribute significantly to the idiotypic determinant (idiotope), the binding of E225 to a genetically engineered variant of D1.3 was tested. This "reshaped" antibody consists of the CDR loops of the murine D1.3 antibody transplanted onto the β-sheet framework regions of the human immunoglobulin New (Reichmann *et al.*, 1988). Since the majority of solvent-accessible FR residues differ between the mouse and human antibodies, E225 would be expected to recognize reshaped D1.3 only if the idiotope were composed primarily of CDR residues. That this is the case is suggested by a gel filtration assay in which formation of a specific E225-reshaped D1.3 complex could be demonstrated.

Consistent with this result, a preliminary, low-resolution X-ray study of the E225–D1.3 complex shows the two Fabs to be in a "head-to-head" orientation with nearly all the intermolecular contacts made by CDR loops.

IV. Solubilization of the T Cell Antigen Receptor

The molecular basis for dual recognition by the T cell receptor of processed antigen plus MHC is unknown despite the recent isolation of the genes encoding the α and β polypeptide chains of the T cell receptor heterodimer (Hood *et al.*, 1985) and crystal structure determination of a human MHC class I molecule (Bjorkman *et al.*, 1987). The principal difficulty resides in the fact that it is not feasible to purify sufficient quantities of this membrane protein directly from T cell lines or clones to permit the study of its structure and binding properties by conventional biochemical and X-ray crystallographic techniques.

To circumvent this difficulty, we have attempted to engineer secreted hybrid proteins containing the V (and presumably antigen-binding) domains of the T cell receptor linked to immunoglobulin C domains. The structural rationale for this approach is the likelihood that the V_α and V_β chain segments fold into immunoglobulin-like domains (Novotny *et al.*, 1986b) and the fact that in the three-dimensional structure of antibody molecules, the V and C domains are essentially independent (Saul *et al.*, 1978). Rearranged "genomic" V_α and V_β sequences were prepared from the corresponding cDNA clones derived from a human T cell clone (PH28) specific for diphtheria toxin and the DR7 MHC molecule (Triebel *et al.*, 1988) to allow their substitution into the immunoglobulin heavy chain expression vector of Jones *et al.* (1986). The entire heavy chain leader peptide, including the signal peptidase cleavage site, was retained, as were residues 119–120, which form part of the "switch" region between the V_H and C_H1 domains. The 5'- and 3'- V_H splice junctions are consequently preserved. The hybrid coding sequences were cloned into the expression vectors $pSV_{gpt}V_{NP}$ or $pSV_{neo}V_{NP}$ (Neuberger *et al.*, 1985) and the immunoglobulin constant regions $C_\gamma2$ or C_κ introduced to complete the constructions. In other constructions, the first T-cell receptor C region exons ($C1_\alpha$ and $C1_\beta2$), which encode the immunoglobulin-like domain of the C regions of T cell receptor α and β chains, were inserted into the intronic region between the T cell receptor V regions and immunoglobulin C regions.

When the plasmid $pSV_{gpt}L_HV_\alpha C_\kappa$ was transfected into myeloma cells, the corresponding $V_\alpha C_\kappa$ chimeric protein was secreted into the culture supernatant at levels of 400–500 μg/liter (Mariuzza and Winter, 1989). The $V_\alpha C_\kappa$ chimera is extensively glycosylated, and its secretion is glycosylation-dependent. Gel filtration of affinity-purified material showed that secreted $V_\alpha C_\kappa$ chimeric protein exists largely as a noncovalent homodimer in solution although in equilibrium with significant amounts of monomer.

In order to attempt to obtain assembly and secretion of α/β heterodimer, plasmids encoding the following combinations of chimeric proteins were cotransfected into myeloma cells: (1) $V_\alpha C_\kappa$ and $V_\beta C_\kappa$; (2) $V_\alpha C_\kappa$ and $V_\beta C_\gamma2$; (3) $V_\alpha C1_\alpha C_\kappa$ and $V_\beta C1_\beta2C_\kappa$; and (4) $V_\alpha C1_\alpha C_\kappa$ and $V_\beta C1_\beta2C_\gamma2$. In none of these cases could secretion of a V_β-containing chimeric protein be demonstrated, even though the intracellular levels of expression of V_β-containing chimeras were comparable to that of $V_\alpha C_\kappa$.

The ability of a $V_\alpha C_\kappa$ chimera to form Bence–Jones protein-like homodimers and of a $V_\alpha C_\gamma2a$ chimera to assemble with a λ light chain (Gascoigne *et al.*, 1987) suggests that T cell receptor V_α domains and immunoglobulin V are sufficiently similar structurally to be interchangeable. In contrast, V_β–IgC chimeras were not secreted either individually or in conjunction with potential partners. These data suggest that the problem is V_β-specific, either because of an error in the design of the V_β-containing chimeras themselves, or because of the absence in the myeloma host of vital T cell specific accessory molecules required for translocating the β chain from one cell compartment to the next. To test the first hypothesis, we asked whether an isolated V_β domain, unattached to an immunoglobulin C domain, could be secreted. A termination codon was therefore introduced at the J_β–C_β junction of the cDNA encoding the β chain of the human T cell line Jurkat. Upon infection of SF9 insect cells with a recombinant baculovirus carrying the truncated β gene, secretion of the V_β domain into the culture medium was demonstrated by immunoprecipitation using several V_β-specific monoclonal antibodies (O. Acuto and R. A. Mariuzza, unpublished observations).

The fact that an isolated V_β domain can be secreted, while all the V_β–IgC chimeras tested remain in the cytoplasm, opens the possibility of obtaining V_α–V_β heterodimers by coexpressing the corresponding V_α domain. It is also worth noting that V_α–V_β fragments should be more amenable to crystallization than V_α, V_β–IgC chimeras due to their significantly smaller size.

References

Amit, A. G., Mariuzza, R. A., Phillips, S. E. V., and Poljak, R. J. (1986). *Science* **233**, 747–753.

Amzel, L. M., and Poljak, R. J. (1979). *Annu. Rev. Biochem.* **48**, 961–997.

Amzel, L. M., Poljak, R. J., Saul, F., Varga, J. M., and Richards, F. F. (1974). *Proc. Natl. Acad. Sci. USA* **71**, 1427–1430.

Barlow, D. J., Edwards, M. S., and Thornton, J. M. (1986). *Nature (London)* **322**, 747–748.

Benjamin, D. C., Berzofsky, J. A., East, I. J., Gurd, F. R. N., Hannum, C., Leach, S. J., Margoliash, E., Michael, J. G., Miller, A., Prager, E. M., Reichlin, M., Sercarz, E. E., Smith-Gill, S. J., Todd, P. E., and Wilson, A. (1984). *Annu. Rev. Immunol.* **2**, 67–101.

Berzofsky, J. A. (1985). *Science* **229**, 932–940.

Bjorkman, P. L., Saper, M. A., Samraoui, B., Bennett, W. S., Strominger, J. L., and Wiley, D. C. (1987). *Nature (London)* **329**, 506–512.

Blake, C. C. F., Koenig, D. F., Mair, G. A., North, A. C. T., Phillips, D. C., and Sarma, V. R. (1965). *Nature (London)* **206**, 757–761.

Boulot, G., Rojas, C., Bentley, G. A., Poljak, R. J., Barbier, E., Le Guern, C., and Cazenave, P. A. (1987). *J. Mol. Biol.* **194**, 577–579.

Chothia, C., Lesk, A. M., Levitt, M., Amit, A. G., Mariuzza, R. A., Phillips, S. E. V., and Poljak, R. J. (1986). *Science* **233**, 755–757.

Colman, P. M., Laver, W. G., Varghese, J. N., Baker, A. T., Tulloch, P. A., Air, G. M., and Webster, R. G. (1987). *Nature (London)* **326**, 358–363.

Davies, D. R., Sheriff, S., and Padlan, E. A. (1988). *J. Biol. Chem.* **263**, 10541–10544.

Fischmann, T., Souchon, H., Riottot, M.-M., Tello, D., and Poljak, R. J. (1988). *J. Mol. Biol.* **203**, 527–529.

Gascoigne, N. R. J., Goodnow, G. C., Dudzik, K. I., Oi, V. T., and Davis, M. M. (1987). *Proc. Natl. Acad. Sci. USA* **84**, 2936–2940.

Guillon, V., Alzari, P. M., and Poljak, R. J. (1987). *J. Mol. Biol.* **197**, 375–376.

Hood, L., Kronenberg, M., and Hunkapiller, T. (1985). *Cell* **40**, 225–229.

Jones, P. T., Dear, P. H., Foote, J., Neuberger, M., and Winter, G. (1986). *Nature (London)* **321**, 522–525.

Kabat, E. A., Wu, T. T., Bilofsky, H., Reid-Miller, M., and Perry, H. (1983). "Sequences of Proteins of Immunological Interest," pp. 396–847. U.S. Dep. Health Hum. Serv., Public Health Serv., Natl. Inst. Health, Bethesda, Maryland.

Lascombe, M.-B., Alzari, P. M., Boulot, G., Saludjian, P., Tougard, P., Berek, C., Haba, S., Rosen, E. M., Nisonoff, A., and Poljak, R. J. (1989). *Proc. Natl. Acad. Sci. USA* **86**, 607–611.

Mariuzza, R. A., and Winter, G. (1989). *J. Biol. Chem.* **264**, 7310–7316.

Mariuzza, R. A., Philips, S. E. V., and Poljak, R. J. (1987). *Annu. Rev. Biophys. Biophys. Chem.* **16**, 139–159.

Neuberger, M. S., Williams, G. T., Mitchell, E. B., Jouhal, S., Flanagan, J. G., and Rabbits, T. H. (1985). *Nature (London)* **314**, 268–270.

Novotny, J., Handschumacher, M., Haber, E., Bruccoleri, R. E., Carlson, W. B., et al. (1986a). Proc. Natl. Acad. Sci. USA **83**, 226–230.

Novotny, J., Tonegawa, S., Saito, H., Kranz, D. M., and Eisen, H. N. (1986b). Proc. Natl. Acad. Sci. USA **83**, 742–746.

Reichmann, L., Clark, M., Waldmann, H., and Winter, G. (1988). Nature (London) **332**, 323–327.

Saul, F. A., Amzel, L. M., and Poljak, R. J. (1978). J. Biol. Chem. **253**, 585–597.

Segal, D. M., Padlan, E. A., Cohen, G. H., Rudikoff, S., Potter, M., and Davies, D. R. (1974). Proc. Natl. Acad. Sci. U.S.A. **71**, 4298–4302.

Sela, M., Schechter, B., Schechter, I., and Borek, F. (1967). Cold Spring Harbor Symp. Quant. Biol. **32**, 537–545.

Sheriff, S., Silverton, E. W., Padlan, E. A., Cohen, G. H., Smith-Gill, S. J., Finzel, B. C., and Davies, D. R. (1987). Proc. Natl. Acad. Sci. USA **84**, 8075–8079.

Tainer, J. A., Getzoff, E. D., Patterson, Y., Olson, A. J., and Lerner, R. A. (1985). Annu. Rev. Immunol. **3**, 501–536.

Triebel, F., Breathnach, R., Graziani, M., Hercend, T., and Debré, P. (1988). J. Immunol. **140**, 300–304.

17 Analysis of Antibody–Protein Interactions Utilizing Site-Directed Mutagenesis and a New Evolutionary Variant of Lysozyme

Thomas B. Lavoie

Laboratory of Genetics, National Cancer Institute, National Institutes of Health, Bethesda, Maryland 20892, and Department of Zoology, University of Maryland, College Park, Maryland 20742

Lauren N. W. Kam-Morgan

Department of Biochemistry, University of California, Berkeley, California 94720

Corey P. Mallett

Laboratory of Genetics, National Cancer Institute, National Institutes of Health, Bethesda, Maryland 20892

James W. Schilling

California Biotechnology, Inc., Mountain View, California 94043

Ellen M. Prager and Allan C. Wilson

Department of Biochemistry, University of California, Berkeley, California 94720

Sandra J. Smith-Gill

Laboratory of Genetics, National Cancer Institute, National Institutes of Health, Bethesda, Maryland 20892

I. Introduction

In order to define as precisely as possible the molecular interactions underlying antibody specificity and affinity for a protein antigen, we have utilized as a model system high-affinity monoclonal antibodies specific for hen egg-white lysozyme c from the chicken (HEL)[1]. We chose HEL

---

[1] Abbreviations. HEL, hen (chicken) egg white lysozyme c; BWQ, bobwhite quail; JQ, Japanese quail; ELISA, enzyme linked immunosorbent assay; PCFIA, particle concentration fluorescence immunoassay; IC_{50}, amount of inhibitor required for 50% inhibition in a competitive inhibition assay; DNP, dinitrophenyl; rHEL and rHyHEL-10, wild-type (unmutated) HEL and HyHEL-10 produced by recombinant DNA methods; $R68_{HEL}K$, mutant recombinant of HEL made with lysine replacing the naturally occurring arginine at position 68, with analogous abbreviations for other mutants of HEL and of the V_H chain of HyHEL-10; $G100\backslash TF/D101_{VH}$, mutant recombinant of HyHEL-10 V_H, made by inserting codons for amino acids T and F between the codons for $G100_{VH}$ and $D101_{VH}$; MAb, monoclonal antibody; Fab, Fv, antigen-binding fragment and variable domain, respectively, of antibody;

Use of X-Ray Crystallography in the Design of Antiviral Agents
213

because there is extensive structural and functional information on bird lysozymes, and lysozyme has long served as a prototype protein for investigating the immune response (reviewed in Benjamin *et al.*, 1984; Smith-Gill and Sercarz, 1989). HEL is a small globular protein (129 amino acid residues) whose three-dimensional structure and catalytic function have been extensively studied (reviewed in Osserman *et al.*, 1974; Smith-Gill and Sercarz, 1989). Complexes of HEL with antibody Fab fragments have proven to be particularly amenable to crystallization, and to date three X-ray structures of Fab–HEL complexes have been refined (Amit *et al.*, 1986; Sheriff *et al.*, 1987, 1988; Padlan *et al.*, 1989). We summarize here recent experiments analyzing the interactions of two Fab–HEL complexes for which the structures have been determined (Table I) (Sheriff *et al.*, 1987; Padlan *et al.*, 1989; see also Davies *et al.*, 1988, 1989, and Chapter 15 in this volume). Some of these results have been summarized previously (Lavoie *et al.*, 1989).

II. Interactions of HyHEL-5 with HEL

A. Evolutionary Variants Point to the Importance of HEL Residues 68 and 45

Utilizing evolutionary variants of lysozyme, we were able to predict critical residues in the epitope recognized by the MAb HyHEL-5 (Smith-Gill *et al.*, 1982). In particular, significantly reduced binding of HyHEL-5 to bobwhite quail but not to other quail lysozymes allowed us to identify $R68_{HEL}$ as a critical residue (see Fig. 1). At the time of these studies, we were utilizing the X-ray structure of tetragonal lysozyme (Phillips, 1967; Diamond, 1974) for uncomplexed chicken lysozyme, in which $R68_{HEL}$ and $R45_{HEL}$ are within hydrogen-bonding distance of each other, and we hypothesized that $R45_{HEL}$ would also be an important contact residue in the complex.

V_H and V_L, variable regions of immunoglobulin (Ig) heavy (H) and light (L) chains, respectively; CDR, complementarity-determining region; H1, H2, and H3, CDR1, 2, and 3 of H chain, with analogous terminology for L chain; J_H and J_K, J-encoded regions of H and L chains; D_H, D-encoded region of H chain. The Kabat amino acid numbering system for immunoglobulin H and L chains is used throughout this paper (Kabat *et al.*, 1987). For reconstituted and chimeric antibody molecules, the H and L chains are designated with numerical subscripts to H and L which refer to the numbers of the parental antibody used for that chain.

Table I. Characteristics of the HyHEL-5 and HyHEL-10 Fab–Lysozyme[a] Complexes

Characteristic	HyHEL-5 Complex[b]	HyHEL-10 Complex[c]
Association constant (K_a)	$\sim 1.4 \times 10^{10}\ M^{-1}$	$\sim 4 \times 10^9\ M^{-1}$
Buried surface area		
Lysozyme	750 Å^2	770 Å^2
Antibody	750 Å^2	720 Å^2
Lysozyme epitope		
Contacting residues[d]	14	15
Buried residues[e]	10 at least partly buried	3 at least partly buried
Epitope topography	3 discrete segments, relatively flat with protruding ridge	Exposed surface of α helix, with surrounding residues
Antibody paratope		
Regions in contact	All 6 CDRs, 1 framework residue	All 6 CDRs, 1 framework residue
Predominant regions	H2, L3	H2
Contacting residues[d]	17	19
Buried residues[e]	15 at least partly buried	2 at least partly buried
Aromatic residues	5 of 17 contacting residues	6 of 19 contacting residues
Paratope topography	Relatively flat, with groove	Relatively flat, with protrusions
Antibody–lysozyme contacts		
Hydrogen bonds	10	14
van der Waals contacts	~ 75	~ 111
Salt bridges	3 in center	1 on periphery

[a] Chicken lysozyme.
[b] Summarized from Sheriff *et al.* (1987); see Davies *et al.* (1989, and Chapter 15 in this volume).
[c] Summarized from Padlan *et al.* (1989); see Davies *et al.* (1989, and Chapter 15 in this volume).
[d] Forms hydrogen bond or van der Waals contact.
[e] Inaccessible in complex, but does not make any actual contacts.

B. X-Ray Structure and Site-Directed Mutants Confirm the Importance of R68 and R45

The X-ray structure of the HyHEL-5 Fab–HEL complex (Sheriff *et al.*, 1987) shows that R68$_{HEL}$ and R45$_{HEL}$ are similarly within hydrogen-bonding distance of each other in the complex, forming a prominent ridge in the center of the epitope. The two Arg residues sit in a complementary groove in the antibody combining site, forming three salt bridges with Glu 35 and Glu 50 (E35$_{VH}$ and E50$_{VH}$) of the antibody heavy chain.

We have produced two site-specific mutants of HEL, R68$_{HEL}$K and

LYSOZYME	AMINO ACID				HyHEL-5 Reactivity (IC$_{50}$)
	40	55	68	91	
Chicken	T	I	R	S	1
Bobwhite quail	S	V	K	T	100
California quail	S	V	-	T	1

Figure 1. Identification of Arg 68 of chicken lysozyme as a critical residue for recognition by HyHEL-5, summarized from Smith-Gill et al. (1982). California quail, which bound HyHEL-5 as well as chicken did, differs from bobwhite quail only at amino acids 68 and 121, while bobwhite quail is identical to chicken at position 121. The reduced affinity of bobwhite quail lysozyme for HyHEL-5 can therefore be attributed to the R68$_{HEL}$→K substitution. Reactivity here is defined as the IC$_{50}$ in a competitive inhibition ELISA (see Figs. 2 and 3), with the IC$_{50}$ of chicken set equal to one, and all other IC$_{50}$ values expressed as a ratio relative to that value. Dashes in the sequences here and in Figs. 3 and 4 and Tables III and IV indicate identity at that position to the sequence shown in the first line.

R45$_{HEL}$K, to measure directly the contributions of R68$_{HEL}$ and R45$_{HEL}$ to HyHEL-5 binding (Lavoie et al., 1989). The wild type and site-directed mutants of a cDNA clone of chicken lysozyme were expressed as secreted proteins in yeast (Malcolm et al., 1989). A fluorescence immunoassay (PCFIA) (Jolley et al., 1984; Hartman et al., 1989) was used to measure competitive inhibition of HyHEL-5 binding to HEL by native and expressed lysozymes and to estimate association constants (Lavoie et al., 1989).

Figure 2A demonstrates that native and recombinant HEL behaved equivalently in a competitive inhibition PCFIA. An excess of over 5000-fold of BWQ compared to chicken lysozyme was necessary to inhibit competitively HyHEL-5 binding to HEL (Fig. 2A, Table II). These results are consistent with our previous ELISA results (Smith-Gill et al., 1982) which allowed identification of R68$_{HEL}$ as a critical residue. The reduced binding of BWQ lysozyme is also reflected in the association constant, which was 10^3 lower than that of HEL (Table II). The site-specific mutant R68$_{HEL}$K was used in parallel competitive inhibition and affinity assays in order to test the hypothesis that the reduced affinity of BWQ lysozyme for HyHEL-5 is attributable to the R68$_{HEL}$→K (Fig. 1). The results (Figs. 2A, Table II) confirmed the importance of R68$_{HEL}$→K to HyHEL-5 binding, but the affinity of R68$_{HEL}$K was significantly lower than that of BWQ lysozyme in both assays. The meaning of this difference between R68$_{HEL}$K and native BWQ lysozyme is currently under investigation.

In order to test the hypothesis that the salt bridge between R45$_{HEL}$ and

A

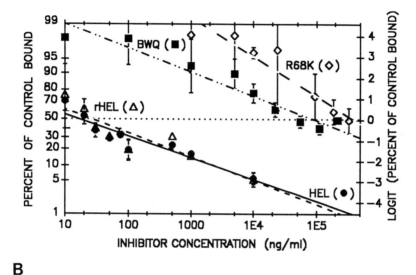

B

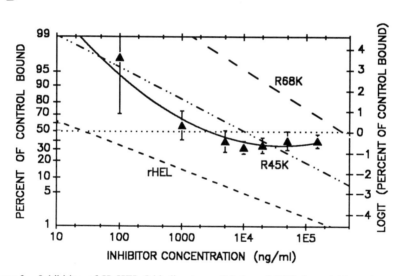

Figure 2. Inhibition of HyHEL-5 binding to particle-bound HEL by soluble native and recombinant lysozymes in a competitive inhibition PCFIA. All data points represent the mean of at least three independent determinations, with the 95% confidence limits indicated by error bars; where no error bars are shown, the 95% confidence interval is contained within (i.e., is equal to or less than) the width of the data point symbol. Panel (A) shows that the inhibition profiles of recombinant (wild-type) and native HEL are equivalent and compares the wild-type profile to those of native BWQ and recombinant R68K lysozyme. Panel (B) compares the inhibition curves of the three recombinant lysozymes; the linear regression lines for rHEL and R68K (dashed lines, means not shown) are the same as in (A), and the linear (dash-dot-dot) and curvilinear (solid) regressions are shown for R45K. The IC_{50} (Table II) for each lysozyme is the calculated intersection of each regression with the dotted line, which represents 50% control bound. A preliminary analysis of these data has been summarized previously (Lavoie *et al.*, 1989).

Table II. Affinities of Native and Mutant
Lysozymes for the Monoclonal Antibody
HyHEL-5

Lysozyme	$IC_{50}^{a,b}$	$\sim K_a^{b,c}$ (M^{-1})
Chicken	1	1.4×10^{10}
Recombinant chicken	1.6	3.6×10^{10}
Bobwhite quail	5300	1.5×10^{7}
$R68_{HEL}K$	23,300	2.8×10^{6}
$R45_{HEL}K$	140	1.5×10^{9}

[a] The amount of each lysozyme required for 50% inhibition of antibody to particle-bound HEL, calculated from linear regressions (except for $R45_{HEL}K$, where the second-order curvilinear regression was utilized), and corresponding to the intersection of each inhibition curve with the 50% control bound line (dotted line) in Fig. 2. The value for chicken lysozyme has been set equal to one, and all other values expressed as a ratio relative to that control value. A preliminary analysis of these data has been presented in Lavoie *et al.* (1989); some values presented here differ slightly from those reported previously due to the addition of several replicate assays and to the use of the curvilinear rather than linear regression for $R45_{HEL}K$.

[b] Details of covalent coupling of lysozyme to beads and conditions of PCFIA are given in Hartman *et al.* (1989).

[c] The association constants were measured by PCFIA, utilizing the method of Friguet *et al.* (1985) with the correction of Stevens (1987) applied as appropriate. A preliminary summary of these data was presented in Lavoie *et al.* (1989).

$E50_{VH}$ also contributes significantly to HyHEL-5 binding, we performed immunoassays with the site-directed mutant $R45_{HEL}K$. More than a 100-fold excess of $R45_{HEL}K$ was required to inhibit binding of HyHEL-5 to HEL (Fig. 2B), and the measured association constant was 10-fold lower than that of HEL (Table II). Notably, the log-logit transformed competitive inhibition data fit a second-order curvilinear model better ($r = .99$) than the expected linear model ($r = .92$) (Fig. 2B), although the binding curve for calculation of the association curve did not indicate significant second-order interactions. The meaning of the apparent second-order inhibition curve will require further investigation to interpret; the shape of the curve may suggest enhanced antibody binding at higher concentrations of antigen (see below).

C. The Role of Salt Bridges and Hydrophobic Interactions

The results just described are directly relevant to understanding viral "escape" mutants, for R→K substitutions in two adjacent contact residues have dramatically different effects on antibody binding; a substitution such as $R68_{HEL}$→K, which reduces antibody affinity by approximately 10^4, might be sufficient to escape neutralization, while the smaller quantitative effect of the $R45_{HEL}$→K substitution probably would not.

Free energy calculations from energy-minimized X-ray coordinates of the HyHEL-5 Fab–HEL complex predicted a greater contribution of $R68_{HEL}$ than $R45_{HEL}$ to binding energy (Novotny *et al.*, 1989). An examination of the X-ray structure of the HyHEL-5 Fab–HEL complex suggests a possible testable hypothesis. In the complex with HyHEL-5, $R68_{HEL}$ forms two salt links to $E35_{VH}$ and $E50_{VH}$ of the antibody heavy chain, while $R45_{HEL}$ forms a single salt link to $E50_{VH}$ (Sheriff *et al.*, 1987). If a Lys side chain is modeled into the $R68_{HEL}$ position, the substituted side chain in principle still could form one of the salt links (S. Sheriff, personal communication), but the substitution would result in the loss of at least one salt link, due to the chemical difference between Arg and Lys residues, and thus would leave unneutralized charges at the interface. On the other hand, a Lys at position 45 in principle still could form a salt link, but perhaps a weaker one due to the smaller size of the Lys side chain. We examined the X-ray structure to ask whether there was an obvious structural barrier to the $R45_{HEL}$ residue forming a second salt bridge to replace that lost by the $R68_{HEL}$→K substitution; R45 in the complex is surrounded by three Trp side chains from the antibody H and L chains, which would likely interfere with the side chain moving close enough to form a salt bridge with $E35_{VH}$. The strongly hydrophobic environment created by the preponderance of aromatic side chains probably increases the strength of the Arg–Glu salt bridges at the HyHEL-5 Fab–HEL interface.

We have recently partially characterized several other MAbs from BALB/c mice immunized with HEL that appear to be sensitive to the $R68_{HEL}$→K replacement found in BWQ lysozyme (i.e., low binding to BWQ but not to other quail lysozymes) and also appear to be of lower affinity than HyHEL-5 (Mallett *et al.*, 1989b). Fischmann *et al.* (1988; see also Mariuzza *et al.*, 1989, and Chapter 16 in this volume) have reported preliminary X-ray diffraction studies of Fab–HEL complexes of two similar MAbs. A structural comparison of these antibodies with HyHEL-5 and with each other should be of interest with respect to understanding specificity and affinity for a specific protein epitope.

III. Interactions of HyHEL-10 and HyHEL-8 with HEL

A. Similar Affinities and Epitopes but Different Fine Specificities

The interactions of HyHEL-10 and the structurally related HyHEL-8 with lysozyme provide an interesting model for the study of specificity and affinity. The serologically defined epitopes recognized by HyHEL-8 and -10 are very similar (Smith-Gill *et al.*, 1984a,b) (Fig. 3), and the affinities of the two antibodies are comparable (Lavoie *et al.*, 1989c). However, while HyHEL-10 is very sensitive to a variety of amino acid substitutions in HEL, HyHEL-8 is relatively insensitive to the same substitutions (Table III, Fig. 3). Experiments with chimeric antibodies, produced by recombining H and L chains of the immunoglobulin proteins, have demonstrated that both the H and L chains contribute to these differing fine specificities, that is, the $H_{10}L_8$ and H_8L_{10} chimeric antibodies have intermediate sensitivities to a variety of amino acid substitutions compared to the reconstituted parental antibodies (Smith-Gill *et al.*, 1987b). Understanding the structural differences underlying these specificity requirements is vital for understanding viral "escape" mutants: While a single mutation in a viral protein could very likely allow escape from neutralization by an antibody like HyHEL-10, several independent mutations might be required to abro-

LYSOZYME	*RESIDUE*						*REACTIVITY*			
	3	*19*	*21*	*102*	*103*	*121*	*HH8*	*HH10*	*HH12*	*HH15*
Chicken	F	N	R	G	N	Q	1	1	1	1
Japanese quail	Y	K	Q	V	H	N	20	800	16	»20
California quail	-	-	-	-	-	H	1	2	3	1
Turkey	Y	-	-	-	-	H	1	4	1	1
Ring-necked pheasant	Y	-	-	-	-	N	1	2	0.2	0.4

Figure 3. Identification of the "Japanese quail-defined" residues N19$_{HEL}$, R21$_{HEL}$, G102$_{HEL}$, and N103$_{HEL}$ as potential critical residues for recognition of chicken lysozyme by HyHEL-8, -10, -12, and -15. When compared to the sequences of the other lysozymes, which bound all the antibodies nearly equivalently to (or better than) chicken, these four residues are uniquely substituted in Japanese quail lysozyme. Reactivity for each antibody is defined as in Fig. 1 and Table II. The data for HyHEL-8, -10, and -12 are based on competitive inhibition ELISA, summarized from Smith-Gill *et al.* (1984a). The data for HyHEL-15 are summarized in part from Lavoie *et al.* (1989).

Table III. Lysozymes Having Substitutions at Positions That Are Contact Residues within the Crystallographically Defined Epitope Recognized by the Antibody HyHEL-10

Lysozyme	Contact residue[a]									Reactivity[b]	
	15	21	73	75	89	93	97	101	102	HyHEL-10	HyHEL-8
Chicken	H	R	R	L	T	N	K	D	G	1	1
Japanese quail	—	Q	—	—	—	—	—	—	V	800	20
Montezuma quail[c]	—	W	—	—	—	—	—	—	—	300–500	3–6
Turkey	L	—	K	—	—	—	—	G	—	4	1
Ring-necked pheasant	M	—	K	—	—	—	—	—	—	2	1
Chachalaca	Y	—	K	—	A	R	R	—	—	70	4
Duck	L	—	K	A	—	R	R	—	—	>4000	10
D101$_{HEL}$G[d]	—	—	—	—	—	—	—	G	—	1–2	1

[a] Amino acids that form van der Waals contact(s) and/or hydrogen bond(s) with HyHEL-10, based on X-ray structure of HyHEL-10 Fab–HEL complex (Padlan *et al.*, 1989; see Davies *et al.*, 1989, and Chapter 15 in this volume).

[b] IC$_{50}$ in a competitive inhibition assay. All values are based on ELISA, summarized from Smith-Gill *et al.* (1984a), with the exception of Montezuma quail and D101$_{HEL}$G, which are based on PCFIA performed as described in Lavoie *et al.* (1989a).

[c] Values for Montezuma quail are preliminary determinations reported for the first time here.

[d] The values for D101$_{HEL}$G were reported previously in Lavoie *et al.* (1989).

gate binding by HyHEL-8. Of course, it is possible that mutations in residues for which we have no variants could abrogate HyHEL-8 binding, although to date we have not identified a substitution to which HyHEL-8 is more sensitive than HyHEL-10.

Four antibodies (HyHEL-8, -10, -12, and -15) were characterized by significantly reduced binding to Japanese quail but not to other quail lysozymes, allowing the identification of four HEL residues (Asn 19, Arg 21, Gly 102, and Asn 103) as potential contact residues in the respective epitopes (see Fig. 3); other potential contact residues were eliminated by binding results with turkey and ring-necked pheasant lysozymes (Smith-Gill *et al.*, 1984a,b; Lavoie *et al.*, 1989).

HyHEL-8 and HyHEL-10 are in different complementation groups from HyHEL-12 and HyHEL-15 as defined by two types of assays (Smith-Gill *et al.*, 1984a,b; Lavoie *et al.*, 1989):

1. HyHEL-8 and -10 strongly competed with each other for binding to HEL but not with HyHEL-12 and -15, both of which competed with each other.

2. HyHEL-8 and -10 would not co-bind with each other but each would co-bind with HyHEL-12 or -15, and similarly HyHEL-12 and -15 would not co-bind with each other.

We therefore reasoned that HyHEL-8 and -10 must each recognize subsets of the four residues defined by JQ lysozyme, which were nonoverlapping with the subset(s) of residues recognized by HyHEL-12 and -15. The patterns of interactions of these antibodies with each other, with other antibodies (including HyHEL-5), and with oligosaccharide substrates led us to hypothesize that HyHEL-8 and HyHEL-10 recognized $R21_{HEL}$ and/ or $N19_{HEL}$, while HyHEL-12 and -15 recognized $G102_{HEL}$ and/or $N103_{HEL}$ (Smith-Gill *et al.*, 1984b; Lavoie *et al.*, 1989).

B. A New Lysozyme Variant and Antibody Interactions with HEL Residue 21

We have obtained the complete amino acid sequence (which will be described in detail elsewhere) of Montezuma quail (*Cyrtonyx montezumae*) lysozyme, whose purification and partial characterization were described by Jollès *et al.* (1979). This lysozyme proved to be a novel evolutionary variant, consistent with its reactivity with polyclonal antisera (White, 1976). At position 21 Montezuma quail has a Trp, an amino acid residue which has not been seen at that position in any other of the more than 50 lysozymes *c* sequenced in this region (E. M. Prager, unpublished observations). The $R21_{HEL} \rightarrow W$ substitution is the only replacement within the HyHEL-10 epitope (Table III).

We examined the binding of HyHEL-8 and -10 to Montezuma quail lysozyme in order to assess the importance of HEL residue 21 to the binding of these two antibodies. The apparent affinity of Montezuma quail lysozyme for HyHEL-10 is more than two orders of magnitude lower than that of chicken lysozyme, providing serological confirmation of the importance of $R21_{HEL}$ to HyHEL-10 binding (Table III). In contrast, the effect of the $R21_{HEL} \rightarrow W$ substitution on HyHEL-8 binding is minimal, although statistically significant (Table III). We are currently producing site-directed mutants of $R21_{HEL}$ to assess directly the contribution of the $R21_{HEL} \rightarrow Q$ substitution to the reduced affinity of JQ lysozyme of HyHEL-10.

C. Molecular Modeling of HyHEL-10

Based upon primary sequence data and our epitope mapping data, we built a molecular model of the HyHEL-10 Fv (Smith-Gill *et al.*, 1987a) and used molecular models to dock it with HEL (using coordinates of tetragonal

HEL) to form a Fv–HEL complex (Mainhart *et al.*, 1989). In our modeled complex, the epitope included both $R21_{HEL}$ and $N19_{HEL}$, but neither $G102_{HEL}$ nor $N103_{HEL}$ (these latter two residues had previously been defined as "border" residues) (Smith-Gill *et al.*, 1984b). After molecular dynamics, the final modeled complex, which will be described in detail elsewhere (Mainhart *et al.*, 1989), provided ready interpretation of most, although not all, of our serological data but did not accurately predict the crystallographically defined epitope.

D. Comparison of the HyHEL-10 Fab–HEL X-Ray Structure and Serological Data

Serological mapping was less accurate in predicting the epitope recognized by the antibody HyHEL-10 than it was for HyHEL-5. The solution of the HyHEL-10 Fab–HEL complex (Padlan *et al.*, 1989) showed that both $R21_{HEL}$ and $G102_{HEL}$ were in contact with the antibody, while both $N19_{HEL}$ and $N103_{HEL}$ were at least partially buried. This result presents an apparent dilemma, because all four of these residues would appear to be inaccessible to HyHEL-12 and -15, although the serological data discussed above suggest that they both are sensitive to substitution in at least one of them (Fig. 3). Furthermore, when HyHEL-8, -10, and -12 were tested for binding to a series of site-specific mutants at $D101_{HEL}$, all three antibodies were found to be sensitive to several substitutions at this position (Lavoie *et al.*, 1989).

The apparent recognition of the same amino acid residue(s) by nonoverlapping antibodies raises some important questions with respect to the functional definition of an epitope, and also with respect to the assumptions commonly made in serological epitope mapping (reviewed in Benjamin *et al.*, 1984). Of particular importance is whether noncontact residues can significantly influence binding, that is, whether there are long-range influences of amino acid substitutions. The bulk of serological evidence to date suggests that the effects of most surface amino acid substitutions on antibody binding are local, and a significant effect of a substitution on binding has therefore been interpreted as evidence that a particular amino acid is likely to be a contact residue (Benjamin *et al.*, 1984). However, one possible explanation of the reduced affinity of HyHEL-12 and -15 for JQ lysozyme is that noncontact residues are influencing binding. Another important question which these observations raise is how closely opposed the epitopes of two serologically "nonoverlapping" (i.e., noncompeting) antibodies can be. It will be of particular interest to analyze the interactions between HyHEL-10 and D1.3 (Amit *et al.*, 1986), whose epitopes overlap by only a few residues on the edge of each respective epitope (Davies *et al.*, 1988, 1989, and Chapter 15 in this volume).

Contacts in the HyHEL-10 Fab–HEL complex also include $H15_{HEL}$ and $R73_{HEL}$, both of which are substituted in turkey and ring-necked pheasant lysozymes, and $D101_{HEL}$, which is substituted in turkey lysozyme (Table III). Yet the binding of turkey and ring-necked pheasant lysozymes to HyHEL-10 is only slightly reduced compared to chicken lysozyme (Table III), which suggests that changes of some contact residues may yield minimal functional differences. Experiments with the $D101_{HEL}G$ mutant of HEL confirm that although $D101_{HEL}$ is a hydrogen-bonding contact residue (Padlan *et al.*, 1989), this substitution and elimination of the hydrogen bond has little measurable effect on HyHEL-10 binding (Table III).

E. Relationship of Primary Structures to Specificity

HyHEL-8 and HyHEL-10 are also of interest because, although independently derived, they are structurally very similar to each other and to the IgAκ DNP-binding myeloma protein XRPC-25 (Smith-Gill *et al.*, 1987b; Lavoie *et al.*, 1989). Several mechanisms contribute to structural immunoglobulin diversity, including combinatorial diversity deriving from the joining of the gene segments comprising the antibody V_H chain (V_H, D_H, J_H) and the V_κ chain (V_κ, J_κ) (see Berek and Milstein, 1988, for a review). Although this process apparently contributes stochastically to the preimmune repertoire, with little restraint on V-gene assembly during differentiation (Kaushik *et al.*, 1989), restricted V_H–V_L pairing has been observed for a number of antigen specificities (Potter, 1984).

It is therefore notable that although HyHEL-8 and -10 represent one antigen-binding specificity and XRPC-25 represents a distinct, nonoverlapping specificity, the variable regions of all three antibodies are derived from V_H36–60 and V_κ23 family genes and share greater than 90% identity in the V-gene encoded portions of the variable regions (Lavoie 1989). The most notable difference between the two HEL-binding MAbs and XRPC-25 is in the length and sequence of the H3 region. The H3 of XRPC-25 utilizes a different D_H gene and is two amino acids longer than those of HyHEL-8 and -10 (Fig. 4), while HyHEL-8 and -10 share highly conserved H3 and L3 structures, although they were independently derived (from different mice and different fusions performed on different days). In L3, they utilize the same J_κ gene. In H3 both utilize the same D_{Q52} gene-encoded segment, putative N addition sequence[2] and the same J_H gene, conserving both sequence (with the exception of a single somatic

---

[2]Randomly added nucleotides at the sites of V_H–D_H and D_H–J_H joining (Alt and Baltimore, 1982; Desiderio *et al.*, 1984).

mutation-derived difference) and length of H3 of these two antibodies (Fig. 4).

Experiments with chimeric antibodies produced by recombination of the H and L polypeptide chains had demonstrated that the major specificity for DNP versus HEL is determined absolutely by the heavy chains, and that with respect to this major specificity the three light chains are essentially interchangeable, leading to the hypothesis that H3 is a likely candidate in determining the major specificity differences (Smith-Gill *et al.*, 1987b; Lavoie, 1989). Spin-labeling studies with three DNP-binding antibodies, XRPC-25, MOPC-315, and MOPC-460, had previously suggested a correlation between the size of the binding pocket and affinity for DNP (Willan *et al.*, 1977). Since there is not an apparent pocket in the HyHEL-10 combining site (Padlan *et al.*, 1989), we proposed that the H3 of HyHEL-8 and -10 are too short to create a binding pocket.

Both HyHEL-8 and -10 have a somatic mutation of the J_H2-encoded $A101_{VH}$, to $T101_{VH}$ in HyHEL-8 and $D101_{VH}$ in HyHEL-10 (Fig. 4). From this observation we hypothesized that a polar or charged residue is required at this position to obtain high-affinity binding to HEL and that this somatic mutation-derived difference might contribute to the fine specificity differences between the two antibodies. [In our modeled complex we had actually placed $D101_{VH}$ of HyHEL-10 in contact with $R21_{HEL}$, al-

ANTIBODY	R E S I D U E										
	92					100			101		103
HyHEL-10	C	A	N	W	D	G	*	*	D	Y	W
HyHEL-8	-	-	-	-	-	-	*	*	T	-	-
H3V2 (D101$_{vH}$A)	-	-	-	-	-	-	*	*	A	-	-
H3V3 (D101$_{vH}$T)	-	-	-	-	-	-	*	*	T	-	-
H3V6 (G100\TF/D101)	-	-	-	-	-	-	T	F	-	-	-
XRPC-25	-	-	R	Y	R	T	T	F	-	-	-

Figure 4. Diagram of H3 of the H chains of the native and site-directed mutant antibodies described in this study. The genetically defined H3 begins at the V_H–D_H junction, usually after residue 95, and ends at the first J_H-encoded residue (Kabat *et al.*, 1987), which is 101_{VH} for the antibodies shown here. For purposes of comparing the total length of H3, we are following the convention introduced by Potter (1984) of counting the total number of residues between the invariant $C92_{VH}$ and $W103_{VH}$ residues, since it is not always possible to specify the precise spans of the V_H and D_H gene segments when all germ line gene sequences of any given family are not known. Dashes indicate identity to HyHEL-10; stars indicate the absence of a residue at that position. Summarized from Lavoie (1989).

though the X-ray structure (Padlan *et al.*, 1989) subsequently showed that those two amino acid residues are not in contact.]

F. Site-Directed Mutants of HyHEL-10 and the Role of 101_{VH}

Site-directed mutants of HyHEL-10 were made in order to test the above hypotheses about the role of H3 in determining major and fine specificity of the three V_H36–60 antibodies. The genomic rearranged heavy and light chain genes of HyHEL-10 were cloned in EMBL phage λ vectors and utilized to construct eukaryotic expression vectors, which were used to transfect SP2/0 Ag14 nonproducing cells by protoplast fusion, as described in detail elsewhere (Lavoie, 1989). The product of wild-type rHyHEL-10 secreting cell lines was characterized by PCFIA for affinity and fine specificity with a panel of evolutionary variants of bird lysozymes and was found to be indistinguishable from native HyHEL-10.

Three changes in the H3 region of HyHEL-10 were produced by oligo-nucleotide-directed mutagenesis in phage M13 (Lavoie, 1989). The H3V3 mutant[3] $D101_{VH} \rightarrow T$ (to the residue found at position 101_{VH} in HyHEL-8) was produced to test the hypothesis that the $A101_{VH} \rightarrow T$ and $A101_{VH} \rightarrow D$ somatic mutations in HyHEL-8 and -10 contributed to the fine specificities of these two antibodies and that the Asp and Thr were interchangeable with respect to affinity; the prediction was that the fine specificity of the $D101_{VH}$ T mutant would be more similar to that of HyHEL-8 and less sensitive to some of the amino acid changes within the HEL epitope (based on the modeled complex). The H3V2 mutant $D101_{VH} \rightarrow A$ (to the germline-encoded residue) was produced in order to examine the effect of this residue on the fine specificity and affinity of the B cell clone which eventually mutated to the HyHEL-10 sequence. Finally, in the H3V6 mutant, $G100\backslash TF/D101_{VH}$, two codons were inserted into the H3 region in order to examine the role of the length of H3 in determining DNP versus HEL major specificity; the two residues $T100a_{VH}$ and $F100b_{VH}$ were placed into their cognate position in the XRPC-25 structure, between $G100_{VH}$ and $D101_{VH}$ (Fig. 4).

All three expressed mutant antibodies were found by PCFIA to bind HEL but with differing affinities (Lavoie, 1989). The $D101_{VH}T$ mutant bound HEL with specific activity roughly equivalent to that of recombinant and native HyHEL-10, while the $D101_{VH}A$ and $G100\backslash TF/D101_{VH}$ mutants bound with 40–60-fold and 20-fold lower specific activity, respectively, as measured by titration on HEL-coupled polystyrene particles. The affinities of HyHEL-10, rHyHEL-10, and the $D101_{VH}T$ mutant

---

[3] We have adopted the following nomenclature for site-directed mutants of HyHEL-10 (Lavoie, 1989): H3V*n* refers to variant *n* of the H3 region. The specific variants discussed in this paper are H3V2, $D101_{VH} \rightarrow A$; H3V3, $D101_{VH} \rightarrow T$; H3V6, $G100\backslash TF/D101_{VH}$.

Table IV. Summary of PCFIA Competitive Inhibition Experiments with HyHEL-10, H3V3, and HyHEL-8, Suggesting an Effect of $D101_{VH}$ on Sensitivity to Substitutions Found in Japanese and Montezuma Quails

Quail Lysozyme	Epitope residue[a]				Reactivity[b]		
	19	21	102	103	HyHEL-10 $(D101_{VH})$	H3V3 $(D101_{VH}T)$	HyHEL-8 $(T101_{VH})$
California	N	R	G	N	1	1	1
Japanese	K	Q	V	H	~500	~200	40
Montezuma	—	W	—	—	~500	~350	5

[a] Residues which are within the crystallographically defined epitope; $R21_{HEL}$ and $G102_{HEL}$ are contact residues (van der Waals contact(s) and/or hydrogen bond(s) with an antibody residue), while $N19_{HEL}$ and $N103_{HEL}$ are at least partially buried (made inaccessible) by the Fv but do not form van der Waals contacts or hydrogen bonds with the antibody (Padlan *et al.*, 1989). California quail is identical to chicken lysozyme at these four positions; all other substitutions in these three quails relative to chicken lysozyme are neither contact nor buried residues in the HyHEL-10 epitope.

[b] Relative ability to inhibit binding of the antibody to particle-bound HEL in PCFIA. The IC_{50} of California quail lysozyme has been set equal to one for each antibody (in all cases, these values were not significantly different from those for chicken, which was also assayed in parallel), and the other values for that antibody expressed as a ratio relative to this value. The reactivities shown represent a single (replicated) data set from a series of experiments (Lavoie, 1989) that illustrates the relative relationships between the inhibitors and these antibodies.

for HEL were all of the same order of magnitude, $\sim$2–4 $\times$ 10^9 M^{-1}, whereas the affinities of the $D101_{VH}A$ and $G100\backslash TF/D101_{VH}$ mutants were $\sim$5 $\times$ 10^6 M^{-1} and $\sim$2 $\times$ 10^7 M^{-1}, respectively. The 1000-fold decrease in affinity of the $D101_{VH}A$ mutant indicates that this position is important for high-affinity HEL binding by both HyHEL-8 and -10, since at this position either a Thr or Asp, but not the germ line Ala, supports high affinity binding.

The fine specificity of the $D101_{VH}T$ mutant was examined by competitive inhibition PCFIA (Table IV). California quail behaves identically to chicken lysozyme in competitive inhibition with all three antibodies and is used as a control in this assay because California quail shares with Montezuma quail lysozyme substitutions at residues 40_{HEL}, 55_{HEL}, 91_{HEL}, and 121_{HEL} (all not in the HyHEL-10 epitope)[4], and is like JQ lysozyme in being different from chicken at position 121 (i.e., JQ has a substitution at residue 121_{HEL}, although not the same one found in California and Montezuma quails). As demonstrated previously (see Fig. 3), HyHEL-10 is much more sensitive to the substitutions presented by Japanese and Montezuma quail lysozymes than is HyHEL-8. Binding of both these quail lysozymes by the $D101_{VH}T$ mutant is intermediate to HyHEL-10 and HyHEL-8

[4] Montezuma quail also differs from chicken lysozyme at position 47, which is also not in the HyHEL-10 epitope, and thus has a total of six substitutions relative to the chicken sequence, with only $R21_{HEL}{\rightarrow}W$ occurring within the HyHEL-10 epitope.

(Table IV). These results clearly demonstrate that the single somatic mutation-derived difference at 101_{VH} contributes to the fine-specificity differences between HyHEL-8 and -10 with respect to recognition of the four residues that are defined by JQ (or a subset thereof), and specifically with respect to recognition of $Arg21_{HEL}$.

G. Possible Long-Range Interactions

The results with both $D101_{VH}T$ and $D101_{VH}A$ are perplexing, because in the crystal structure of the HyHEL-10 Fab–HEL complex, $D101_{VH}$ does not contact HEL, nor does it interact directly with any residues that do contact HEL (Padlan *et al.*, 1989). In fact, $D101_{VH}$ is near HEL not in the vicinity of $R21_{HEL}$, but rather at least 15 Å away on the opposite side of the catalytic cleft, with the nearest HEL residue to $D101_{VH}$ being Asn 77_{HEL}. One possible explanation of this result is that single mutations at this position in H3 directly alter the conformation of H3 and/or possibly other hypervariable regions through long-range interactions, either by movement of side chains or of the protein backbone. An additional formal possibility is that HyHEL-10 interacts with its epitope in more than one way, one of which is captured in the crystal form and another which is detectable in binding assays. Additional structural and functional studies are necessary to examine these and other possibilities.

H. Multispecific Antibodies and Multivalent Recognition

Although the $G100\backslash TF/D101_{VH}$ mutant of HyHEL-10 showed variable binding to polystyrene particles coated with DNP coupled to bovine serum albumin, it was retained on a DNP-Affigel-10 column and could be eluted with DNP (10^{-3} M), HEL (7×10^{-4} M), or Glycine-HCl (5×10^{-2} M, pH 2); the antibody that was eluted from the DNP column bound HEL in a PCFIA. Binding to HEL was partially inhibitable by DNP-lysine, whereas the HEL binding by neither HyHEL-10 nor the single site $D101_{VH}$ mutants was inhibitable by DNP-lysine, nor were they retained on a DNP affinity column. The $G100\backslash TF/D101_{VH}$ mutant antibody thus appears to have gained a low affinity for DNP, but it also retains HEL binding which is of reasonable affinity. This antibody therefore can be considered a multispecific antibody (Lavoie, 1989). Presumably a B cell with this antibody as its cell surface receptor could gain affinity and specificity for either antigen by somatic mutation.

The $G100\backslash TF/D101_{VH}$ and $D101_{VH}A$ mutants share another unusual property that we believe has not been previously described (Lavoie, 1989). Soluble HEL enhances rather than inhibits binding of these MAbs to HEL-coated particles. The enhancement is not nonspecific: other proteins (e.g., cytochrome *c*) do not produce it and there are differences in the ability of evolutionary variants of lysozyme to produce enhancement. Notably, higher concentrations than chicken lysozyme are required of JQ or ring-necked pheasant to enhance HEL-binding by either mutant. The enhancement can be explained if these two MAbs each bind at least two nonoverlapping sites on HEL, and thus larger antibody–HEL complexes are formed, resulting in more than one molecule of antibody being captured by the particles per molecule of antibody bound directly to the particles. Consistent with this interpretation is the observation that in a true competitive inhibition assay, utilizing fluorescein-conjugated rather than particle-bound HEL, the G100\TF/D101$_{VH}$ mutant is inhibited by chicken but poorly by JQ lysozyme. (The affinity of D101$_{VH}$A for HEL is too low to bind enough of the labeled HEL to perform this assay.) In addition, a series of well-characterized single epitopic MAbs (i.e., HyHEL-8, HyHEL-10, HyHEL-5, HyHEL-9, and D101$_{VH}$T), whose binding to HEL is normally inhibited in a competitive inhibition PCFIA, display enhanced binding under appropriate conditions using as the soluble inhibitor chemically dimerized HEL (Lavoie, 1989). The weaker binding of both mutant antibodies by JQ lysozyme suggests that they bind HEL at one site that is similar to the epitope recognized by the HyHEL-10 parent antibody; in addition to this site they each must recognize at least one additional site on HEL, which may contain residues that vary in pheasant lysozyme, as this lysozyme behaves like HEL in the competitive inhibition assay with fluorescein-labeled HEL but shows reduced enhancement in the particle-binding assay.

The properties of multispecificity and "dual" or "multivalent" recognition in IgG antibodies are not unique to these mutants created *in vitro*. In experiments designed to examine the effect of antigen-specific immunization on the BALB/c B cell repertoire (Mallett *et al.*, 1989a), a significant number of IgG antibodies in both categories were found following immunization with HEL. As many as a third of the HEL-specific IgG MAbs were found to behave as if they recognized HEL as a multivalent antigen (Mallett *et al.*, 1989b), and over a fifth of all antigen-reactive IgG MAbs were multireactive when screened against a panel of 13 complex antigens (Mallett *et al.*, 1989a). Thus, the antibodies created by site-directed mutagenesis of HyHEL-10 exhibit binding properties which are naturally produced during the course of the immune response to HEL.

IV. Summation and Conclusion

High affinity MAbs specific for chicken lzsozyme provide a well-defined system for understanding the molecular basis of antibody recognition. Focusing on two antibodies for which the X-ray structures of the Fab–

lysozyme complexes are known, HyHEL-5 and HyHEL-10, we have manipulated both antibody and the protein antigen to examine a full range of specificity properties. The variety of biochemical and molecular techniques used included (1) comparative studies with a pair of antibodies which are structurally closely related, (2) production of chimeric antibodies by recombining immunoglobulin H and L chains, (3) site-directed mutagenesis of both antibody and lysozyme clones expressed *in vitro*, and (4) comparative studies with a panel of evolutionary variants of lysozyme. The results suggest that antibody specificity and affinity may vary independently, and that a variety of factors may contribute to both properties, including (1) electrostatic interactions and (2) indirect interactions involving antibody residues which contact neither the antigen nor another antibody residue which contacts the antigen. Montezuma quail lysozyme, a novel evolutionary variant at residue 21, was particularly useful in examining the role of lysozyme residue 21 and possible long-range interactions in binding of HyHEL-10. Site-directed mutants of HyHEL-10 have yielded unexpected specificities, including a bispecific antibody and antibodies which appear to recognize lysozyme as a multivalent antigen. An understanding of these interactions has implications for antibody design and vaccine development.

Acknowledgments

We thank Verne Schumaker for providing the dimerized HEL, Steven Sheriff for many hours of discussion and help with computer graphics of the Fab–HEL complexes, and Jack F. Kirsch for many helpful discussions and grant support. A portion of this work was supported by NIH grants GM-35393 to J. F. Kirsch and GM-21509 to A.C.W.

References

Alt, F. W., and Baltimore, D. (1982). Joining of immunoglobulin heavy chain gene segments: Implications from a chromosome with evidence of three $D-J_H$ fusions. *Proc. Natl. Acad. Sci. USA* **79,** 4118–4122.

Amit, A. G., Mariuzza, R. A., Phillips, S. E. V., and Poljak, R. J. (1986). Three-dimensional structure of an antigen–antibody complex at 2.8 Å resolution. *Science* **233,** 747–753.

Benjamin, D. C., Berzofsky, J. A., East, I. J., Gurd, F. R. N., Hannum, C., Leach, S. J., Margoliash, E., Michael, J. G., Miller, A., Prager, E. M., Reichlin, M., Sercarz, E. E., Smith-Gill, S. J., Todd, P. E., and Wilson, A. C. (1984). The antigenic structure of proteins: A reappraisal. *Annu. Rev. Immunol.* **2,** 67–101.

Berek, C., and Milstein, C. (1988). The dynamic nature of the antibody repertoire. *Immunol. Rev.* **105,** 5–26.

Davies, D. R., Sheriff, S., and Padlan, E. A. (1988). Antibody–antigen complexes. *J. Biol. Chem.* **263,** 10541–10544.

Davies, D.R., Sheriff, S., Padlan, E. A., Silverton, E. W., Cohen, G. H., and Smith-Gill, S. J.

(1989). Three-dimensional structures of two Fab complexes with lysozyme. *In* "The Immune Response to Structurally Defined Lysozyme Model" (S. J. Smith-Gill and E. E. Sercarz, eds.), pp. 125–132. Adenine Press, Schenectady, New York.

Desiderio, S. V., Yancopoulos, G. D., Paskind, M., Thomas, E., Boss, M. A., Landau, N., Alt, F. W., and Baltimore, D. (1984). Insertion of N regions into heavy-chain genes is correlated with expression of terminal deoxytransferase in B cells. *Nature (London)* **311,** 752–755.

Diamond, R. (1974). Real-space refinement of the structure of hen egg-white lysozyme. *J. Mol. Biol.* **82,** 371–391.

Fischmann, T., Souchon, H., Riottot, M.-M., Tello, D., and Poljak, R. J. (1988). Crystallization and preliminary X-ray diffraction studies of two new antigen–antibody (lysozyme–Fab) complexes. *J. Mol. Biol.* **203,** 527–529.

Friguet, B., Chaffotte, A. F., Djavadi-Ohaniance, L., and Goldberg, M. E. (1985). Measurements of the true affinity constant in solution of antigen–antibody complexes by enzyme-linked immunosorbent assay. *J. Immunol. Methods* **77,** 305–319.

Hartman, A. B., Mallett, C. P., Srinivasappa, J., Prabahakar, B. S., Notkins, A. L., and Smith-Gill, S. J. (1989). Organ reactive autoantibodies from nonimmunized BALB/c mice are polyreactive and express nonbiased V_H usage. *Mol. Immunol.* **26,** 359–370.

Jollès, J., Ibrahimi, I. M., Prager, E. M., Schoentgen, F., Jollès, P., and Wilson, A. C. (1979). Amino acid sequence of pheasant lysozyme. Evolutionary change affecting processing of prelysozyme. *Biochemistry* **18,** 2744–2752.

Jolley, M. E., Wang, C. J., Ekenberg, S. J., Zvelke, M. S., and Kelsoe, D. M. (1984). Particle concentration fluorescence immunoassay (PCFIA): A new, rapid immunoassay technique with high sensitivity. *J. Immunol. Methods* **67,** 21–35.

Kabat, E. A., Wu, T., Reid-Miller, M., Perry, H. M., and Gottesman, K. S. (1987). "Sequences of Proteins of Immunological Interest," 4th Ed. U.S. Dep. Health Hum. Serv., Public Health Serv., Nat. Inst. Health, Bethesda, Maryland.

Kaushik, A., Schulze, D. H., Bona, C., and Kelsoe, G. (1989). Murine $V\kappa$ does not follow the V_H paradigm. *J. Exp. Med.* **169,** 1859–1864.

Lavoie, T. B. (1989). Structure–function relationships in lysozyme-binding antibodies examined by site-directed mutagenesis. Ph.D. thesis, University of Maryland, College Park, Maryland.

Lavoie, T. B., Kam-Morgan, L. N. W., Hartman, A. B., Mallett, C. P., Sheriff, S., Saroff, D. A., Mainhart, C. R., Hamel, P. A., Kirsch, J. F., Wilson, A. C., and Smith-Gill, S. J. (1989). Structure–function relationships in high affinity antibodies to lysozyme. *In* "The Immune Response to Structurally Defined Lysozyme Model" (S. J. Smith-Gill and E. E. Sercarz, eds.), pp. 151–168. Adenine Press, Schenectady, New York.

Mainhart, C. R., Brooks, B., and Smith-Gill, S. J. (1989). In preparation.

Malcolm, B. A., Rosenberg, S., Corey, M. J., Allen, J. S., de Baetselier, A., and Kirsch, J. F. (1989). Site-directed mutagenesis of the catalytic residues Asp-52 and Glu-35 of chicken egg white lysozyme. *Proc. Natl. Acad. Sci. USA* **86,** 133–137.

Mallett, C. P., Rousseau, P. G., Saroff, D. A., and Smith-Gill, S. J. (1989a). B cell specificity repertoire during the development of the immune response to hen lysozyme. *In* "The Immune Response to Structurally Defined Lysozyme Model" (S. J. Smith-Gill and E. E. Sercarz, eds.), pp. 353–367. Adenine Press, Schenectady, New York.

Mallett, C. P., Newman, M. A., Rousseau, P. G., and Smith-Gill, S. J. (1989b). In preparation.

Mariuzza, R. A., Alzari, P. M., Amit, A. G., Bentley, G. A., Boulot, G., Chitarra-Guillon, V., Fischmann, T., Lascombe, M.-B., Riottot, M.-M., Saludjian, P., Saul, F. A., Souchon, H., Tello, D., and Poljak, R. J. (1989). Structural studies of antibody: lysozyme complexes. *In* "The Immune Response to Structurally Defined Proteins:

The Lysozyme Model" (S. J. Smith-Gill and E. E. Sercarz, eds.), pp. 133–140. Adenine Press, Schenectady, New York.

Novotny, J., Bruccoleri, R. E., and Saul, F. A. (1989). On the attribution of binding energy in antigen–antibody complexes McPC603, D1.3, and HyHEL-5. *Biochemistry* **28,** 4735–4749.

Osserman, E. F., Canfield, R. E., and Beychok, S., eds. (1974). "Lysozyme." Academic Press, New York.

Padlan, E. A., Silverton, E. W., Sheriff, S., Cohen, G. H., Smith-Gill, S. J., and Davies, D. R. (1989). Structure of an antibody : antigen complex: Crystal structure of the HyHEL-10 Fab : lysozyme complex. *Proc. Natl. Acad. Sci. USA,* **86,** 5938–5942.

Phillips, D. C. (1967). The hen egg-white lysozyme molecule. *Proc. Natl. Acad. Sci. USA* **57,** 484–495.

Potter, M. (1984). Immunoglobulins and immunoglobulin genes. *In* "The Mouse in Biomedical Research" (H. L. Foster, J. D. Small, and J. G. Fox, eds.), Vol. 3, pp. 347–380. Academic Press, New York.

Sheriff, S., Silverton, E. W., Padlan, E. A., Cohen, G. H., Smith-Gill, S. J., Finzel, B. C., and Davies, D. R. (1987). Three-dimensional structure of an antibody–antigen complex. *Proc. Natl. Acad. Sci. USA* **84,** 8075–8079.

Sheriff, S., Silverton, E., Padlan, E., Cohen, G., Smith-Gill, S., Finzel, B., and Davies, D. R. (1988). Antibody–antigen complexes: Three dimensional structure and conformational change. *In* "Structure and Expression. From Proteins to Ribosomes" (R. H. Sarma and M. H. Sarma, eds.), Vol. 1, pp. 49–53. Adenine Press, Schenectady, New York.

Smith-Gill, S. J., and Sercarz, E. E., eds. (1989). "The Immune Response to Structurally Defined Proteins: The Lysozyme Model." Adenine Press, Schenectady, New York.

Smith-Gill, S. J., Wilson, A. C., Potter, M., Prager, E. M., Feldmann, R. J., and Mainhart, C. R. (1982). Mapping the antigenic epitope for a monoclonal antibody against lysozyme. *J. Immunol.* **128,** 314–322.

Smith-Gill, S. J., Mainhart, C. R., Lavoie, T. B., Rudikoff, S., and Potter, M. (1984a). V_L–V_H expression by monoclonal antibodies recognizing avian lysozyme. *J. Immunol.* **132,** 963–967.

Smith-Gill, S. J., Lavoie, T. B., and Mainhart, C. R. (1984b). Antigenic regions defined by monoclonal antibodies correspond to structural domains of avian lysozyme. *J. Immunol.* **133,** 384–393.

Smith-Gill, S. J., Mainhart, C., Lavoie, T. B., Feldmann, R. J., Drohan, W., and Brooks, B. R. (1987a). A three-dimensional model of an anti-lysozyme antibody. *J. Mol. Biol.* **194,** 713–724.

Smith-Gill, S. J., Hamel, P. A., Lavoie, T. B., and Dorrington, K. J. (1987b). Contributions of immunoglobulin heavy and light chains to antibody specificity for lysozyme and two haptens. *J. Immunol.* **139,** 4135–4144.

Stevens, F. J. (1987). Modification of an ELISA-based procedure for affinity determination: Correction necessary for use with bivalent antibody. *Mol. Immunol.* **24,** 1055–1060.

White, T. J. (1976). Protein evolution in rodents. Ph.D. Thesis, Appendix II, pp. 207–225. Univ. of California, Berkeley.

Willan, K. J., Marsh, D., Sunderland, C. A., Sutton, B. J., Wain-Hobson, S., Dwek, R. A., and Givol, D. (1977). Comparison of the dimensions of the combining sites of the dinitrophenyl-binding immunoglobulin A myeloma proteins MOPC 315, MOPC 460 and XRPC 25 by spin-label mapping. *Biochem. J.* **165,** 199–206.

18 Approaches toward the Design of Proteins of Enhanced Thermostability

J. A. Bell, S. Dao-pin, R. Faber, R. Jacobson, M. Karpusas,
M. Matsumura*, H. Nicholson, P. E. Pjura, D. E. Tronrud,
L. H. Weaver, K. P. Wilson, J. A. Wozniak, X-J. Zhang,
T. Alber,† and B. W. Matthews
Institute of Molecular Biology, University of Oregon, Eugene, Oregon 97403

I. Introduction

The advent of directed mutagenesis has made it possible to alter protein structures at will. For the first time it is possible to design and to introduce modifications into a protein that are intended to change its behavior in predictable ways.

We have been using the lysozyme from bacteriophage T4 (Color Plate 24) as a model system to test ways in which the stability of a protein might be improved. Such studies also provide quantitative information on the contributions that different types of interactions (H-bonds, hydrophobic interactions, salt bridges, etc.) make to the stability of proteins (Matthews, 1987). As such, these studies are also relevant to the contributions that these different interactions can make in enzyme–inhibitor and drug–receptor complexes.

In this report we review some of the approaches that are being explored in an attempt to increase the thermostability of T4 lysozyme.

II. Hydrophobic Interactions

It has long been known that the hydrophobic effect plays a very important role in stabilizing protein structures (Kauzmann, 1959). In order to quantitate the contribution of the hydrophobic effect at a specific site in a protein, isoleucine 3 in T4 lysozyme was replaced with 13 different amino acids (Matsumura *et al.*, 1988). It was found that the contributions of different

* Present address: Department of Immunology, Research Institute of Scripps Clinic, La Jolla, California 92037
† Present address: Department of Biochemistry, School of Medicine, University of Utah, Salt Lake City, Utah 84132

residues at position 3 to the overall stability of the protein were directly proportional to the hydrophobicity of the substituted residue (Fig. 1). Phenylalanine, tyrosine, and tryptophan were exceptions because, as inferred from the crystal structure of the Tyr variant, the side chains of these residues are too large to be accommodated within the interior of the protein.

It was of interest to find that two substitutions at position 3 yielded mutant proteins with thermostability greater than that of wild-type lysozyme. The first thermostable variant has cysteine at position 3, and the enhanced stability can be attributed to the formation of a nonnative disulfide bridge (Perry and Wetzel, 1984) (see below). The second variant more stable than the wild type has leucine at position 3, and the enhanced stability is presumed to be due to increased hydrophobic stabilization. Because crystals could not be obtained, the structure of the Leu 3 variant has not been determined. Model building suggests that the Leu 3 side chain is completely buried within the protein, whereas the side chain of the Ile 3 in wild-type lysozyme is 15% exposed to solvent and therefore does not manifest its full hydrophobic potential (Matsumura *et al.*, 1988, 1989b). This result suggests that there may be other sites within the T4 lysozyme

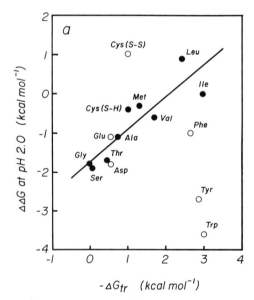

Figure 1. Stabilities of lysozymes with 13 different amino acid substitutions at Ile 3. The free energy of stabilization ($\Delta\Delta G$) of each mutant lysozyme at pH 2.0 is plotted against the free energy of transfer ($-\Delta G_{tr}$) of the individual amino acid from water to ethanol. The protein free energies are plotted relative to the wild type (Ile). (Reprinted by permission from *Nature* **334**, 406–410, Copyright (c) 1988 Macmillan Magazines Ltd.)

molecule (or within proteins in general) at which enhanced hydrophobic stabilization might be achieved by appropriately chosen amino acid replacements. In an attempt to find such sites a systematic search was made for cavities within the T4 lysozyme structure that might permit the replacement of smaller hydrophobic amino acids by larger ones, thus achieving greater hydrophobic stabilization (Karpusas *et al.*, 1989). There are such cavities, and one of these appears to allow the replacement of Leu 133 by a phenylalanine. This mutant protein has been constructed and high-resolution crystallographic analysis confirms that the Phe 133 side chain does occupy the cavity as expected. The crystallographic analysis also shows, however, that the side chain rotational angle, χ^1, is not optimal, and the strain energy associated with this distortion appears to offset the hydrophobic stabilization expected from the Leu→Phe replacement. As a result, the measured thermostability of the mutant protein is essentially identical with the wild type.

III. Hydrogen Bonding

The importance of hydrogen bonding in determining the folded structures of proteins does not need to be emphasized. The energetic contributions of individual hydrogen bonds, however, are not well understood. In folded proteins, as in protein–ligand complexes, the consideration of hydrogen bonding is complicated by the absolute requirement to take into account the role of solvent in stabilizing both the folded and the unfolded state (or, in the case of protein–ligand complexes, the associated and dissociated states).

The role of hydrogen bonding at one site in T4 lysozyme has been analyzed by a series of substitutions of Thr 157. In wild-type lysozyme the γ hydroxyl of this threonine participates in a network of hydrogen bonds (Fig. 2). Early studies of randomly generated temperature-sensitive mutant lysozymes had shown that the replacement of threonine with isoleucine at position 157 substantially destabilizes the protein, apparently because it results in the disruption of the hydrogen bond network (Fig. 2) (Alber *et al.*, 1986, 1987; Grütter *et al.*, 1987). Other changes between the two structures could, however, also contribute to instability. To determine how Thr 157 contributes to the stability of T4 lysozyme, 13 different amino acids were substituted at this site. The structures of these modified lysozymes have been determined and their stabilities measured (Alber *et al.*, 1987). The results show that the main way in which Thr 157 contributes to stability is through its hydrogen-bonding interactions. An interesting situation occurs when glycine is substituted at position 157. The lack of a side

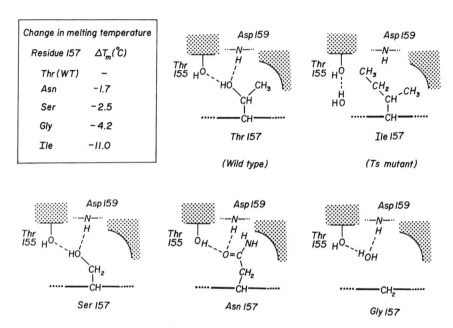

Figure 2. Schematic illustration showing the interactions displayed by five representative amino acids at position 157 of T4 lysozyme. In wild-type lysozyme the γ-oxygen of Thr 157 participates in a network of hydrogen bonds that is conserved in the Ser 157 and Asn 157 structures. In the case of the Gly 157 structure, X-ray analysis shows that a water molecule serves to retain the H-bond network, giving a relatively stable protein. The insert shows the relative melting temperatures at pH 2.0 of the mutant lysozymes illustrated in the figure. (Adapted from Alber *et al.*, 1987.)

chain allows a water molecule to bind at the site previously occupied by the γ hydroxyl of the threonine and to restore the hydrogen bond network, giving a protein whose stability is close to that of the wild type (Fig. 2).

In the case of the 13 different substitutions that have been made at position 157 in T4 lysozyme, no mutant is more stable than the wild type. Also, no engineered mutant protein is of lower stability than Ile 157, the variant that was obtained as a temperature-sensitive mutant after random chemical mutagenesis. This tends to suggest that proteins are tolerant of change and, within reason, relatively resistant to destabilization by amino acid replacements.

The above study suggests that one way in which proteins might be stabilized would be to locate any hydrogen bond donors or acceptors in the folded structure of the protein that do not participate in hydrogen bonding and to satisfy their H-bonding potential by appropriate site-directed substitutions. A detailed analysis of the T4 lysozyme structure, using the program of Baker and Hubbard (1984), did not reveal a single candidate for such a substitution. Within the accuracy of the X-ray structural analysis it appears that every hydrogen bond donor and acceptor in T4 lysozyme participates in at least one hydrogen bond (although not necessarily with good geometry). Unsatisfied H-bonding groups in folded proteins seem to be very rare, perhaps because they would tend to be very destabilizing, and the protein structure therefore relaxes to alternative conformations which permit hydrogen bonds to occur.

IV. Helix–Dipole Interactions

Recent evidence has shown that the stabilities of proteins can be enhanced by the introduction of appropriately charged groups at the ends of α helices (Mitchinson and Baldwin, 1986; Nicholson *et al.*, 1988).

In the case of T4 lysozyme initial experiments have focused on the introduction of aspartic acids at or near the amino termini of α helices. Two such substitutions, Ser 38→Asp and Asn 144→Asp (Fig. 3), were both found to increase the melting temperature of the protein by about 2°C at pH values where the introduced aspartates were negatively charged. The double mutant was found to increase the melting temperature by about 4°C (Nicholson *et al.*, 1988). A related substitution, Asn 144→Glu has also been constructed and found to yield essentially the same increase in stability as the replacement with aspartic acid (H. Nicholson *et al.*, unpublished observations). This therefore seems to be a rather general way to increase protein stability.

Structural studies of the wild-type and mutant lysozymes indicate that the stabilization is due to generalized electrostatic interaction of the introduced aspartic acid side chain with the positive charge at the end of the α-helix and does not require precise hydrogen bonding to the terminal amino groups. In the case of the Asn 144→Asp substitution, for example, neither the Asn or Asp side chain makes any hydrogen bonds to the end of the helix (Fig. 4). Because precise hydrogen bonding is not required, it greatly simplifies the design of stabilizing substitutions.

V. Substitutions That Decrease the Entropy of Unfolding

It has been proposed that selected substitutions of the form Xaa→Pro and Gly→Xaa can be used to decrease the configurational entropy of unfolding of a protein and so increase its thermostability (Matthews *et al.*, 1987). The

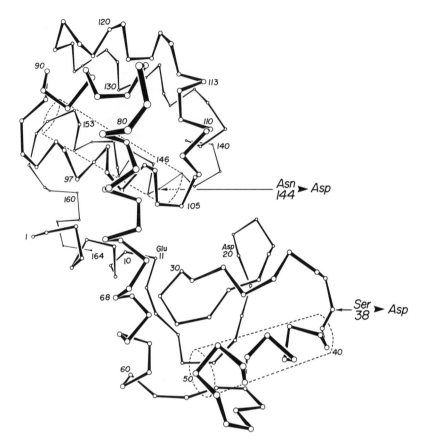

Figure 3. Backbone of T4 lysozyme showing the locations of the substitutions that were used to introduce negatively charged residues close to the amino termini of α-helices. The respective helices are outlined with broken lines. (Reprinted by permission from *Nature* **336,** 651–656, Copyright © 1988 Macmillan Magazines Ltd.)

basic idea is that glycine has greater conformational flexibility than a residue with a β carbon and so requires greater free energy to change from the unfolded to the folded state. Conversely, proline has a very restricted conformation and so requires a relatively low expenditure of free energy to retain in the folded state. The sites in a protein at which glycines are to be replaced or prolines substituted must, of course, be chosen in such a way that the native structure of the protein is not perturbed.

One such substitution, Ala 82→Pro was found to increase the melting temperature to T4 lysozyme by 2°C (Matthews *et al.,* 1987). In this case a crystal structure analysis confirmed that the structure of the mutant protein was virtually identical with that of the wild type, apart from the

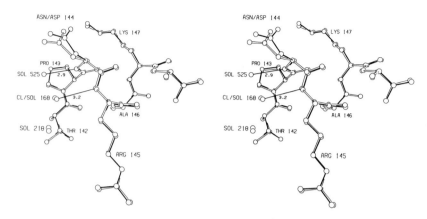

Figure 4. Superposition of the crystal structures of Asn 144 (wild-type) lysozyme (solid bonds) and Asp 144 (mutant) lysozyme (open bonds) showing that neither the wild-type nor the mutant structure displays any hydrogen bonds to the peptide nitrogens within the last turn of the α helix. (Reprinted by permission from *Nature* **336**, 651–656 Copyright © 1988 Macmillan Magazines Ltd.)

addition of the pyrrolidine ring of the proline. A second substitution, Ala 93→Pro, enhanced the stability of T4 lysozyme toward irreversible denaturation but increased the reversible melting temperature at pH 6.5 by only 0.2°C (H. Nicholson *et al.*, unpublished observations).

The replacement of Gly 77 with alanine was found to increase the melting temperature by 1°C at pH 6.5 (although not at pH 2.0) (Matthews *et al.*, 1987). A second replacement, Gly 113→Ala increased the melting temperature by 0.6°C at pH 5.0 and 0.3°C at pH 2.0 (H. Nicholson *et al.*, 1989).

Thus the removal of glycine residues and the insertion of prolines can enhance thermostability, although the increase in melting temperature is only marginal in some cases.

The above approach is not restricted to substitutions involving glycines and prolines. Replacements of the form Ser→Thr and Ala→Val, among others, can also be considered. For example, the variant Ala 41→Val has been constructed and found to increase the melting temperature by 1.5°C (Dao-pin *et al.*, 1989).

VI. Removal of Strain

A possible approach to stabilizing proteins might be to identify parts of the protein structure that are under strain and to relieve such strain by appropriate amino acid replacements. One difficulty in such an approach is that strain is likely to be distributed over the whole protein and not localized at a few sites. Also, the distortions due to strain that might be anticipated in a typical protein structure are likely to be difficult to detect, given the relatively limited accuracy that can be achieved even from highly refined protein crystal structures.

One approach along these lines that has been tested in T4 lysozyme is to replace so-called left-handed helical residues with glycine (Fig. 5). The rationale is that non-glycine residues are rarely observed to have conformations on the right-hand side of the Ramachandran diagram, whereas

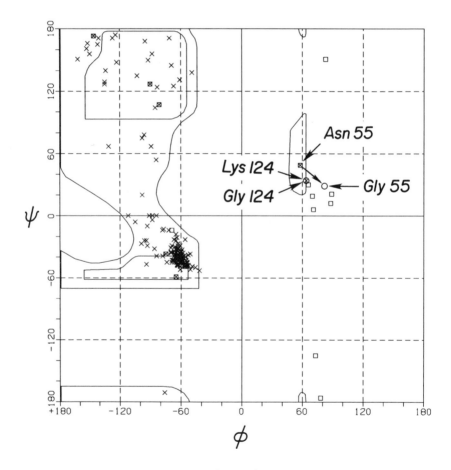

Figure 5. Ramachandran diagram of T4 lysozyme showing the (ϕ,ψ) conformations of the two left-handed-helical residues, Asn 55 and Lys 124, that were, respectively, replaced with glycine. In neither case is the stability of the protein significantly altered.

such conformations for glycines are common and are expected to be of lower energy.

Replacements of both Lys 124 and Asn 55 with glycine were constructed and found to have free energies of folding virtually identical with that of wild-type lysozyme (Nicholson *et al.*, 1989). This indicates that there is actually very little difference between the energy of a glycine and a non-glycine in the left-handed helical conformation.

Another approach to the removal of strain is to look for residues that have values of the side chain rotation angle χ^1 that are displaced from the expected energy minima. One possible candidate in T4 lysozyme is Val 131, for which the χ^1 value is 18° from the low-energy value. Val 131 is within an α-helix and the rotation of the side chain is presumably due to a close contact with atoms in the next turn of the helix. Replacement of Val 131 with alanine increased the melting temperature of the protein by 0.7°C (Dao-pin *et al.*, 1989). In this case the design was successful although the X-ray crystallographic structure analysis suggests that the expected stabilization due to relaxation of the χ^1 angle is offset somewhat by loss of hydrophobic stabilization and by entropic stabilization associated with the Val→Ala replacement.

VII. Disulfide Bridges

A number of attempts have been made to increase the stability of proteins by the introduction of nonnative disulfide bridges (Villafranca *et al.*, 1987; Perry and Wetzel, 1984; Sauer *et al.*, 1986; Pantoliano *et al.*, 1987; Wells and Powers, 1986). Surveys of disulfide bridges in known protein structures show that the geometry of such bridges is very restricted (Thornton, 1981; Richardson, 1981; Pabo and Suchanek, 1986). For this reason it is often difficult to find suitable pairs of residues in a protein that can be linked by a disulfide bridge without concomitant introduction of strain. Although polymer theory indicates that genetically engineered disulfide bridges can increase protein stability, experience to date shows that in many cases the engineered protein is not more stable than the wild type.

In an attempt to elucidate general principles relevant to the design of disulfide linkages, five different bridges have been introduced into phage lysozyme (Fig. 6) (Perry and Wetzel, 1984; Matsumura *et al.*, 1989a). Three of these bridges increase the melting temperature of the protein by 5–11°C at pH 2; the other two bridges destabilize the protein relative to wild-type lysozyme (Fig. 7). In each case the oxidized (cross-linked) form of the protein is more stable than the reduced (non-cross-linked) form.

It is anticipated that stabilization from disulfide bridges arises from the

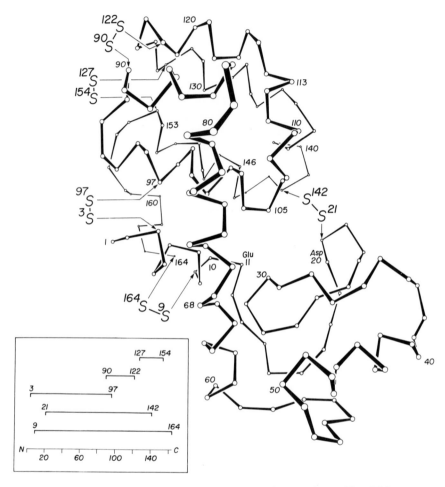

Figure 6. Locations of five disulfide bridges that have been engineered into T4 lysozyme. The lengths of the loops formed by these bridges are shown schematically in the insert. (Adapted from Matsumura *et al.*, 1989a.)

reduction in entropy of the unfolded protein and increases logarithmically with the size of the loop that is formed (see, e.g., Pace *et al.*, 1988). In practice, this contribution to stability will be offset by loss of preexisting favorable interactions resulting from the replacement of residues with cysteines and/or by strain associated with the formation of the S—S bridge.

As seen in Fig. 7 the results obtained with T4 lysozyme are consistent with these expectations. In each case the reduced mutant proteins are

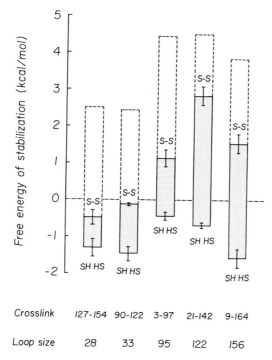

Figure 7.   Schematic diagram showing the free energies of stabilization of the reduced (SH) and oxidized (SS) forms of the disulfide-bridge mutant lysozymes shown in Fig. 6. The dotted line illustrates the theoretical stabilization to be expected from the reduction in entropy of the unfolded form, calculated using the formula of Pace *et al.* (1988). This theoretical contribution to the free energy of stabilization is added to the observed free energy of the reduced form.

about 0.5–1.5 kcal/mol less stable than the wild type, suggesting that the introduction of each single cysteine decreased the stability of the protein by about 0.75 kcal/mole on average. In no case is the stabilization expected from the entropic effect fully realized. Rather, the formation of each disulfide bridge appears to introduce strain into the molecule that reduces the net stability of the oxidized form of the mutant protein. This inference is supported by the observation that the disulfide bridges that appear to be under greatest strain (127–154 and 90–122) have measured stabilities substantially lower than the theoretical upper limit (Matsumura *et al.*, 1989a).

   The three disulfide bridges that are most effective in stabilizing T4 lysozyme have the largest loop sizes. Consistent with theoretical expectation, this suggests that a large loop size is a desirable attribute of engineered disulfide bridges that are intended to maximize protein stability.

When possible, disulfide bridges should also be introduced at sites at which the conformation of the native protein is geometrically compatible with the known requirements for formation of an unstrained disulfide bridge. Bearing in mind, however, that such sites are rare, it may be necessary to choose a site that is less than ideal. In the case of T4 lysozyme, the two bridges that are most effective in stabilizing the protein (9–164 and 21–142) are introduced into a flexible part of the structure. The use of such flexible sites could be another general attribute that is desirable in designing disulfide bridges in general (Matsumura *et al.*, 1989a).

VIII. Conclusions

There are many possible ways in which the thermostability of proteins might be increased.

Tests to date with T4 lysozyme suggest that the most effective approaches are (1) use of disulfide bridges, (2) stabilization of α-helix dipoles, and (3) entropic stabilization via the introduction of prolines and/or removal of glycines.

It also appears that the effects of independent mutations are additive. The combination of different mutations, each of which may provide only a modest increase in thermostability, therefore holds the promise of engineering proteins of substantially enhanced thermostability.

Acknowledgments

We are most grateful to a number of collaborators in the Institute of Molecular Biology including W. Baase, W. J. Becktel, F. W. Dahlquist, M. Lindorfer, D. C. Muchmore, S. J. Remington, and J. A. Schellman.

The work was supported in part by grants from the National Institutes of Health (GM21967; GM20066), the National Science Foundation (DMB8611084), and the Lucille P. Markey Charitable Trust.

References

Alber, T., Grütter, M. G., Gray, T. M., Wozniak, J., Weaver, L. H., Chen, B.-L., Baker, E. N., and Matthews, B. W. (1986). *UCLA Symp. Mol. Cell. Biol.* **39,** 307–318.
Alber, T., Dao-pin, S., Wilson, K., Wozniak, J. A., Cook, S. P., and Matthews, B. W. (1987). *Nature (London)* **330,** 41–46.
Baker, E. N., and Hubbard, R. E. (1984). *Prog. Biophys. Mol. Biol.* **44,** 97–179.
Dao-pin, S., Baase, W., and Matthews, B. W. (1989). *Proteins: Struct. Funct. Genet.* Submitted for publication.

Grütter, M. G., Gray, T. M., Weaver, L. H., Alber, T., Wilson, K., and Matthews, B. W. (1987). *J. Mol. Biol.* **197**, 315–329.

Karpusas, M., Baase, W. A., Matsumura, M., and Matthews, B. W. (1989). *Proc. Natl. Acad. Sci. U.S.A.* In press.

Kauzmann, W. (1959). *Adv. Protein Chem.* **14**, 1–63.

Matsumura, M., Becktel, W. J., and Matthews, B. W. (1988). *Nature (London)* **334**, 406–410.

Matsumura, M., Becktel, W. J., Levitt, M., and Matthews, B. W. (1989a). *Proc. Natl. Acad. Sci. U.S.A.* In press.

Matsumura, M., Wozniak, J. A., Daopin, S., and Matthews, B. W. (1989b). *J. Biol. Chem.* Submitted for publication.

Matthews, B. W. (1987). *Biochemistry* **26**, 6885–6888.

Matthews, B. W., Nicholson, H., and Becktel, W. J. (1987). *Proc. Natl. Acad. Sci. USA* **84**, 6663–6667.

Mitchinson, C., and Baldwin, R. L. (1986). *Proteins: Struct. Funct. Genet.* **1**, 23–33.

Nicholson, H., Becktel, W. J., and Matthews, B. W. (1988). *Nature (London)* **336**, 651–656.

Nicholson, H., Söderlind, E., Tronrud, D. E., and Matthews, B. W. (1989). *J. Mol. Biol.* In press.

Pabo, C. O., and Suchanek, E. G. (1986). *Biochemistry* **25**, 5987–5991.

Pace, C. N., Grimsley, G. R., Thomson, J. A., and Barnett, B. J. (1988). *J. Biol. Chem.* **263**, 11820–11825.

Pantoliano, M. W., Ladner, R. C., Bryan, P. N., Rollence, M. L., Wood, J. F., and Poulos, T. L. (1987). *Biochemistry* **26**, 2077–2082.

Perry, L. J., and Wetzel, R. (1984). *Science* **226**, 555–557.

Richardson, J. S. (1981). *Adv. Protein Chem.* **34**, 167–339.

Sauer, R. T., Hehir, K., Stearman, R. S., Weiss, M. A., Jeitler-Nilsson, A., Suchanek, E. G., and Pabo, C. O. (1986). *Biochemistry* **25**, 5992–5998.

Thornton, J. M. (1981). *J. Mol. Biol.* **151**, 261–287.

Villafranca, J. E., Howell, E. E., Oatley, S. J., Xuong, N.-H., and Kraut, J. (1987). *Biochemistry* **26**, 2182–2189.

Wells, J. A., and Powers, D. B. (1986). *J. Biol. Chem.* **261**, 6564–6570.

19 Interplay among Enzyme Mechanism, Protein Structure, and the Design of Serine Protease Inhibitors

Paul A. Bartlett, Nicole S. Sampson, Siegfried H. Reich,
David H. Drewry, and Lawrence A. Lamden
Department of Chemistry, University of California, Berkeley, California 94720

I. Introduction

It is still a challenge to design a potent enzyme inhibitor. Nevertheless, a number of effective strategies for doing so have emerged in recent decades. Among the first was the concept of "affinity labels" (Wofsy *et al.,* 1962) or "active-site directed irreversible inhibitors" (Baker, 1967); of greater generality are "suicide inhibitors" (Walsh, 1980) (often referred to as "mechanism-based inhibitors") and "transition state analogs" (Wolfenden, 1976). Although each of these strategies reflects a different approach to rational inhibitor design, they nevertheless have a common element in that their implementation requires knowledge of the enzyme substrate and mechanism. These inhibitors can all be characterized as *mechanism-derived inhibitors*[1], to indicate the source of information which went into their design. These strategies have matured to the extent that for some classes of enzymes, for example, the peptidases or those utilizing the cofactor pyridoxal phosphate, one can, given only the structure of a specific substrate, propose a molecule which will almost certainly be an effective inhibitor. However, even with this degree of success, there remain some enzymes for which a general solution to the design of inhibitors has not proved straightforward; moreover, mechanism-derived approaches are for the most part inapplicable to the design of ligands for noncatalytic protein binding sites such as receptors or allosteric sites.

With the accelerating pace at which structures of enzymes and other biologically interesting proteins are being determined in the crystalline state by X-ray and in solution by NMR, it is clear that another approach to

---

[1] The term *mechanism-derived inhibitors,* especially in the context of inhibitor design, is intended to apply to the broader subset of inhibitors than that indicated by the term *mechanism-based* inhibitors, which are a subset of the former.

Use of X-Ray Crystallography in the Design of Antiviral Agents

inhibitor design is emerging. From a knowledge of protein structure alone, inhibitors will be designed without reference to the mechanism of an enzyme or its normal substrate. We denote such compounds *structure-derived inhibitors* in order to distinguish them from the more traditional kinds. The types of compounds that one can imagine arising from such a design approach would be mimics of natural, perhaps peptidic, ligands for the target binding site, or de novo inventions devised to complement the binding cavity sterically and electronically.

In this presentation, the serine proteases are used as examples of enzymes for which inhibitors of both types have been devised. Of the four archetypal mechanistic motifs within the peptidase class, that involving acyl transfer to an activated serine hydroxyl (Scheme 1) is found in enzymes which are both important and ubiquitous. The serine peptidase class embodies digestive enzymes such as trypsin and chymotrypsin (Kossiakoff, 1987; Appel, 1986; Blow, 1976), those involved in response cascades such as blood clotting (Davie, 1986; Davie *et al.,* 1979) and the complement activation process (Colomb *et al.,* 1987; Reid and Porter, 1981), and defensive enzymes such as leukocyte elastase (Stockley, 1987; Groutas, 1987; Barrett, 1981), to name a few. They have been well characterized mechanistically (Kossiakoff, 1987; Blow, 1976) and structurally

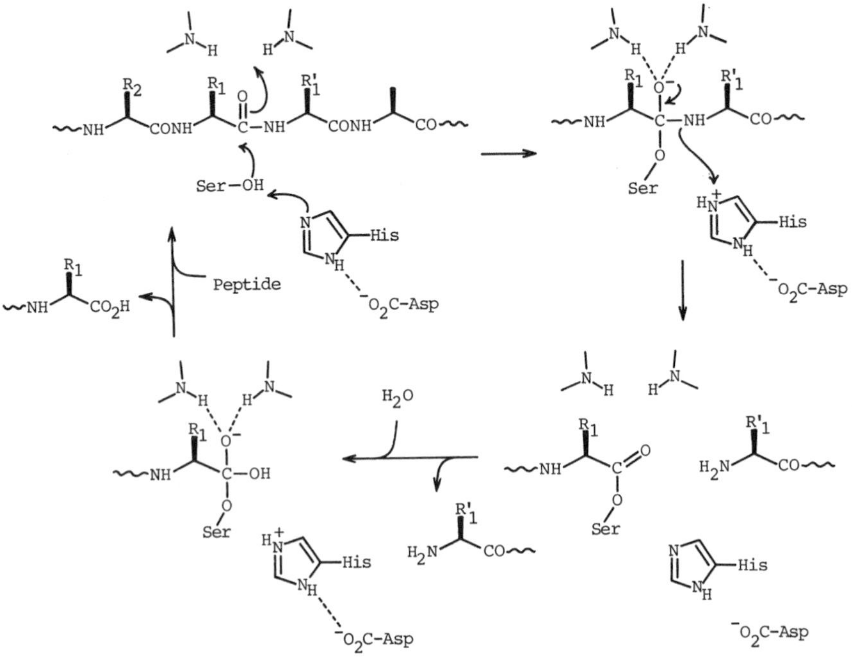

Scheme 1

(Kraut, 1977), and a variety of naturally occurring, macromolecular inhibitors (Laskowski, 1986; Gebhard and Fritz, 1986; Carrell and Travis, 1985), as well as small, synthetic inactivators are known (Shaw, 1970).

II. Mechanism-Derived Inhibitors

Although most of the small-molecule inhibitors of the serine proteases are covalently bound, both reversible as well as irreversible members of this group have been devised. The classic inactivators are the phosphorylating agent diisopropyl fluorophosphate (DFP) (Mounter *et al.*, 1963; Ooms and van Dijk, 1966) and peptidyl chloromethyl ketones (e.g., compound 2 of Fig. 1) (Wei *et al.*, 1988; Shaw *et al.*, 1981; Larsen and Shaw, 1976). Both of these bond irreversibly to residues in the active site, in the case of DFP by phosphorylating the serine hydroxyl and for the chloromethyl ketones by alkylating the histidine residue. Inhibitors which are freely reversible, such as the peptidyl aldehydes (e.g., compound 3 of Fig. 1) (Thompson and Bauer, 1979; Brayer *et al.*, 1979; Hunkapiller *et al.*, 1975), di- and trifluoromethylene analogs (e.g., compound 4 of Fig. 1) (Imperiali and Abeles, 1986), and peptidyl boronic acids (e.g., compound 5 of Fig. 1) (Kettner and Shenvi, 1984; Bachovchin *et al.*, 1988; Bone *et al.*, 1987) also bind covalently to the serine hydroxyl, but via a reversibly formed hemiacetal or boronate adduct. These inhibitors enjoy greater or lesser specificity among

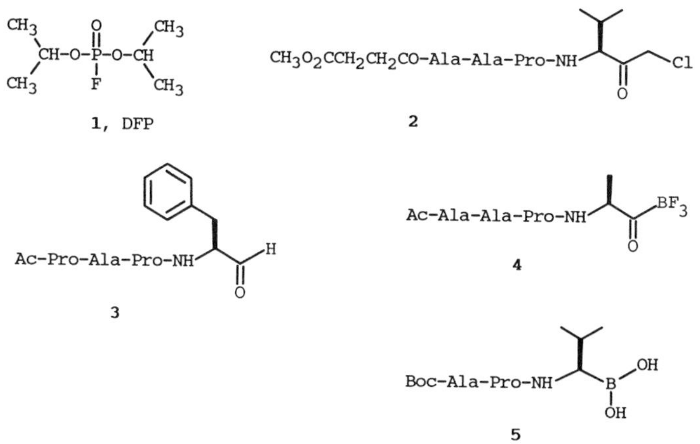

Figure 1. Inhibitors of serine proteases. 1, Diisopropyl fluorophosphate (DFP); 2, peptidyl chloromethyl ketone; 3, peptidyl aldehyde; 4, trifluoromethylene analog; 5, peptidyl boronic acid.

the various serine proteases depending on the extent to which they incorporate structural elements which allow them to mimic specific substrates. All serine proteases are inactivated by DFP, for example, because they all have the nucleophilic serine hydroxyl group necessary for "recognition" by the phosphorylating agent. The longer peptide derivatives in contrast, show considerable selectivity because they occupy the binding subsites which determine the enzyme substrate specificity.

In spite of the nonspecific character of DFP, it is an appealing design because the covalent complex between enzyme and inactivator can be viewed as a model for a tetrahedral intermediate along the acylation–deacylation pathway (see Scheme 1) (Bernhard and Orgel, 1959; Stroud *et al.*, 1974; Porubcan *et al.*, 1979). We reasoned that replacement of the nonspecific isopropoxy groups with peptidyl moieties would, on the one hand, improve this analogy, and, on the other, might confer useful specificity on this class of inactivator (see also Becker, 1967; Boone *et al.*, 1964; Mounter *et al.*, 1963; Nayak and Bender, 1978; Silipo *et al.*, 1979).

Our initial investigations focused on the phosphonofluoridates which we demonstrated are exceedingly rapid inactivators of chymotrypsin and pancreatic elastase (Table I) (Lamden and Bartlett, 1983; Bartlett and Lamden, 1986). Replacement of one of the isopropoxy substituents with a Cbz-aminoalkyl moiety not only increases the rate of inactivation of these enzymes, it also confers a measure of selectivity between chymotrypsin and pancreatic elastase.

The high reactivity of the phosphorus–fluorine bond presents an obstacle toward improvement in the design of the peptidyl phosphonofluo-

Table I. Inactivation of Elastase and Chymotrypsin by Fluorophosphorus Derivatives

Inhibitor	Pancreatic elastase	Chymotrypsin
	K_i $(M^{-1} \, sec^{-1})^a$	
DFP	—	250^b
R = CH$_3$, YR' = OCH$_3$	2000	—
R = CH$_3$, YR' = OCH(CH$_3$)$_2$	1280	8,800
R = CH$_3$, YR' = NHCH(CH$_3$)$_2$	160	—
R = CH$_2$Ph, YR' = OCH(CH$_3$)$_2$	160	180,000

[a] Determined at 25°C, pH 6.5, 0.05 *M* phosphate (Bartlett and Lamden, 1986).
[b] Determined at pH 7.7, 0.067 *M* Veronal buffer (Ooms and van Dijk, 1966).

ridates as practical inhibitors. They are not only hydrolyzed rapidly but are also reactive toward the peptide linkage itself. For example, the half-lives of these compounds are around 5 min at 25°C, 0.05 M phosphate buffer, pH 6.5. We expected that replacement of the fluoride leaving group with a phenyl or thiophenyl ester would reduce this sensitivity, permitting oligo-peptides to be constructed with the phosphonylating moiety embedded in the middle of the chain and yet retaining enough activity for them to inactivate the target enzymes effectively (Bartlett and Lamden, 1986; P. A. Bartlett and N. S. Sampson, unpublished observations).

A critical step in the development of these compounds as versatile inhibitors was the discovery of a convenient method for incorporating a reactive phosphonate ester in the heart of an oligopeptide (Sampson and Bartlett, 1988). Phenyl and thiophenyl phosphonate esters, while more stable than the corresponding fluoridates, are nonetheless sensitive to many of the synthetic procedures employed in assembling a polypeptide chain. The conventional methods of activating a phosphonic acid via the phosphonochloridate, however, are too harsh to be carried out after as-sembly of the polypeptide. This dilemma is resolved by carrying the phosphorus moiety through the polypeptide assembly process at the phos-phinate oxidation level, and, as the last step in the synthesis, activating it in the presence of the leaving group by reaction of the phosphonite tauto-mer with carbon tetrachloride (Scheme 2).

A variety of hexapeptide phosphonylating agents have been assembled in this manner and evaluated as inactivators of chymotrypsin, subtilisin,

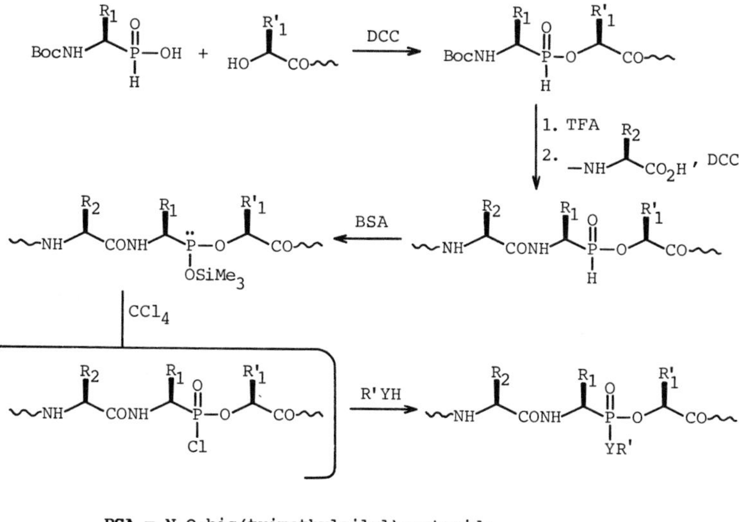

BSA = N,O-bis(trimethylsilyl)acetamide

Scheme 2

pancreatic elastase, and human leukocyte elastase (Table II). Although phosphorus amino acid analogs of the correct α-configuration are available (Baylis *et al.*, 1984), the activation step is not stereospecific and two diastereomers are produced as a result of the stereocenter at phosphorus. While these isomers (designated A and B) can be separated chromatographically, we do not yet have a method for assigning their configurations. Although we expected otherwise, the diastereomeric inhibitors do not differ significantly in their rates of inactivation of the target enzymes, which suggests that phosphonylation of the enzyme does not take place from an inhibitor–enzyme complex in which both the Boc-Ala-Ala-Pro- and -Ala-Ala-OCH_3 segments are occupying their respective subsites.

We anticipate that the major utility of inhibitors of this sort will be to provide inhibited complexes which are models for the tetrahedral species leading to formation of the acyl enzyme intermediate, either for NMR or crystallographic studies. The ^{31}P nucleus is advantageous for NMR studies, and the fact that residues extending in both the amino and carboxyl directions can be incorporated distinguishes them from many of the reversible inhibitors. Taking as a starting point the crystal structure of the

Table II.   Inactivation of Serine Proteases by Hexapeptide Phosphonylating Agents[a]

Inhibitor	Enzyme:	HLE[b]	SUB[c]	CHY[c]	PPE[b]
BocAAPV^P (SPh) (O)AA-OMe[d]	Isomer A	470	360	42	160
(R_1 = $CH(CH_3)_2$, Y = S)	Isomer B	540	440	75	340
BocAAPV^P (OPh) (O)AA-OMe	Isomer A	320			
(R_1 = $CH(CH_3)_2$, Y = O)	Isomer B	630			
BocAAPF^P (OPh) (O)AA-OMe	Isomer A		3000[e]		
(R_1 = CH_2Ph, Y = O)	Isomer B		3500[e]		

(K_i values are in units of $M^{-1}\,min^{-1}$)

[a] All rates measured at 25°C.
[b] HLE, human leukocyte elastase; PPE, porcine pancreatic elastase. Inactivation rates measured in 0.1 M citrate, 0.5 M NaCl, pH 6.0.
[c] SUB, subtilisin; CHY, chymotrypsin. Inactivation rates measured in 50 mM Tris, pH 8.6.
[d] Pseudo-first-order rate of hydrolysis = 0.063 min^{-1}, pH 6.0.
[e] Inactivation rates measured in 50 mM TAPS buffer, pH 8.6.

complex formed on inactivation of human leukocyte elastase (HLE) with methoxysuccinyl-Ala-Ala-Pro-Val-chloromethyl ketone (Wei *et al.*, 1988), we have constructed a model of the adduct we envisage is formed on inactivation of this enzyme with the hexapeptide phosphonates (Fig. 2). Simply overlapping the residues corresponding to Ala-Ala-Pro-Val allows the phosphorus moiety to be positioned within covalent bonding distance of the activated serine hydroxyl and allows the Ala-Ala-OCH₃ terminus to occupy the S_1' and S_2' sites as well. In this representation, the phosphoryl oxygen occupies the "anionic hole" which stabilizes the negative charge that develops on the carbonyl oxygen as the tetrahedral species is generated. We look forward to experimental confirmation (or refutation) of this model, and in particular to a comparison of the structure of this complex with those of other, "substrate-like" complexes.

III. Structure-Derived Inhibitors

Because of the strong evolutionary relationships among most of the serine peptidases, it is not surprising that the gross structures of their binding clefts should be similar. Nor is it surprising that the proteinaceous mole-

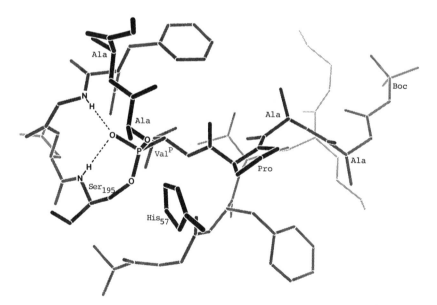

Figure 2. Proposed structure of covalent complex formed on inactivation of human leukocyte elastase with activated Boc-Ala-Ala-Pro-Val^P-(O)Ala-Ala-OMe.

cules that have evolved to inhibit these enzymes should adopt similar conformations in the regions that complement the peptidase active sites. However, except for these coincident active loops, the protease inhibitors have evolved quite dissimilar overall structures. Although these molecules are large in comparison to synthetic inhibitors, only a fraction of their residues interact directly with the target enzymes; the rest of the peptide is there to provide appropriate scaffolding for the active loop, as well as for ancillary reasons of biosynthesis, regulation, etc. Could a different, non-peptidic scaffolding do a similar job? Could the critical interactions of the inhibitor active loops with the enzyme active sites be sustained in a small molecule mimic? The successful resolution to these questions could provide a new type of peptidase inhibitor, and, more importantly, demonstrate the soundness of a "structure-derived" approach to the design of biologically active molecules.

Our design process began with the crystal structure of the complex between bovine pancreatic trypsin inhibitor (BPTI) and trypsin (Marquart *et al.*, 1983). We focused on the three inhibitor residues in the P_3, P2, and P1 positions and tried to devise cyclic structures which would position the three carbonyl groups of these units in the mutually perpendicular orientation which they adopt in BPTI. The scissile peptide was replaced with a ketone moiety, with the hope that a hemiketal adduct would be formed with the catalytic serine residue. Candidate structures were built (as the ketone hydrates), minimized (starting from the desired conformation) by molecular mechanics calculations (Allinger, 1976), and then compared with the BPTI–trypsin complex on a computer graphics system (Ferrin *et al.*, 1988). Those candidates which appeared to match the backbone conformation of BPTI best were subjected to further energy refinement to ascertain if the desired conformation was likely to be a global minimum. A selection of the core structures of some of the candidates is illustrated in Fig. 3.

The various candidates were evaluated on a subjective basis, with the recognition that the validity of the minimization routines and the accuracy of a protein crystal structure place a limit on the exactness with which any match can be predicted. Moreover, ease of synthesis was an important consideration at each stage of the design process and was in fact responsible for our decision to pursue a *chymotrypsin* inhibitor (compound 6 in Scheme 3 and 4) as our first target.

The highlights of the synthetic route to compound 6 are illustrated in Scheme 3. Of the four stereocenters in the target, three were obtained in the starting materials phenylalanine and *allo*-hydroxyproline; the fourth was introduced using an asymmetric aldol condensation (Evans, 1982). Assembly of the acyclic precursor was straightforward, and the macrocy-

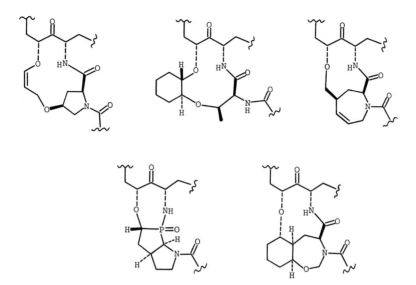

Figure 3. A selection of structures considered as potential mimics of the active loop of serine peptidase inhibitors.

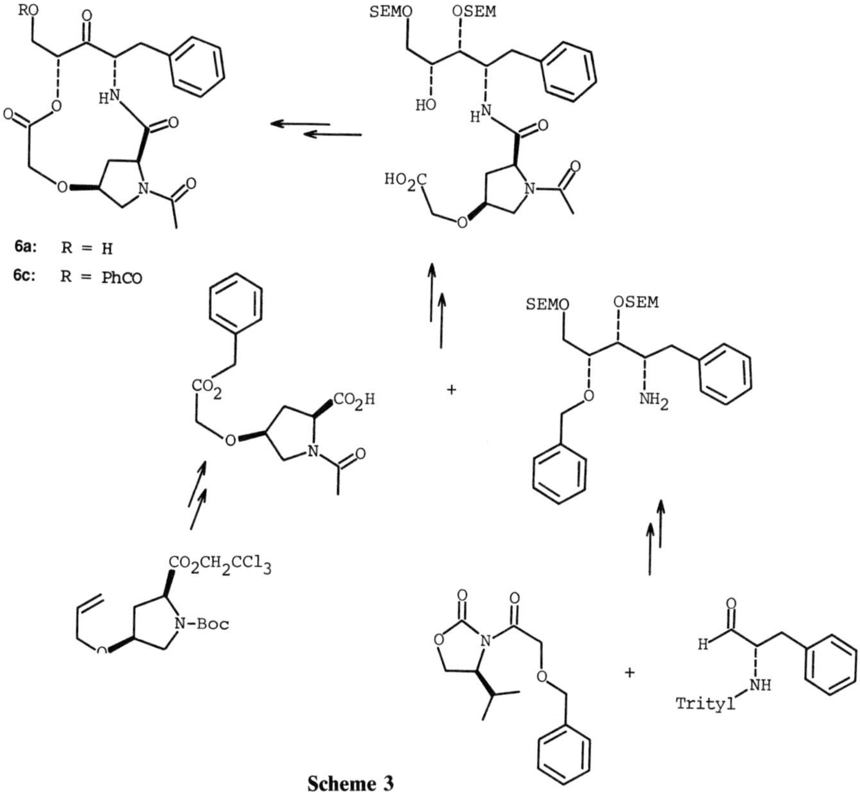

6a: R = H
6c: R = PhCO

Scheme 3

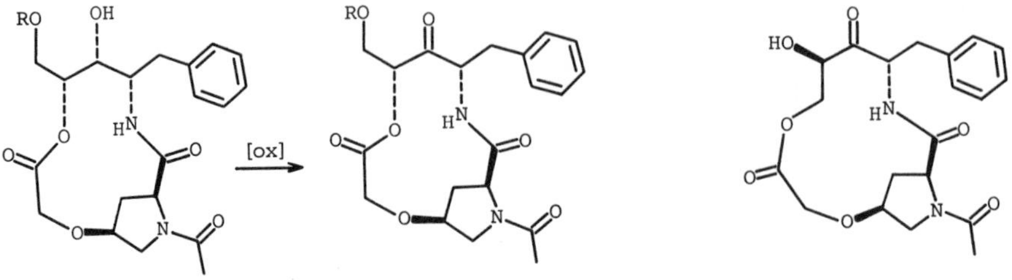

7a: R = H
7b: R = Ph₃C
7c: R = PhCO

6b: R = Ph₃C
6c: R = PhCO

8

Scheme 4

clization was carried out in good yield using Keck's procedure (Boden and Keck, 1985). After deprotection of the cyclic diol (compound 7a, Scheme 4), the primary and secondary hydroxyls could be distinguished by formation of the primary trityl ether (compound 7b, Scheme 4). Unfortunately, removal of this protecting group after oxidation to the ketone was accompanied by ring expansion to the 13-membered lactone (compound 8, Scheme 4). Selective benzoylation of the diol and retention of the ester in the final product (compound 6c, Scheme 4) avoided this complication.

Both the final product and the ring-expanded material have undergone preliminary evaluation as inhibitors of chymotrypsin; at pH 7.5, they show K_i values of 10 μM and 73 μM, respectively, and they are purely competitive against the substrate succinyl-Ala-Ala-Pro-Phe nitroanilide (K_m = 53 μM). We are encouraged by these initial results, which demonstrate that moderate inhibition can be obtained even with a stripped-down structural mimic. Too, we are gratified that the "correct" inhibitor 6c is bound more tightly than the considerably larger tetrapeptide substrate, as well as better than the "incorrect" inhibitor, although we find in hindsight that the 13-membered ring system can also adopt a conformation which overlaps with the active loop of the desired one. A more complete evaluation of the success of our design will depend upon confirmation of the postulated hemiketal adduct, as well as our ability to improve the binding affinity by appropriate substitution at the extremities of the molecule.

Acknowledgment

Support for this research was provided by fellowships from the National Science Foundation (to N.S.S. and L.A.L.) and the National Institutes of Health (to S.H.R.) and by research grants from the National Institutes of Health (grants no. CA-22747 and GM-30759).

References

Allinger, N. L. (1976). Calculation of molecular structure and energy by force-field methods. *Adv. Phys. Org. Chem.* **13**, 1–82.

Appel, W. (1986). Chymotrypsin: Molecular and catalytic properties. *Clin. Biochem. (Ottawa)* **19**, 317–322.

Bachovchin, W. W., Wong, W. Y. L., Farr-Jones, S., Shenvi, A. B., and Kettner, C. A. (1988). Nitrogen-15 NMR spectroscopy of the catalytic-triad histidine of a serine protease in peptide boronic acid inhibitor complexes. *Biochemistry* **27**, 7689–7697.

Baker, B. R. (1967). "Design of Active-Site-Directed Irreversible Enzyme Inhibitors." Wiley (Interscience), New York.

Barrett, A. J. (1981). Leukocyte elastase. *Methods Enzymol.* **80**, 581–588.

Bartlett, P. A., and Lamden, L. A. (1986). Inhibition of chymotrypsin by phosphonate and phosphonamidate peptide analogs. *Bioorg. Chem.* **14**, 356–377.

Baylis, E. K., Campbell, C. D., and Dingwall, J. G. (1984). 1-Aminoalkyl phosphonous acids. Part 1. Isosteres of the protein amino acids. *J.C.S. Perkin I* pp. 2845–2853.

Becker, E. (1967). The relationship of the structure of phosphonate esters to their ability to inhibit chymotrypsin, trypsin, acetyl cholinesterase and C'_{1a}. *Biochim. Biophys. Acta* **147**, 289–296.

Bernhard, S. A., and Orgel, L. E. (1959). Mechanism of enzyme inhibition by phosphate esters. *Science* **130**, 625–626.

Blow, D. M. (1976). Structure and mechanism of chymotrypsin. *Acc. Chem. Res.* **9**, 145–152.

Boden, E. P., and Keck, G. E. (1985). Proton-transfer steps in Steglich esterification: A very practical new method for macrolactonization. *J. Org. Chem.* **50**, 2394–2395.

Bone, R., Shenvi, A. B., Kettner, C. A., and Agard, D. A. (1987). Serine protease mechanism: Structure of an inhibitory complex of α-lytic protease and a tightly bound peptide boronic acid. *Biochemistry* **26**, 7609–7614.

Boone, B. J., Becker, E. L., and Canham, D. H. (1964). Enzyme inhibitory activity of certain phosphonate esters against chymotrypsin, trypsin, and acetylcholinesterase. *Biochim. Biophys. Acta* **85**, 441–445.

Brayer, G. D., Delbaere, L. T. J., James, M. N. G., Bauer, C. A., and Thompson, R. C. (1979). Crystallographic and kinetic investigations of the covalent complex formed by a specific tetrapeptide aldehyde and the serine protease from *Streptomyces griseus*. *Proc. Natl. Acad. Sci. USA* **76**, 96–100.

Carrell, R., and Travis, J. (1985). α-1-Antitrypsin and the serpins: Variation and countervariation. *Trends Biochem. Sci.* **10**, 20–24.

Colomb, M. G., Arlaud, G. J., Aude, C., and Journet, A. (1987). The role of proteolytic enzymes in the complement system. *Adv. Biosci.* **65**, 11–18.

Davie, E. W. (1986). Introduction to the blood coagulation cascade and cloning of blood coagulation factors. *J. Protein Chem.* **5**, 247–253.

Davie, E. W., Fujikawa, K., Kurachi, K., and Kisiel, W. (1979). The role of serine proteases in the blood coagulation cascade. *Adv. Enzymol. Relat. Areas Mol. Biol.* **48**, 277–318.

Evans, D. A. (1982). Studies in asymmetric synthesis. The development of practical chiral enolate synthons. *Aldrichimica Acta* **15**, 23–32.

Ferrin, T. E., Huang, C. C., Jarvis, L. E., and Langridge, R. (1988). The MIDAS display system. *J. Mol. Graphics* **6**, 13–27.

Gebhard, W. H., and Fritz, H. (1986). Biochemistry of aprotinin and aprotinin-like inhibitors. *Res. Monogr. Cell Tissue Physiol.* **12**, 375–388.

Groutas, W. C. (1987). Inhibitors of leukocyte elastase and leukocyte cathepsin G. Agents for the treatment of emphysema and related ailments. *Med. Res. Rev.* **7**, 227–241.

Hunkapiller, M. W., Smallcombe, S. H., and Richards, J. H. (1975). Mechanism of serine protease action. Ionization of tetrahedral adduct between α-lytic protease and tripeptide aldehyde studied by carbon-13 magnetic resonance. *Org. Magn. Reson.* **7**, 262–265.

Imperiali, B., and Abeles, R. H. (1986). Inhibition of serine proteases by peptidyl fluoromethyl ketones. *Biochemistry* **25**, 3760–3767.

Kettner, C. A., and Shenvi, A. B. (1984). Inhibition of the serine proteases leukocyte elastase, pancreatic elastase, cathepsin G, and chymotrypsin by peptide boronic acids. *J. Biol. Chem.* **259**, 15106–15114.

Kossiakoff, A. A. (1987). Catalytic properties of trypsin. *Biol. Macromol. Assem.* **3**, 369–412.

Kraut, J. (1977). Serine proteases: Structure and mechanism of catalysis. *Annu. Rev. Biochem.* **46**, 331–358.

Lamden, L. A., and Bartlett, P. A. (1983). Aminoalkylphosphonofluoridate derivatives: Rapid and potentially selective inactivators of serine peptidases. *Biochem. Biophys. Res. Commun.* **112**, 1085–1090.

Larsen, D., and Shaw, E. (1976). Active-site-directed alkylation of chymotrypsin by reagents utilizing various departing groups. *J. Med. Chem.* **19**, 1284–1286.

Laskowski, M., Jr. (1986). Protein inhibitors of serine proteinases—mechanism and classification. *Adv. Exp. Med. Biol.* **199**, 1–17.

Marquart, M., Walter, J., Deisenhofer, J., Bode, W., and Huber, R. (1983). The geometry of the reaction site and of the peptide groups in trypsin, trypsinogen and its complexes with inhibitors. *Acta Crystallogr., Sect. B* **B39**, 480–490.

Mounter, L. A., Shipley, B. A., and Mounter, M. E. (1963). The inhibition of hydrolytic enzymes by organophosphorus compounds. *J. Biol. Chem.* **238**, 1979–1983.

Nayak, P. L., and Bender, M. L. (1978). Organophosphorus compounds as active site-directed inhibitors of elastase. *Biochem. Biophys. Res. Commun.* **83**, 1178–1182.

Ooms, A. J. J., and van Dijk, C. (1966). The reaction of organophosphorus compounds with hydrolytic enzymes. III. The inhibition of chymotrypsin and trypsin. *Biochem. Pharmacol.* **15**, 1361–1377.

Porubcan, M. A., Westler, W. M., Ibañez, I. B., and Markley, J. L. (1979). (Diisopropylphosphoryl)serine proteinases. Proton and phosphorus-13 nuclear magnetic resonance-pH titration studies. *Biochemistry* **18**, 4108–4116.

Reid, K. B. M., and Porter, R. R. (1981). The proteolytic activation systems of complement. *Annu. Rev. Biochem.* **50**, 433–464.

Sampson, N. S., and Bartlett, P. A. (1988). Synthesis of phosphonic acid derivatives by oxidative activation of phosphinate esters. *J. Org. Chem.* **53**, 4500–4503.

Shaw, E. (1970). Chemical modification by active-site-directed reagents. *In* "The Enzymes" (P. D. Boyer, ed.), 3rd Ed., Vol. 1, pp. 91–146. Academic Press, New York.

Shaw, E., Kettner, C., and Green, G. D. J. (1981). Peptidyl chloromethyl ketones and diazomethyl ketones as protease affinity labels. *Pept.: Synth., Struct., Funct., Proc. Am. Pept. Symp., 7th* (D. H. Rich and E. Gross, eds.), pp. 401–409. Pierce Chem. Co., Rockford, Illinois.

Silipo, C., Hansch, C., Grieco, C., and Vittoria, A. (1979). Inhibition of chymotrypsin by alkyl phosphonates: A quantitative structure–activity analysis. *Arch. Biochem. Biophys.* **194**, 552–557.

Stockley, R. A. (1987). Alpha-1-antitrypsin and the pathogenesis of emphysema. *Lung* **165**, 61–77.

Stroud, R. M., Kay, L. M., and Dickerson, R. E. (1974). Structure of bovine trypsin. Electron density maps of the inhibited enzyme at 5 Å and 2.7 Å resolution. *J. Mol. Biol.* **83**, 185–208.

Thompson, R. C., and Bauer, C. A. (1979). Reaction of peptide aldehydes with serine proteases. Implications for the entropy changes associated with enzymic catalysis. *Biochemistry* **18**, 1552–1558.

Walsh, C. (1980). Suicide substrates: Mechanism-based inactivators of specific target enzymes. *Tetrahedron* **38**, 871–909.

Wei, A. Z., Mayr, I., and Bode, W. (1988). The refined 2.3 Å crystal structure of human leukocyte elastase in a complex with a valine chloromethyl ketone inhibitor. *FEBS Lett.* **234**, 367–373.

Wofsy, L., Metzger, H., and Singer, S. J. (1962). Affinity labeling—a general method for labeling the active sites of antibody and enzyme molecules. *Biochemistry* **1**, 1031–1039.

Wolfenden, R. (1976). Transition state analog inhibitors and enzyme catalysis. *Annu. Rev. Biophys. Bioeng.* **5**, 217–306.

20 Applications of Crystallographic Databases in Molecular Design

R. Scott Rowland
Department of Biochemistry, University of Alabama at Birmingham,
Birmingham, Alabama 35294

W. Michael Carson
Center for Macromolecular Crystallography, University of Albama at
Birmingham, Birmingham, Alabama 35294

Charles E. Bugg
Department of Biochemsitry, University of Alabama at Birmingham,
Birmingham, Alabama 35294

I. Introduction

Advances in techniques of macromolecular crystallography have significantly increased the number of known structures of biological macromolecules. This wealth of structural information has spurred molecular design initiatives such as protein engineering and rational drug design. Knowledge of the three-dimensional structure of a molecule can be invaluable in molecular design efforts. In drug design, the structure of a target site serves as a chemical map to guide the medicinal chemist in the development of a potential inhibitor or activator. In the absence of such structural information, development would normally proceed by tedious trial-and-error methods.

Molecular design usually proceeds in two phases. The first phase is conceptualization and formalization of a specific design: for example, the addition of a substituent to a lead compound or the proposal of a completely new lead compound. The next phase is evaluation of the design by various modeling techniques such as graphical visualization or molecular dynamics calculations. A major concern at this stage is the reliability of these calculations, since most modeling procedures do not explicitly consider hydrophobicity or polarizability of atoms.

Both stages of molecular design can be improved by utilizing the vast amount of structural information in crystallographic databases. Analysis of databases can provide tabulations of the preferential association of different chemical groups. This type of information can be extremely valuable in the first phase when deciding which chemical groups are to be

Use of X-Ray Crystallography in the Design of Antiviral Agents

included in the design. Instead of relying solely on chemical intuition, a more quantitative approach can be used to select substituents. Since crystallographic databases contain the coordinates of structures and information for reconstructing crystal packing, the geometry of preferential associations can be determined from these databases. Results from molecular mechanics calculations can then be compared with observed patterns of interaction to test and improve modeling methods. One of the chief advantages in using this type of information from databases is that they are based upon actual experimental observations: crystallographically determined structures. This paper discusses the investigation of preferred interaction patterns within crystallographic databases and how knowledge of these patterns can be utilized in molecular design.

II. Description of Crystallographic Databases

The two crystallographic databases discussed in this paper are the Cambridge Structural Database and the Brookhaven Protein Database, hereafter referred to as CSD and PDB respectively. Although other crystallographic databases exist, these two are the primary depositories of three-dimensional structural information. Below is a brief overview of CSD and PDB. A more detailed description of these and other crystallographic databases is given in a review by Allen et al. (1987a).

A. Cambridge Structural Database

The Cambridge Structural Database (Allen et al., 1979) is the depository of small organic molecule structures determined by X-ray and neutron diffraction analysis. The current 1989 release contains coordinates for approximately 71,000 small molecule structures. Even more impressive is the growth rate of the database. A plot of the number of new entries added each year for the last 20 years indicates an exponential increase in the annual number of new entries (Allen et al., 1987b). Although many factors have contributed to this massive increase in structural analyses, two of the most important were the advent of automatic diffractometers and direct methods.

In addition to the coordinate file, an excellent suite of programs is provided for searching the database and calculating molecular geometry. A wide range of parameters (e.g., R-factor, unit cell parameters, presence of a particular element) can be used to search the database for crystal structures of interest. The software also provides the ability to locate user-defined chemical substructures, an option that has contributed in large part to the success of the CSD. Once a set of structures is located, the geometry analysis program (GSTAT89) can be used to calculate a variety of geometrical parameters (e.g., bond lengths, sugar pucker, least squares planes, plus user-defined geometry). The rapid growth in the number of entries, the accuracy of the data, and the versatile search and analysis software make the CSD a powerful tool for the molecular designer.

B. Brookhaven Protein Database

The Brookhaven Protein Database (Bernstein *et al.*, 1977) is the depository of atomic coordinates for biological macromolecules including proteins, nucleic acids, and polysaccharides. Although the majority of structures were experimentally determined by X-ray diffraction, the database does contain some model structures. The current 1989 release of the PDB contains approximately 420 entries, including complexes and homologous proteins from different sources. However, the advent of area detectors and more powerful computers is likely to cause an increase in growth similar to the CSD. In fact, the PDB currently has 35 proteins in the processing stage. The PDB is distributed with some software, but for the most part users must write their own programs to utilize the database effectively.

III. Molecular Design and Databases

A. Choice of Database and Structures

Before choosing which database to use for an analysis one must be aware of the particular limitations of each database under consideration. One of the most important considerations is the reliability or accuracy of the structures within the database. In general, small molecule structures from the CSD are much more accurate than protein structures from the PDB, due to the inherent difficulties and complexities of protein crystallography. Flexibility or crystal disorder can cause regions of a protein to be poorly defined, and thus the quality of a structure varies from region to region. In contrast, most small molecule diffraction studies can accurately locate all atoms, including hydrogens. Therefore, in an analysis of the PDB it is desirable to use structures that are refined to relatively high resolution (<2.0 Å). If hydrogens are required for the analysis they must be explicitly added at their predicted positions. Although neutron diffraction studies can locate hydrogens in protein structures, the vast majority of PDB entries are the results of X-ray analysis. For the CSD, it is desirable to use

structures that have no disorder and R-factors < 7.5%. Most such structures contain hydrogen coordinates but they must be normalized in the
case of X-ray analysis due to systematic shortness of covalent hydrogen
bonds (Taylor and Kennard, 1984).

In a survey of the PDB one must decide whether sets of homologous
proteins (e.g., hemoglobins) are to be used. If members of such sets have a
high degree of both structural and sequence homology, their inclusion
could result in redundant measurements. However, some workers (Bryant
and Amzel, 1987) have found that the presence of homologous proteins did
not signifcantly affect their results. The presence of identical compounds
is usually not a problem with the CSD.

One final caution should be noted about any study that uses crystallographic databases. Crystallographic databases only contain structures of
molecules that are amenable to diffraction analysis. For the PDB, this
means mostly soluble, globular proteins, although new techniques are
allowing structural analyses of membrane proteins. It also implies that
only molecules for which single crystals were obtainable are present in
these databases. However, the rapid growth of both databases makes this
less of a consideration.

B. Preferential Association

One of the most valuable pieces of information obtainable from databases
is an estimation of the preferential association of one group of atoms with
another. This information may be obtained by constructing frequency
tables of the various types of interactions. A variety of methods can be
used to construct these tables depending on the type of interactions one
wants to study. Such tables are then normalized to correct for the differen-
tial number of groups present in the database. For these tables to have any
meaning, they must also be shown to be different from a random distribution at some level of statistical significance.

C. Directionality of Association

After discovering some affinity of one group for another, the next step is
analysis of the geometry of the interaction to see if there is a pattern. The
reason there may be a pattern is that atoms do not interact in a spherically
symmetric fashion. The anisotropic nature of an interaction may be due to
the nonsphericity of the electron distribution and density (Ramasubbu et
al., 1986). The real power of crystallographic databases is that they contain
the three-dimensional coordinates of interacting groups, allowing one to
determine the preferred directionality of an interaction. If an observed

pattern of interactions is based on chemical preference, it can be statistically distinguished from a random distribution.

IV. Examples of Database Utilization

A. Preferences of Amino Acids Based on Shared Surface Area

Considerable information about amino acid interaction patterns can be obtained from a tabulation of the preferential association of the 20 amino acids using the PDB. Although tables of preferential association have been compiled by other workers (Narayana and Argos, 1984; Warme and Morgan, 1978a,b), the method that we use is unique in that association is not based on contact frequencies but instead on shared surface area between nonbonded neighbors. This method has several advantages over methods based on contact frequencies (i.e., strict distance cutoffs). First, if a distance cutoff between centroids of amino acids is used to define a contact, distortions occur because amino acids are not spherical in shape. The next level of sophistication, atom-to-atom distance cutoffs, also presents difficulties. In this method, the contact frequencies will be affected by the surface area of the residue, since the number of contacts is dependent on surface area (Warme and Morgan, 1978a). Lastly, by using shared surface area as a definition for association, a degree of physical meaning is given to the tabulated values. Hydrophobic interactions are thought to play an important role in the folding and stability of proteins and in the binding of small ligands to protein sites (Kauzmann, 1959). Although hydrophobicity is a difficult interaction to quantitate and model, Chothia (1974) demonstrated that there is an approximate linear relationship between surface area and the free energy of transfer of amino acid residues from an aqueous to a nonpolar environment. By using shared surface area for defining association preferences, a measure of hydrophobicity is implicitly included to some degree.

1. Methods

Surface areas were calculated by using van der Waals dot surfaces constructed around each amino acid residue. The surface is constructed by treating each atom as a sphere with the appropriate radius (see Table I). Since hydrogens were not considered, the "united atom" concept was used by slightly increasing the van der Waals radii. A 42-point polyhedron is then mapped onto the sphere to generate the dots that comprise the surface. Color Plate 25 illustrates such a surface for the interior of bovine pancreatic trypsin inhibitor. The surface area of a given region on an atom,

Table I. Radii Used to Generate van der Waals
Surfaces of Amino Acids[a]

Atom	Radius (Å)
O(carbonyl)	1.82
O(hydroxyl)	1.93
N(amide)	1.94
S	2.04
C(carbonyl, trivalent aromatic)	2.06
CH(aromatic)	2.17
C(methyl)	2.18
C(methylene)	2.23
C_α	2.31

[a] Radii were taken from the CEDAR package (Carson
and Hermans, 1985) and are based on the equilibrium distance given by Lennard-Jones nonbonded force constants.
They are "united atom" radii, which are slightly larger
than standard van der Waals radii to account for hydrogens
when they are not explicitly included.

s, can then be approximated by

$$s = N^a S^r / 42 \tag{1}$$

where N^a is the number of dots in the region and S^r is the surface area of
the atom of radius r. As a test of the sampling size, areas were recalculated
with 92 dots per atom. The largest change in any residue's area was only
1%. Calculation of surface area by this method has been shown (Frömmel,
1984) to approximate closely the values calculated by the method of Lee
and Richards (1971). The percentage of buried surface area calculated by
this method is virtually identical to the values reported by Rose *et al.*
(1985).

Once the dot surface is constructed for the protein chain, each dot is
assigned to the nearest noncovalently bonded atom in a neighboring residue or a calculated water molecule as shown in Color Plate 26. The waters
are placed tangent to the dot at a distance of 2.0 Å and tested for overlap
with neighboring atoms. If the water is closer than 3.0 Å to a neighboring
atom, it is rejected. If no atom is within 3.75 Å, the dot is assigned to a
water. Inclusion of these "phantom" waters is used to represent the effect
of solvent. No crystallographic waters were used in this analysis.

Sums are then tabulated of the number of dots of atom i assigned to
atom j in a neighboring group and converted to area using Eq. (1). The

shared surface area of a side chain can then be calculated by summing the areas of its constituent atoms. One can then define E_{ij} (Lifson and Sander, 1980), the expected shared area of a side chain of type i with type j,

$$E_{ij} = A_i A_j / A_{\text{tot}} \tag{2}$$

where A_i and A_j are the sums of the areas of side chains of type i and j, respectively. A_{tot} is the total surface area of all residues. The preferential association or shared surface correlation, A_{ij}, of side chain i for j is defined by

$$A_{ij} = S_{ij} / E_{ij} \tag{3}$$

where S_{ij} is the observed shared surface of side chain i for j. Stated another way, A_{ij} is the contact correlation of a given group i to be in contact with group j. When the correlation is favorable, $A_{ij} > 1$, while $A_{ij} < 1$ indicates an unfavorable correlation. In general, $A_{ij} \neq A_{ji}$ due to the unique microenvironment of each amino acid.

2. Results

A plot of the A_{ij} values for each side chain is shown in Fig. 1. Side chains are identified by their standard one-letter codes. The peptide backbone atoms were treated separately, except in the case of glycine and proline, and are indicated by B. Disulfides were distinguished from cysteines and are represented by Cs and Ch, respectively. All heteroatoms present in a protein structure (e.g., ligands, cofactors, ions) were grouped together and are identified by X. Inclusion of heteroatoms was done for completeness even though the diverse chemical nature of their constituent groups makes interpretation of their A_{ij} values difficult. Terminal amino and carboxyl groups were also treated separately and are indicated by Z. The code O represents the calculated water molecules. A total of 23 proteins (see Table II) was used to construct this table of A_{ij} values. This set of globular monomeric proteins is representative of the well-refined structures in the PDB.

Although complex, the plot of A_{ij} values in Fig. 1 contains much valuable information about the preferential association of amino acid side chains. Only the more salient features of the plot will be discussed here (a more detailed analysis will be published elsewhere). One of the most striking trends is the expected increase in preference for hydrophobic side chains as one moves from charged groups to hydrophobic groups. Also expected is the observed decrease in preference of hydrophobic groups for charged side chains. Another intresting trend is the hourglass shape of the plot. Side chains at the extremes of the plot have very strong likes and dislikes while those in the middle will associate with almost any

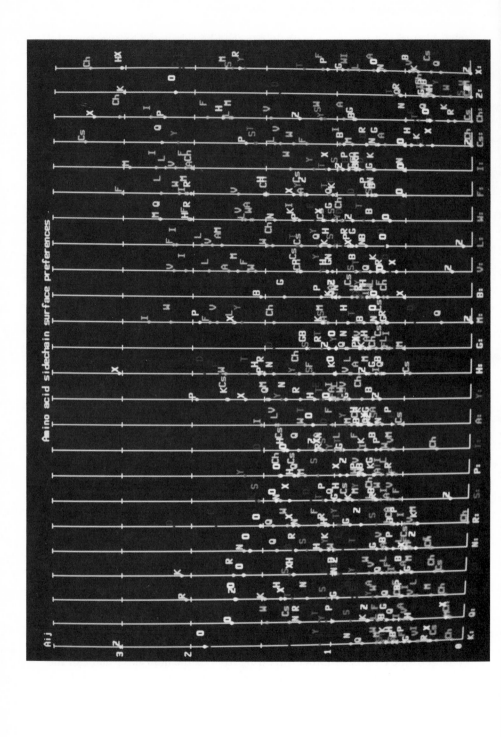

Amino acid sidechain surface preferences

Table II.   The 23 Proteins Used to Calculate Preferential Association of Amino Acid Side Chains Based on Shared Surface Area

Code	Protein	Code	Protein
1EST	Elastase	3C2C	Cytochrome c2
1FDX	Ferredoxin	3CPV	Calcium-binding parvalbumin B
1HIP	High potential iron protein	3GPD	D-Glyceraldehyde-3-phosphate
1RN3	Ribonuclease A		dehydrogenase
1SBT	Subtilisin	3MBN	Myoglobin
1TIM	Triose phosphate isomerase	3RXN	Rubredoxin
1TPO	β-Tryspin	3TLN	Thermolysin
2ACT	Actinidin	4FNX	Flavodoxin
2CNA	Concanavalin A	4LDH	Lactate dehydrogenase
2LYZ	Lysozyme	4TPI	Trypsinogen
2SNS	Staphylococcal nuclease	5CPA	Carboxypeptidase A
2SOD	Superoxide dismutase	8PAP	Papain

residue, suggesting that the amino acids have varying degrees of selectivity.

It is worth noting that most of the interactions with preference values above 2.0 are hydrophobic. In other words, hydrophobic side chains share twice the surface area with each other as would be expected for a random distribution. This is in agreement with the observation of Bryant and Amzel (1987) that hydrophobic residues make twice as many hydrophobic contacts than would be expected by random chance. This implies that our method of calculating association correlations does an adequate job of modeling hydrophobicity. Other workers (Eisenberg and McLachlan, 1986) have also used surface area to model hydrophobicity.

The side chain of phenylalanine should be noted in particular, as it exhibits several interesting features. First, the preference of phenylalanine for itself is slightly larger than 3.0, the highest value of any association correlation on the plot (excluding Z and X). A more surprising feature is

Figure 1.   Plot of amino acid side chain preferential association values, A_{ij}, based on shared surface area. Standard one letter codes are used for the side chains. Backbone atoms are indicated by B; heteroatoms, X; terminal groups, Z; disulfides, Cs; cysteine, Ch; and water, O. The preference of side chain i for j is found by locating i on the x-axis and moving up the y-axis to j at which point the A_{ij} is read off the scale on the left. Note that in general $A_{ij} \neq A_{ji}$. The scale of the y-axis is nonlinear in order to clarify tightly clustered groups. The amino acids on the x-axis are sorted by hydrophilicity with the most hydrophilic at the left.

the high preference of phenylalanine for arginine, greater than 2.0. Both of these observations can be explained in terms of recent discoveries about the character of aromatic side chains. Several workers (Burley and Petsko, 1985; Singh and Thornton, 1985) have shown that Phe-Phe clustering is the most common type of aromatic–aromatic interaction in proteins (note the example in Plate 25). It has been postulated that such interactions may play an important role in the stabilization of protein structure. This type of interaction has also been shown to have a preferred geometry; the aromatic rings tend to be tilted to each other rather than parallel. This "herringbone" pattern is similar to the packing of molecules in the crystal structure of benzene (see Fig. 2).

One possible explanation for the observed Phe-Phe interaction pattern

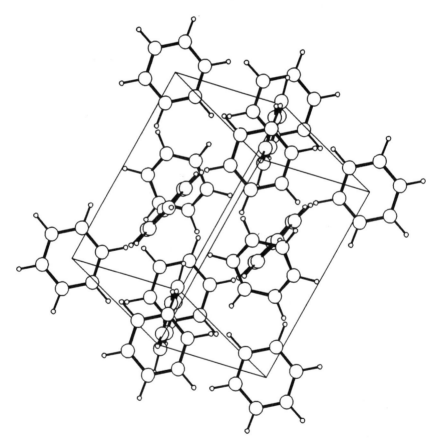

Figure 2. A crystal packing diagram of benzene. Notice the "herringbone" arrangement of the molecules. Coordinates were obtained from the CSD entry BENZEN.

is in terms of an electrostatic model. Since H atoms have a partial positive charge and C atoms a partial negative charge in aromatic molecules, there is a coulombic attraction that favors close $C \cdots H$ approaches and causes the edge-to-face interaction between aromatic molecules. This model has been the basis of suggestions that amino–aromatic contacts would be favorable (Burley and Petsko, 1986) and could possibly explain the high preference of Phe for Arg. However, this observation could be an artifact due to the low occurrence of both Phe and Arg in proteins. In order to further examine the role of electrostatic effects in Phe interaction patterns, one needs many accurate structures, which makes the CSD an ideal choice.

B. Phenylalanine–Carbonyl Interactions

The results of side chain packing patterns in proteins prompted a geometric analysis of interactions between aromatic rings and polar groups. Several years ago, examination of oxygen distributions around phenylalanines in the PDB led to the first proposal of this type of interaction (Thomas *et al.*, 1982). The PDB was relatively small at the time of that survey so we have reperformed the analysis with a larger set of proteins. We also utilized the CSD to characterize these interactions with very accurate small molecule structures (the previous group only used the PDB).

1. Cambridge Structural Database

a. Methods The CSD was searched for structures that contained both a carbonyl and an unsubstituted phenyl ring attached by a methylene group. The initial screen located 1218 such structures. Restricting the search to entries with no disorder, R-factor < 7.5%, organic class, intensity data measured by diffractometer, and containing no elements with $Z > 18$ reduced the set to 785 entries. The final set retained for analysis was reduced to 90 structures by allowing only those of the amino acid–peptide class with located hydrogens. Only structures from the amino acid–peptide class were used since they more closely approximate protein interactions. All but one of the carbonyls used in the analysis were components of peptide bonds.

These structures were then analyzed for intermolecular contacts involving a phenyl ring and a carbonyl oxygen. Such contacts are part of the interactions that contribute to crystal packing. Previous studies (Rosenfield *et al.*, 1977; Guru Row and Parthasarathy, 1981; Ramasubbu *et al.*, 1986) have shown that the examination of crystal packing can provide information about the directional preferences of two interacting chemical groups. Such information should be transferable to the analysis of protein structure, since the interior of a protein is analogous to a molecular crystal (Schulz and Schirmer, 1979).

The contact criterion was based on van der Waals radii (Bondi, 1964). If the distance between a carbonyl oxygen and a phenyl carbon or hydrogen was less than the sum of the two atom's van der Waals radii, plus a 1.0 Å tolerance, then the two groups were considered to be in contact. A total of 248 carbonyl–phenyl contacts were located, with only 9 of the crystal structures having no such contact. The set of 81 structures containing such contacts is listed in Table III by their six-character REFCOD identifier. The reason that the total number of contacts is larger than the number of crystal structures included is that many structures contained more than one phenyl or carbonyl group. Redundant contacts due to symmetry were not allowed. Reexamination of the structures with no tolerance included in the contact criterion located 59 contacts. This subset consists of contacts shorter than optimal van der Waals contacts.

For all of the contacts located, a complete set of geometrical parameters

Table III.   Structures Used from the CSD for Analysis of Phenyl–Carbonyl Contacts

ALPALC10	ALPRAL10	APALAM	APALTY
APHAMA	BAHNED	BAMGOL10	BCMEGL
BCPSBZ	BERVOJ	BGPLGQ	BHMTDC
BIBDOF	BIBDUL	BIDKOO	BIHTUH10
BIHXUL10	BIPVUR10	BOCAPR	BOCPLB
BOFRIX	BOGPBZ	BOHLUF	BOLWEE
BONROL10	BOPPOL10	BOPZDT10	BOTDOD
BXABPA	BXGLPR	BZCPLE	BZCPRO11
BZGPRO10	BZOAZA10	BZXGPL10	CAHWEN
CALFOK	CAPZIC	CBBLPB10	CEFGID
CEWCIQ10	CEYGAO	CEYGES	CIHLAG
CITNEY10	CITNIC10	CITNOI10	CITXAE10
CIYNIH	COYRIR	DEBHEX	DESXOO
DHCMYD10	DIFNAH	DIRPUP	DOZFED
DUMCET	DUPKEE	DUYTIA	DUZDUX
FABLUP10	FAJBUN	FAJKUW	FAVLOD
FAXLOF	FAXLUL	FESRIE	FESTIG
FETROL	FEYZUE	FIDLAF	FIFMEM
FIKFUA	FIMNUK	FIMVOM	HTENTX10
MLDPHE10	NMLPHE10	PAGLAL	PEPCYC10
PTHPAL			

was calculated. A spherical coordinate system was used to describe the location of the carbonyl oxygens with respect to the phenyl ring. The xy plane coincides with the plane of the aromatic ring with the origin located in the center of the ring. The x axis points towards the methylene linker and the z-axis is normal to the aromatic plane. Thus, for the spherical coordinate parameter θ, a value of $0°$ would be directly over the center of the ring while a value of $90°$ would be in the plane of the ring. All contacts were reduced to one quadrant by the mm symmetry of the aromatic ring. In addition, hydrogen bonding geometry was calculated with the closest phenyl C—H as the donor.

b. Results The θ distribution for the contacts using the 1.0 Å tolerance is shown in Fig. 3a. Approximately 60% of the carbonyl oxygens are in the range $70° < \theta < 90°$ (i.e., are close to lying in the plane of the aromatic ring). For the subset of very close contacts, approximately 70% were in the same range. However, in a spherical coordinate system incremental sections along θ do not have the same volume. For this reason, both of these distributions were tested to determine if they were statistically distinguishable from a random distribution given by $\sin \theta$ in a spherical coordinate system (Singh and Thornton, 1985). The probability of these distributions occurring by chance is less than 0.01, as calculated by a χ^2 analysis.

The subset containing the very close contacts was also used to determine if any of the oxygens were hydrogen bonding to a phenyl C—H. In addition to the distance criterion imposed by the definition of this subset, two angles were required to be between $90°$ and $180°$ before the contact was considered to be a hydrogen bond. The first angle, ANG1, is the standard hydrogen bond angle (C—H$\cdots$O) and the second, ANG2, is about the acceptor atom (H$\cdots$O=C). By this definition, 54 of the contacts (almost the entire subset) were determined to satisfy the geometry expected for hydrogen bonds. For the phenyl–carbonyl interactions that met these geometric constraints the means of ANG1 and ANG2 are $149.3°$ and $128.7°$, respectively. Figure 4a illustrates the distribution of carbonyl oxygens around a reference phenyl ring using pseudodensity contours at the 50 and 25% levels for this subset (Rosenfield *et al.*, 1984). The clustering of the oxygens around the hydrogen atoms of the phenyl ring is quite apparent.

2. Brookhaven Protein Database

a. Methods The PDB was then searched for similar phenyl–carbonyl contacts. In this case the search was confined to intramolecular contacts. The CSD geometry analysis software was modified to handle proteins in order to perform this search. A set of 36 proteins whose structures had

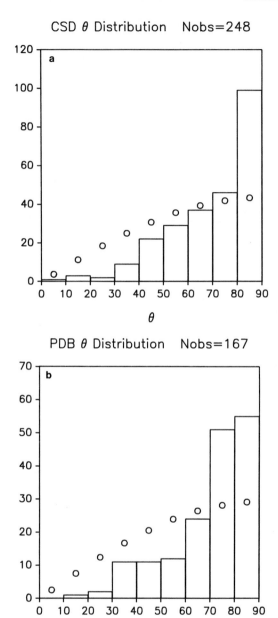

Figure 3. The θ distribution of carbonyl oxygen contacts with phenyl rings. The angle formed by the normal of the aromatic plane and the vector from the center of the ring to the oxygen is θ. The distribution expected by random chance (○) is given by sin θ. (a) The distribution observed in the CSD. (b) The distribution observed in the PDB.

been determined to 2.0 Å resolution or better was used for this analysis and is listed in Table IV. The contact criterion was the same as used for the CSD. This required the calculation of expected hydrogen coordinates for the phenylalanine residues. The 1.0 Å tolerance was useful in this analysis to allow for coordinate errors. In addition, only backbone carbonyls were used so that the results would be directly comparable with the CSD analysis. A subsequent search allowing other types of oxygens showed that backbone carbonyls accounted for 85% of all carbonyl–phenyl contacts. The carbonyl group in the phenylalanine and its adjacent residues were not considered. For this set of proteins, a total of 167 contacts was found.

b. Results Analysis of the geometry was carried out as stated above. Figure 3b shows the θ distribution for the contacts using the 1.0 Å tolerance. The range $70° < \theta < 90°$ contains 66% of the carbonyl oxygens. A χ^2 analysis verified that the probability of this distribution occurring by chance is less than 0.01. The subset of very close contacts contained 34 contacts of which 30 satisfied the conditions stated above for being considered a hydrogen bond. For the hydrogen bonds, the means of ANG1 and ANG2 are 142.6° and 124.4°, respectively. A pseudodensity map at the 50 and 25% levels illustrating the distribution of carbonyl oxygens in this subset is shown in Figure 4b. Although the predominant clustering occurs around the *ortho* position, the other hydrogen positions also show clustering.

2. Relationship Between CSD and PDB

Since contacts of carbonyl oxygens with phenyl groups were defined and located identically in both databases, the results can be compared statistically. The θ distributions were subjected to a χ^2 analysis to determine if they were dependent on the database from which they were derived. The analysis showed that at the 0.01 level of significance the θ distribution of carbonyls around phenyl rings is independent of the database used. This supports the hypothesis that observed directional preferences in small molecule crystals also occur in the interior of proteins. One must caution that this analysis used only one of the geometric parameters that describe a three-dimensional distribution. In fact, Figure 4 shows that there may be differences in hydrogen position preferences between the two databases. (A more complete analysis of the phenyl–carbonyl interaction will be published elsewhere.)

The results from both databases do support the claim that when aromatic rings interact with carbonyl oxygens there is a preferred orientation. The preferred orientation is for carbonyl oxygens to be near the edge of the

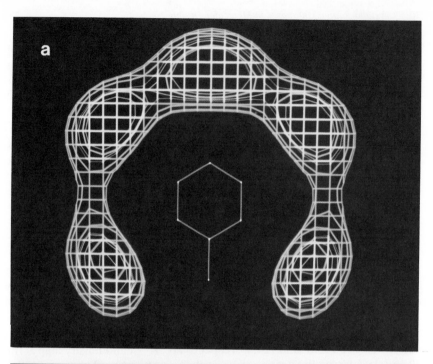

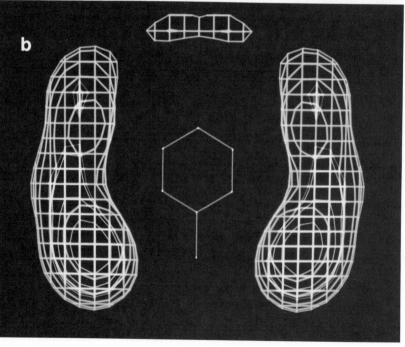

Table IV.   The 36 Proteins Used to Analyze Phenylalanine–Backbone
Carbonyl Interactions

Code	Protein	Code	Protein
1ACX	Actinoxanthin	2APP	Penicillopepsin
1AZA	Azurin	2B5C	Cytochrome *b*5
1BP2	Phospholipase A2	2CDV	Cytochrome *c*3
1CAC	Carbonic anhydrase form C	2CTS	Citrate synthetase
1CRN	Crambin	2GRS	Glutathione reductase
1ECD	Erythrocruorin	2LYZ	Lysozyme
1FB4	Immunoglobulin Fab	2OVO	Ovomucoid third domain
1FDX	Ferredoxin	2PAB	Prealbumin
1HIP	High potential iron protein	2SNS	Staphylococcal nuclease
1HMQ	Hemerythrin	2SOD	Superoxide dismutase
1INS	Insulin	3CPV	Calcium-binding parvalbumin B
1LH1	Leghemoglobin	3TLN	Thermolysin
1NXB	Neurotoxin	4CYT	Cytochrome *c*
1PCY	Plastocyanin	4DFR	Dihydrofolate reductase
1PPD	Papain	5CHA	α-Chymotrypsin
1REI	Bence–Jones rei protein	5CPA	Carboxypeptidase A
1RN3	Ribonuclease A	5PTI	Trypsin inhibitor
2ACT	Actinidin	5RXN	Rubredoxin

aromatic ring. In fact, analysis of C—H···O contacts that are shorter than van der Waals distance reveals restricted angular ranges that are indicative of hydrogen bonding.

V. Application to Design

Interaction preferences and directionality observed in crystallographic databases can be useful information in designing compounds that are tailored to fit selected protein sites. For example, if the target site contains

Figure 4.   Pseudo-density contours illustrating the distribution of carbonyl oxygens around a reference phenyl ring. Each oxygen is represented by an isotropic Gaussian smearing function. A density map is then generated by summing over all the atoms at each point. The thicker contours contain 50% of the map's total integrated density, and the thinner contours 25%. Both maps were generated using the subset of oxygen contacts that are closer than optimal van der Waals contacts and normalized for the number of contacts. (a) Contours for the CSD distribution. (b) Contours for the PDB distribution.

an exposed phenylalanine residue, the plot of preferential associations (see Fig. 1) quite clearly shows that a phenyl group would be an excellent substituent to place next to the phenylalanine. Analysis of aromatic–aromatic interactions in databases indicates there is an optimal orientation of the rings. Therefore, an attempt would be made to position the phenyl ring perpendicular to the phenylalanine to achieve a favorable interaction. One would expect further enhancement if the phenyl ring were positioned to form favorable contacts with a peptide carbonyl group. The rules governing favorable carbonyl–phenyl contacts can be obtained by analyzing crystallographic databases. If the carbonyl is close to the edge of the phenyl, or (even better) one of the hydrogens, the interaction is especially favorable.

The method of design illustrated above can be facilitated by using interactive computer graphics to display database-derived preferences and patterns of interaction in the binding site. A procedure developed by Kuntz *et al.* (1982) uses a set of spheres to describe a receptor binding site. This negative image is quite useful since it represents the available volume of the binding site. A molecular volume potential can be used to contour the set of spheres, thus generating an available volume map suitable for graphical display. Such maps can be color coded by the preferences of the groups forming the active site. This provides not only a steric map but a chemical map as well. Pseudodensity contours can be displayed for exposed groups in the binding site to define the directionality of an interaction further. Our lab is in the process of developing a database for use with such graphical techniques that encapsulates chemical group preferences and directionality observed from databases. One of the advantages of this method of design is that it is based solely on experimental observations.

VI. Conclusions

There are two major crystallographic databases, the Cambridge Structural Database and the Brookhaven Protein Database. Both contain a wealth of structural information that can be used in molecular design. The two main types of information obtainable are the preferential associations and geometrical directionality of chemical groups for one another.

The PDB was used to tabulate the preferential association of amino acid side chains for one another. Our method for tabulating these values is novel in that it is based on shared surface area. Not only does this eliminate the problems inherent in using strict distance cutoffs but it also allows better modeling of hydrophobic interactions. In fact, hydrophobic interactions were the most favorable of all. Hydrophobic residues were found to share twice as much surface area as expected by random chance. Preferential association values that correctly model hydrophobicity are extremely important in molecular design since hydrophobicity is a dominant force in many processes, such as the binding of a ligand.

In this paper, we have provided examples to show how the databases can be used to predict preferred interactions with phenylalanine residues of proteins. The most favorable hydrophobic interaction found within proteins is Phe-Phe pairing. This type of interaction has been explained as the result of coulombic attraction between the partially positive hydrogens and the partially negative carbons. In order to test the extent to which aromatic rings can participate in polar interactions, we examined phenyl–carbonyl contacts in both databases.

The results from both databases clearly showed that there is a preference for carbonyl oxygen atoms to interact with Phe residues. The carbonyl oxygen atoms tend to be in the plane of the aromatic ring and the observed distributions are statistically different from the distribution expected by random chance. Further analysis revealed subsets of contacts from both the CSD and PDB that satisfy the geometry expected for hydrogen bonds. Finally, statistical analysis showed that the θ distribution of carbonyls around phenyls was independent of the database used. This suggests that observed patterns of interaction in crystal packing are similar to the intramolecular interactions that stabilize proteins.

Information about preferences and directionality is of great value in molecular design by making design more quantitative instead of relying solely on chemical intuition. The now widely accepted rules that apply to such fundamental interactions as hydrogen bonding and metal–ligand contacts have been derived largely by analysis of crystallographic results. It is expected that crystallographic databases will permit the preferred interaction patterns to be established for a variety of chemical groups, and that this information will play an increasingly important role in molecular design processes.

References

Allen, F. H., Bellard, S., Brice, M. D., Cartwright, B. A., Doubleday, A., Higgs, H., Hummelink, T., Hummelink-Peters, B. G., Kennard, O., Motherwell, W. D. S., Rodgers, J. R., and Watson, D. G. (1979). The Cambridge Crystallographic Data Centre: computer-based search, retrieval, analysis and display of information. *Acta Crystallogr., Sect. B* **B35,** 2331–2339.

Allen, F. H., Bergerhoff, G., and Sievers, R. (1987a). "Crystallographic Databases." Polycrystal Book Serv., Dayton, Ohio.

Allen, F. H., Kennard, O., Watson, D. G., Brammer, L., Orpen, A. G., and Taylor, R. (1987b). Tables of bond lengths determined by X-ray and neutron diffraction. Part 1. Bond lengths in organic compounds. *J.C.S. Perkin II* pp. S1–S19.

Bernstein, F. C., Koetzle, T. F., Williams, G. J. B., Meyer, E. F., Jr., Brice, M. D., Rodgers, J. R., Kennard, O., Shimanouchi, T., and Tasumi, M. (1977). The Protein Data Bank: A computer-based archival file for macrmolecular structures. *J. Mol. Biol.* **112**, 535–542.

Bondi, A. (1964). van der Waals volumes and radii. *J. Phys. Chem.* **68**, 441–451.

Bryant, S. H., and Amzel, L. M. (1987). Correctly folded proteins make twice as many hydrophobic contacts. *Int. J. Pep. Protein Res.* **29**, 46–52.

Burley, S. K., and Petsko, G. A. (1985). Aromatic–aromatic interaction: A mechanism of protein structure stabilization. *Science* **229**, 23–28.

Burley, S. K., and Petsko, G. A. (1986). Amino–aromatic interactions in proteins. *FEBS Lett.* **203**, 139–143.

Carson, M., and Hermans, J. (1985). Molecular dynamics workshop laboratory. *In* "Molecular Dynamics and Protein Structure" (J. Hermans, ed.), pp. 165–166. Polycrystal Book Serv., Dayton, Ohio.

Chothia, C. H. (1974). Hydrophobic bonding and accessible surface area in proteins. *Nature (London)* **248**, 338–339.

Eisenberg, D., and McLachlan, A. D. (1986). Solvation energy in protein folding and binding. *Nature (London)* **319**, 199–203.

Frömmel, C. (1984). The apolar surface area of amino acids and its empirical correlation with hydrophobic free energy. *J. Theor. Biol.* **111**, 247–260.

Guru Row, T. N., and Parthasarathy, R. (1981). Directional preferences of nonbonded atomic contacts with divalent sulfur in terms of its orbital orientations. 2. S···S interactions and nonspherical shape of sulfur in crystal. *J. Am. Chem. Soc.* **103**, 477–479.

Kauzmann, W. (1959). Some factors in the interpretation of protein denaturation. *Adv. Protein Chem.* **14**, 1–63.

Kuntz, I. D., Blaney, J. M., Oatley, S. J., Langridge, R., and Ferrin, T. E. (1982). A geoemtric approach to macrmolecule–ligand interactions. *J. Mol. Biol.* **161**, 269–288.

Lee, B., and Richards, F. M. (1971). The interpretation of protein structures: Estimation of static accessibility. *J. Mol. Biol.* **55**, 379–400.

Lifson, S., and Sander, C. (1980). Specific recognition in the tertiary structure of β-sheets of proteins. *J. Mol. Biol.* **139**, 627–639.

Narayana, S. V. L., and Argos, P. (1984). Residue contacts in protein structures and implications for protein folding. *Int. J. Pep. Protein Res.* **24**, 25–39.

Ramasubbu, N., Parthasarathy, R., and Murray-Rust, P. (1986). Angular preferences of intermolecular forces around halogen centers: Preferred directions of approach of electrophiles and nucleophiles around the carbon–halogen bond. *J. Am. Chem. Soc.* **108**, 4308–4314.

Rose, G. D., Geselowitz, A. R., Lesser, G. J., Lee, R., and Zehfus, M. H. (1985). Hydrophobicity of amino acid residues in globular proteins. *Science* **229**, 834–838.

Rosenfield, R. E., Jr., Parthasarathy, R., and Dunitz, J. D. (1977). Directional preferences of nonbonded atomic contacts with divalent sulfur. 1. Electrophiles and nucleophiles. *J. Am. Chem. Soc.* **99**, 4860–4862.

Rosenfield, R. E., Jr., Swanson, S. M., Meyer, E. F., Jr., Carrell, H. L., and Murray-Rust, P. (1984). Mapping the atomic environment of functional groups: Turning 3D scatter plots into pseudo-density contours. *J. Mol. Graphics* **2**, 43–46.

Schulz, G. E., and Schirmer, R. H. (1979). "Principles of Protein Structure." Springer-Verlag, New York.

Singh, J., and Thornton, J. M. (1985). The interaction between phenylalanine rings in proteins. *FEBS Lett.* **191**, 1–6.

Taylor, R., and Kennard, O. (1984). Hydrogen-bond geometry in organic crystals. *Acc. Chem. Res.* **17**, 320–326.

Thomas, K. A., Smith, G. M., Thomas, T. B., and Feldmann, R. J. (1982). Electronic distributions within protein phenylalanine aromatic rings are reflected by the three-dimensional oxygen atom environments. *Proc. Natl. Acad. Sci. USA* **79,** 4843–4847.

Warme, P. K., and Morgan, R. S. (1978a). A survey of atomic interactions in 21 proteins. *J. Mol. Biol.* **118,** 273–287.

Warme, P. K., and Morgan, R. S. (1978b). A survey of amino acid side-chain interactions in 21 proteins. *J. Mol. Biol.* **118,** 289–304.

21 Virus Structure and the AIDS Problem: Strategies for Antiviral Design Based on Structure

Edward Arnold and Gail Ferstandig Arnold
Center for Advanced Biotechnology and Medicine (CABM) and Department of Chemistry, Rutgers University, Piscataway, New Jersey 08855-0759

I. Introduction

A detailed knowledge of the three-dimensional structure of the AIDS-causing human immunodeficiency virus (HIV) is not available; in fact, with the exception of the recently determined structure of the aspartyl protease of HIV (Navia *et al.*, 1989), there are no known atomic structures for HIV proteins. Knowledge of these structures will be critical for the development of rationally designed antiviral agents directed against AIDS and related retroviral diseases. Current progress in pharmaceutical design suggests that detailed information about the three-dimensional structures of macromolecular targets can be used to optimize the design of prototype compounds that show therapeutic promise. Long-term prospects hold hope for suggesting the *ab initio* design of therapeutic agents, based solely on the structure of the macromolecular target under consideration.

It is possible that knowledge of the currently available structures can be used to expedite determining unknown structures to high resolution, for example, by providing starting models for the structural solution using the so-called molecular replacement method in crystallography. Even single domains of a multidomain protein can be used successfully to initiate structure solution by the molecular replacement method. The increasing conviction that many different types of viruses have utilized preexisting gene products in their construction lends credence to the hypothesis that our knowledge of other viral structures provides a useful basis for thinking about the details of retroviral structures.

Given the structural complexity and size of the HIV virion as well as the danger that could arise from working with tens of milligrams of virions necessary for biophysical study, the most fruitful approach for determining the structure of HIV at high resolution will likely be to describe the structures of the component viral proteins. This paper examines and com-

Use of X-Ray Crystallography in the Design of Antiviral Agents
283

pares structures and functions of representative viruses and viral proteins as a means to extend our current thinking and hopefully our knowledge of virus structure and function.

II. Overall Structure of HIV and Genomic Arrangement

The HIV genome encodes three major genes: *gag, pol,* and *env.* The *gag* gene codes for the virion structural proteins [abbreviated p17 or MA (matrix), p24 or CA (capsid), and p15 or NC (nucleocapsid)]; the *pol* gene codes for protease [p12 or PR (protease)], polymerase [p66 or RT (reverse transcriptase)], and integrase [p34 or IN (integrase)]; and the *env* gene codes for the surface glycoproteins [gp120 or SU (surface), and gp41 or TM (transmembrane)]. These HIV proteins are digested from polyprotein translation products by a combination of virally coded and cellular proteases. A schematic representation of the intact virion is shown in Fig. 1. As can be seen in the figure, the CA and NC proteins associate with two copies of the (+)-stranded HIV genome, forming a bullet-shaped core particle that contains reverse transcriptase and integrase molecules. Contacts between the nucleoprotein core and the lipid membrane appear to be mediated by the MA protein, forming the icosadeltahedrally symmetric virion shells located at the inner leaflet of the viral membrane (Gelderblom *et al.,* 1987; Marx *et al.,* 1988). The membrane spikes are conglomerates consisting of the outer surface glycoprotein gp120 in association with the hydrophobic membrane-spanning gp41.

In addition to the molecules that make up the infectious virion, HIV encodes and expresses a number of regulatory proteins, some of which appear to be essential for virion production (reviewed in Haseltine and Wong-Staal, 1988). While some of these proteins, such as tat and rev, are likely to be valuable targets for antiviral therapy, we are less knowledgeable about similarities between these proteins and other proteins. Thus, for the purposes of this discussion, the focus will be on the proteins that make up the HIV virion itself.

III. A Detailed Look at the HIV Proteins in Light of Other Viral Structures

A. The *gag* Products: Coat Proteins of HIV

There are compelling reasons to believe that one or more of the gag proteins of HIV shares significant structural homology with a substantial

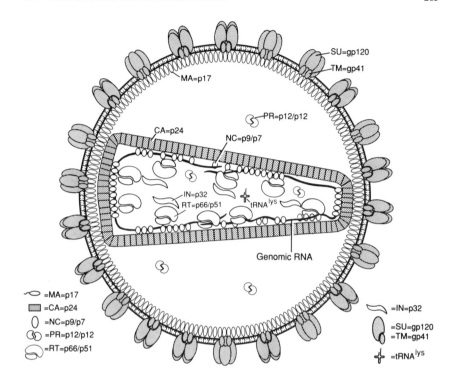

Figure 1. Model of the structure of the HIV virion.

number of other viral coat proteins. Most strikingly, in the presence of tremendous sequence diversity, the coat proteins of viruses including human rhinovirus 14 (HRV14) (Rossmann *et al.*, 1985), poliovirus 1 Mahoney (Hogle *et al.*, 1985), tomato bushy stunt virus (TBSV) (Harrison *et al.*, 1978), southern bean mosaic virus (SBMV) (Abad-Zapatero *et al.*, 1980), and satellite tobacco necrosis virus (STNV) (Liljas *et al.*, 1982) have been seen to share remarkable structural homology (Fig. 2). The RNA virus capsid or RVC proteins (to use the term recommended by Steven Harrison at this workshop) share the common structural motif of an eight-stranded antiparallel β-barrel. In the absence of any reason to believe that RVC proteins need necessarily have this structure in order to form biologically active icosahedral particles, the most likely conclusion is that the coat proteins of these extraordinarily diverse viruses diverged from a common ancestor. The functional and structural roles expected to be shared among the RVC proteins and the gag proteins of HIV have led to the speculation (E. Arnold, unpublished observations; Rossmann, 1988) that one or more of the gag proteins may contain the RVC eight-stranded β-

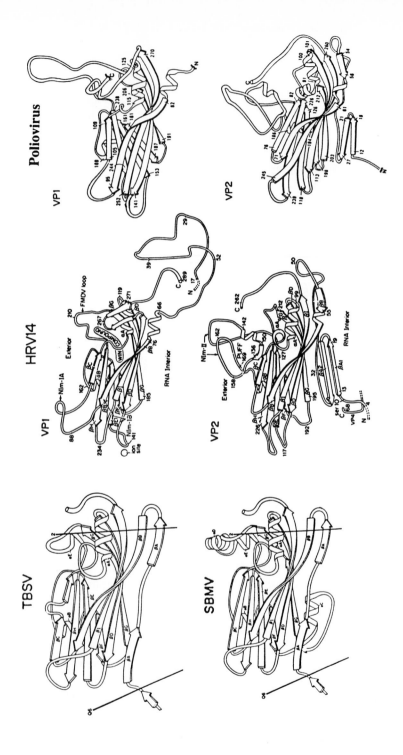

286

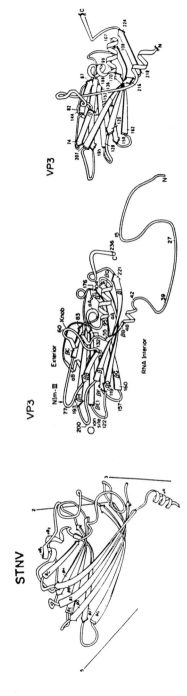

Figure 2. Three-dimensional folding diagrams illustrating the remarkable similarity among coat proteins of a number of representative small RNA viruses. (From Rossmann *et al.*, 1983, Hogle *et al.*, copyright 1985 by AAAS, and Luo *et al.*, copyright 1987 by AAAS.)

barrel. The two most likely candidates for structural homology are the major capsid protein (p24 or CA) that assembles to form the outer layer of the nucleocapsid core and the matrix protein (p17 or MA) that forms the protein shell underneath the lipid bilayer (Gelderblom *et al.*, 1987; Marx *et al.*, 1988). Like other RVC proteins, both p24 and p17 assemble to form capsid structures, and in the case of p24, the protein is also involved in packaging the RNA genome. Thus, it may be possible to model the structure of one or more of the HIV coat proteins starting from available structures of RVC proteins; plausible sequence alignments can be constructed, at least in the case of p24 (E. Arnold and V. G. Thailambal, unpublished observations). Furthermore, given the likely structural similarity of the various viral coat proteins, it might be possible to develop specific classes of antiviral agents that interfere with assembly or disassembly of the HIV virion. For instance, a compound analogous to the conformationally restricting picornavirus-inhibiting WIN compounds (Smith *et al.*, 1986) might insert into a hydrophobic core of one or more of the HIV gag proteins (Rossmann, 1988).

There may be important parallels among many viruses in their means of viral capsid assembly. Among the features expected to unite many viruses in this regard are specific viral protein–nucleic acid interactions. In the case of picornaviruses, the final stage of maturation is characterized by an obligate, apparently autoproteolytic, cleavage of a viral capsid precursor chain into two chains, a step which may involve virion RNA (Arnold *et al.*, 1987). The virion location of the processed termini that were covalently connected prior to proteolytic processing suggests that many of the protein–protein contacts present during the polyprotein stage may remain unchanged in the mature virion. Thus, given that the overall arrangement of the gag polyprotein precursor of HIV, Pr55gag, is (from the amino to carboxyl terminus) MA : CA : NC, it is possible that the majority of contacts among these proteins at the polyprotein stage are maintained in the association between MA and CA on the outside of the bullet-shaped particle of HIV and also in the association of CA and NC on the inside of the virion capsid. That the myristylated amino terminus of MA (Veronese *et al.*, 1988) probably provides an anchor to the viral membrane is consistent with this hypothesis. In fully assembled virions, there are a number of documented examples of protein–nucleic acid interactions. In the cases of the icosahedral viruses HRV14 (Arnold and Rossmann, 1988) and poliovirus (J. M. Hogle, personal communication), there is electron density corresponding to what may be partially ordered RNA near the base of the icosahedral threefold axis, which is also the locus of the autoproteolytic maturation site and some critical interpentamer bonding interactions. More dramatically, another small RNA virus recently determined to share the eight-stranded β-barrel of the other RVC proteins, bean pod mottle virus, shows a rather large portion of the RNA that is in contact with the icosahedral protein shell (on the order of 420 bases in all); this portion is located in precisely the same places as the small hints of RNA in the picornavirus structures (J. E. Johnson, this meeting). For a number of the other small RNA plant viruses, the capsid proteins and RNA appear to interact to a large extent by an attraction through the negatively charged phosphodiester backbone of the RNA and regions of strong positive charge in the coat proteins. Given the highly basic nature of the NC proteins of HIV, recognition of the viral RNA by HIV gag proteins may be, at least in part, mediated by the attraction of opposite charges in a manner that is reminiscent of some of these plant viruses. Other aspects of the viral infectious cycle in which many viral capsid proteins participate, such as receptor recognition and evasion of immune surveillance by rapid variation of antigenic structures, appear to map to non-gag viral proteins of HIV, in particular, to the surface glycoprotein gp120.

B. The Products of *pol:* Protease, Reverse Transcriptase, and Integrase of HIV

The retroviral *pol* gene encodes three proteins essential to the viral infectious cycle: the protease, the reverse transcriptase, and the integrase. The aspartyl protease, which performs proteolytic processing essential to the retroviral cycle, shows significant sequence homology with other aspartyl proteases and is a prime example of how we can extend our knowledge of the detailed three-dimensional structures of proteins through modeling. A three-dimensional model was postulated (Pearl and Taylor, 1987) that has been shown to be markedly similar to the model derived from crystallography (Navia *et al.*, 1989). While it was not tested, it is conceivable that the Pearl and Taylor model could have been used to initiate the structure determination by molecular replacement. With structure in hand, antiviral design can be initiated for disabling the protease.

The integrase, involved in poorly understood ways in catalyzing the integration of the double-stranded DNA form of the retroviral genome into the host chromosomes, is poorly characterized in terms of both structure and function. Furthermore, there are relatively few proteins of known structure that bind to and rearrange the structure of DNA. To date, some of the best examples may be only distantly related and difficult to use for initiating modeling studies.

More opportune structural comparisons might exist for reverse transcriptase (RT), the other *pol* gene product. The RT enzyme, which catalyzes the synthesis of DNA from RNA templates, essentially causes genetic information to flow in "reverse." RT of HIV is an attractive target for antiviral intervention. While essential for retroviral replication, reverse transcription appears to be rare in normal mammalian metabolism. Furthermore, RT appears to be the only nonorganellar polymerase that operates in the cytoplasm of HIV-infected cells, inviting the targeting of compounds against cytoplasmic, and not nuclear, polymerization. In fact, to date, the most promising antiviral compounds used to treat AIDS patients, such as 3'-azido-3'-deoxythymidine (AZT), dideoxycytidine (ddC), and dideoxyadenosine (ddA) (reviewed by Yarchoan *et al.,* 1988), poison the reverse transcription process by terminating chain elongation. In order to optimize the design of RT inhibitors, it would be ideal to have a high-resolution three-dimensional structure of HIV RT.

In addition to the polymerization reactions catalyzed by reverse transcriptase, both RNA-dependent DNA polymerization (RDDP) and DNA-dependent DNA polymerization (DDDP), there is an essential RNase H activity. The RNase H of RT, another potential target for antiviral intervention, is responsible for the degradation of RNA from RNA–DNA heteroduplexes synthesized in the RDDP reaction. Perhaps a nondissociable form of an RNA–DNA heteroduplex could be prepared, for example by cross-linking the strands, which might irreversibly bind to and poison the RNase H active site of the enzyme. Alternatively, a nonhydrolyzable form of the RNA strand in an RNA–DNA heteroduplex might accomplish the same trick.

Another possible strategy for antiviral intervention against HIV takes advantage of the observation that the interaction between retroviral RTs and the tRNAs used for priming reverse transcription is specific, akin to the recognition of tRNAs by tRNA synthetases (see, e.g., Labouze and Bedouelle, 1989). In the case of HIV, it has been inferred from the sequence of the primer binding site that tRNA[lys] is used to prime DNA synthesis. In order to obtain the most useful description of the interaction of RT inhibitors with the enzyme, it will be important to cocrystallize RT with mimics of primer–template complexes. Addition of activated RT inhibitors, such as AZT-triphosphate, to crystals of such complexes would yield important details about recognition that could guide rational inhibitor design.

No three-dimensional structure of any reverse transcriptase is known; however, crystals of HIV reverse transcriptase have been obtained (E. Arnold, G. F. Arnold, A. D. Clark, A. Jacobo-Molina, and R. G. Nanni, unpublished observations; Lowe *et al.,* 1988). Only one three-dimensional structure of any polymerase has been determined to high resolution, that of the Klenow fragment of *Escherichia coli* DNA polymerase I (Ollis *et al.,* 1985). A C_α tracing of the overall structure of Klenow is shown in Color

Plate 27, with the polymerizing and exonucleolytic domains highlighted in different colors. Attempts to determine possible evolutionary and hence structural relatedness between Klenow fragment and retroviral reverse transcriptases on the basis of amino acid sequence alignments have not been convincing (E. Arnold and M. A. Cueto, unpublished observations; Argos, 1988). Given the difficulties in using sequence analysis methods to determine distant evolutionary relationships, it is still possible that the Klenow fragment structure could provide a useful starting model for solution of the HIV RT structure by the molecular replacement method.

C. The Products of *env*: Surface Glycoproteins of HIV

The spikes protruding from the membrane of the HIV particle correspond to the glycoproteins gp120 and gp41 encoded by the *env* gene of HIV. The external surface glycoprotein gp120, associated with the transmembrane glycoprotein gp41, is responsible for recognizing the CD4 receptor and is involved in the fusion of the HIV and cellular membranes. The glycoprotein gp120, the primary antigenic determinant of HIV, displays an impressive amount of antigenic variability; thus, while neutralizing antibodies reacting with gp120 have been described, the high degree of variability of the antigenic portions complicate vaccination strategies based on gp120. Recognition between the CD4 molecule and gp120, critical for initiation of infection by HIV, is strongly influenced by the extent of glycosylation of gp120. Inhibition of gp120 glycosylation, for example by agents such as castanospermine (Gruters *et al.,* 1987), interferes with infectivity of the virus. Completely deglycosylated gp120 is still capable of recognizing the CD4 molecule, although with an affinity 50-fold less than the fully glycosylated molecule (Matthews *et al.,* 1987). This would indicate that structural studies of glycosylated or deglycosylated gp120 would be illuminating with regard to some of the crucial functions of the protein and might also offer insights into antiviral design based on glycosylation inhibition.

Perhaps the best available model for the structure of the HIV envelope glycoproteins is the hemagglutinin (HA) spike of influenza virus. The crystal structure of the bromelain-digested influenza HA trimer has been determined (Wilson *et al.,* 1981; Wiley *et al.,* 1981). Both the HIV envelope glycoproteins and influenza HA are derived from proteolytic processing of a precursor chain into two chains (gp120 and gp41 and HA1 and HA2, respectively) via a host-encoded protease (McCune *et al.,* 1988). In both cases, the proteolysis is obligatory for viral infectivity (McCune *et al.,* 1988). Thus, inhibition of this cleavage might interfere with spread of HIV (and influenza) infection. The relative sizes and mapping of functions to the domains is similar in the two cases; the outermost protrusion of each spike appears to recognize the host cell receptor while the small transmembrane proteins, gp41 and HA2, are important for fusion.

The marked simlarities between the known structure of influenza HA and the unknown structure of the HIV envelope glycoproteins suggests that not only might it be possible to model the structure of gp120 from that of HA, but it might also be possible to extend our appreciation of the biology of the HIV surface glycoproteins by focusing on that of HA and related proteins.

A critical feature of gp120, influenza HA, the viral capsid of HRV14, and many other viral proteins, is the ability to attach to a cellular receptor. In the case of HRV14, one of the most striking features is a narrow groove or canyon (Fig. 3) (Rossmann *et al.*, 1985) that appears to function as the recognition site for the host-cell membrane receptor, now known to be intercellular adhesion molecule 1 (S. Marlin, presented at this meeting). An interesting and probably critical feature of the HRV14 canyon is a marked absence of any residues that are recognized by neutralizing monoclonal antibodies (Rossmann *et al.*, 1985; Sherry and Rueckert, 1985; Sherry *et al.*, 1986). Single amino acid changes in the HRV14 canyon have led to altered receptor binding phenotypes, including tighter binding to the

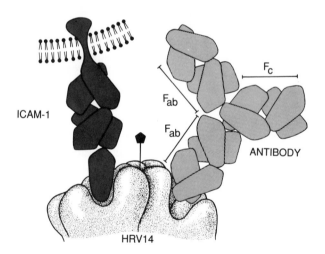

Figure 3.   Model for the means by which the rhinovirus canyon serves as a conserved host cellular receptor attachment site, protecting the recognition site from neutralizing antibodies. (After Luo *et al.*, copyright 1987 by AAAS.) Intercellular adhesion molecule-1 (ICAM-1) has recently been identified (S. Marlin, presented at this meeting) as the cellular receptor for the major group of human rhinoviruses, which includes HRV14 (Abraham and Colonno, 1984). ICAM-1 appears to have five immunoglulin-like domains (Staunton *et al.*, 1988), one of which we hypothesize is capable of binding to the receptor attachment site.

host cellular receptor, providing direct experimental evidence to support the assignment of the canyon as the host cell receptor attachment site (Colonno *et al.,* 1988). Together, this suggests that divergent serotypes of rhinoviruses are able to interact with a common cellular receptor by hiding this site from immune pressure.

Like the capsid shell of HRV14, influenza HA also has a receptor attachment site that maps to a crevice on the protein surface (Color Plate 28). This crevice, like its HRV14 analog, appears to be shielded from recognition by neutralizing antibodies (Weis *et al.,* 1988). HIV, too, is marked by a great serological diversity and perhaps only one or at most a few cellular receptor types, suggesting that the CD4-recognizing portion of gp120 may reside in a surface depression or canyon. This presents another favorable opportunity for antiviral intervention, perhaps through the construction of a small molecular mimic of CD4 to interfere with the critical receptor attachment step of the HIV infectious cycle. Such a compound, which might resemble a "molecular plaster cast" of the gp120 canyon, may be therapeutically effective against a wide range of antigenically distinct variants of the AIDS virus. It is also possible that such a treatment would have fewer side effects than treatments involving the soluble CD4 molecule, as a miniature CD4 mimic could possibly be designed to interfere specifically with gp120 attachment to CD4 but not with the attachment of normal host ligands to CD4.

IV. Conclusion

AIDS will continue to be a major human health problem in the foreseeable future. Detailed information about the three-dimensional structures of HIV components should greatly enhance the current drug design efforts targeted against HIV. In the absence of such information, it is useful to consider the details of functionally and evolutionarily related cellular and viral products. The current base of information about other possibly related structures gives us both a framework for thinking about retroviral protein structures in detail and a possible start for experimental structure determinations by methods such as molecular replacement.

Acknowledgments

We thank Aaron Shatkin and Kevin Ryan for valuable discussions regarding the manuscript, and Hilda Muiños for help with the figures. We also thank the Center for Advanced Biotechnology and Medicine, Marvin Cassman, NIH AI27690, and the New Jersey Commission on Science and Technology for support.

References

Abad-Zapatero, C., Abdel-Meguid, S., Johnson, J. E., Leslie, A. G. W., Rayment, I., Rossmann, M. G., Suck, D., and Tsuikihara, T. (1980). Structure of southern bean mosaic virus at 2.8 Å resolution. *Nature (London)* **286,** 33–39.

Abraham, G., and Colonno, R. J. (1984). Many rhinovirus serotypes share the same cellular receptor. *J. Virol.* **51,** 340–345.

Argos, P. (1988). A sequence motif in many polymerases. *Nucleic Acids Res.* **21,** 9909–9916.

Arnold, E., and Rossmann, M. G. (1988). The use of molecular-replacement phases for the refinement of the human rhinovirus 14 structure. *Acta Crystallogr., Sect. A* **A44,** 270–282.

Arnold, E., Luo, M., Vriend, G., Rossmann, M. G., Palmenberg, A. C., Parks, G. D., Nicklin, M. J. H., and Wimmer, E. (1987). Implications of the picornavirus capsid structure for polyprotein processing. *Proc. Natl. Acad. Sci. USA* **84,** 21–25.

Colonno, R. J., Condra, J. H., Mizutani, S., Callahan, P. L., Davies, M.-E., and Murcko, M. A. (1988). Evidence for the direct involvement of the rhinovirus canyon in receptor binding. *Proc. Natl. Acad. Sci. U.S.A.* **85,** 5449–5453.

Gelderblom, H. R., Hausmann, E. H. S., Ozel, M., Pauli, G., and Koch, M. A. (1987). Fine structure of human immunodeficiency virus (HIV) and immunolocalization of structural proteins. *Virology* **156,** 171–176.

Gruters, R. A., Neefjes, J. J., Tersmette, M., de Goede, R. E. Y., Tulp, A., Huisman, H. G., Miedema, F., and Ploegh, H. L. (1987). Interference with HIV-induced syncitium formation and viral infectivity by inhibitors of trimming glucosidase. *Nature (London)* **330,** 74–77.

Harrison, S. C., Olson, A. J., Winkler, F. K., and Bricogne, G. (1978). Tomato bushy stunt virus at 2.9 Å resolution. *Nature (London)* **276,** 368–373.

Haseltine, W. A., and Wong-Staal, F. (1988). The molecular biology of the AIDS virus. *Sci. Am.* **259,** 52–62.

Hogle, J. M., Chow, M., and Filman, D. J. (1985). Three-dimensional structure of poliovirus at 2.9 Å resolution. *Science* **229,** 1358–1365.

Labouze, E., and Bedouelle, H. (1989). Structural and kinetic bases for the recognition of tRNA^{Tyr} by tyrosyl-tRNA synthetase. *J. Mol. Biol.* **205,** 729–735.

Liljas, L., Unge, T., Jones, T. A., Fridborg, K., Lovgren, S., Skoglund, U., and Strandberg, B. (1982). Structure of satellite tobacco necrosis virus at 3.0 Å rsolution. *J. Mol. Biol.* **159,** 93–108.

Lowe, D. M., Aitken, A., Bradley, C., Darby, G. K., Larder, B. A., Powell, K. L., Purifoy, J. M., Tisdale, M., and Stammers, D. K. (1988). HIV-1 reverse transcriptase: crystallization and analysis of domain structure by limited proteolysis. *Biochemistry* **27,** 8884–8889.

Luo, M., Vriend, G., Kamer, G., Minor, I., Arnold, E., Rossmann, M. G., Boege, U., Scraba, D. G., Duke, G. M., and Palmenberg, A. C. (1987). The atomic structure of Mengo virus at 3.0 Å resolution. *Science* **235,** 182–191.

McCune, J. M., Rabin, L. B., Feinberg, M. B., Lieberman, M., Kosek, J. C., Reyes, G. R., and Weissman, I. L. (1988). Endoproteolytic cleavage of gp160 is required for the activation of human immunodeficiency virus. *Cell* **53,** 55–67.

Marx, P. A., Munn, R. J., and Joy, K. I. (1988). Computer emulation of thin section electron microscopy predicts an envelope-associated icosadeltahedral capsid for human immunodeficiency virus. *Lab. Invest.* **58,** 112–118.

Matthews, T. J., Weinhold, K. J., Lyerly, H. K., Langloise, A. J., Wigzell, H., and Bolognesi, D. P. (1987). Interaction between the human T-cell lymphotropic virus type

IIIB envelope glycoprotein gp120 and the surface antigen CD4: Role of carbohydrate in binding and cell fusion. *Proc. Natl. Acad. Sci. USA* **84**, 5434–5428.

Navia, M. A., Fitzgerald, P. M. D., McKeever, B. M., Leu, C.-T., Heimbach, J. C., Herber, W. K., Sigal, I. S., Darke, P. L., and Springer, J. P. (1989). Three-dimensional structure of aspartyl protease from human immunodeficiency virus HIV-1. *Nature (London)* **337**, 615–620.

Ollis, D. L., Brick, P., Hamlin, R., Xuong, N. G., and Steitz, T. A. (1985). Structure of large fragment of *Escherichia coli* DNA polymerase I complexed with dTMP. *Nature (London)* **313**, 762–766.

Pearl, L. H., and Taylor, W. R., (1987). A structural model for the retroviral proteases. *Nature (London)* **329**, 351–354.

Rossmann, M. G. (1988). Antiviral agents targeted to interact with viral capsid proteins and a possible application to human immunodeficiency virus. *Proc. Natl. Acad. Sci. USA* **85**, 4625–4627.

Rossmann, M. G., Abad-Zapatero, C., Murthy, M. R. N., Liljas, L., Jones, T. A., and Strandberg, B. (1983). Structural comparisons of some small spherical plant viruses. *J. Mol. Biol.* **165**, 711–736.

Rossmann, M. G., Arnold, E., Erickson, J. W., Frankenberger, E. A., Griffith, J. P., Hecht, H.-J., Johnson, J. E., Kamer, G., Luo, M., Mosser, A. G., Rueckert, R. R., Sherry, B., and Vriend, G. (1985). Structure of a human common cold virus and functional relationships to other picornaviruses. *Nature (London)* **317**, 145–153.

Sherry, B., and Rueckert, R. R. (1985). Evidence for at least two dominant neutralization antigens on human rhinovirus 14. *J. Virol.* **53**, 137–143.

Sherry, B., Mosser, A. G., Colonno, R. J., and Rueckert, R. R. (1986). Use of monoclonal antibodies to identify four neutralization immunogens on a common cold picornavirus, human rhinovirus 14. *J. Virol.* **57**, 246–257.

Smith, T. J., Kremer, M. J., Luo, M., Vriend, G., Arnold, E., Kamer, G., Rossmann, M. G., McKinlay, M. A., Diana, G. D., and Otto, M. J. (1986). The site of attachment in human rhinovirus 14 for antiviral agents that inhibit uncoating. *Science* **233**, 1286–1293.

Staunton, D. E., Marlin, S. D., Stratowa, C., Dustin, M. L., and Springer, T. A. (1988). Primary structure of ICAM-1 demonstrates interaction between members of the immunoglobulin and integrin supergene families. *Cell* **52**, 925–933.

Veronese, F. D. M., Copeland, T. D., Oroszlan, S., Gallo, R. C., and Sarngadharan, M. G. (1988). Biochemical and immunological analysis of human immunodeficiency virus *gag* gene products p17 and p24. *J. Virol.* **62**, 795–801.

Weis, W., Brown, J. H., Cusack, S., Paulson, J. C., Skehel, J. J., and Wiley, D. C. (1988). Structure of the influenza virus haemagglutinin complexed with its receptor, sialic acid. *Nature (London)* **333**, 426–431.

Wiley, D. C., Wilson, I. A., and Skehel, J. J. (1981). Structural identification of the antibody-binding sites of Hong Kong influenza hemaegglutinin and their involvement in antigenic variation. *Nature (London)* **289**, 373–378.

Wilson, I. A., Skehel, J. J., and Wiley, D. C. (1981). Structure of the haemagglutinin membrane glycoprotein of influenza virus at 3 Å resolution. *Nature (London)* **289**, 366–373.

Yarchoan, R., Mitsuya, H., and Broder, S. (1988). AIDS therapies. *Sci. Am.* **259**, 110–119.

22 Analysis of the Reverse Transcriptase of Human Immunodeficiency Virus Expressed in *Escherichia coli*

Stephen H. Hughes and Andrea L. Ferris
Bionetics Research Inc.-Basic Research Program, National Cancer Institute,
Frederick Cancer Research Facility, Frederick, Maryland 21701-1013

Amnon Hizi
Sackler School of Medicine, Tel Aviv University, Ramat Aviv, Israel

Retroviral reverse transcriptases are initially synthesized as part of a large polyprotein. During virion maturation reverse transcriptase is cleaved from the polyprotein by a viral protease (Weis *et al.*, 1982, 1985). The relatively small amounts of reverse transcriptase in the virion make large scale purification of the enzyme from virions impractical. Recombinant DNA technology provides an alternate approach: Large amounts of reverse transcriptase can be prepared in genetically engineered cells. The enzyme can be made either with or without the concomitant synthesis of other viral proteins (Kotewicz *et al.*, 1985; Tanese *et al.*, 1985, 1986; Hu *et al.*, 1986; Barr *et al.*, 1987; Farmerie *et al.*, 1987; Larder *et al.*, 1987; LeGrice *et al.*, 1987, 1988; Hizi *et al.*, 1988; Hizi and Hughes, 1988b; Mous *et al.*, 1988). Since reverse transcriptase is part of a polyprotein, the portion of the viral genome that encodes it is not bounded by initiation and termination codons. The region encoding the human immunodeficiency virus (HIV) reverse transcriptase can be modified by introducing initiation and termination codons into the HIV genome at positions that originally encoded the protein segment recognized by the viral protease (Larder *et al.*, 1987; Hizi *et al.*, 1988) (see Figs. 1 and 2). It is possible to introduce a termination codon at precisely the proteolytic recognition site at the C-terminus of the HIV reverse transcriptase. However, this is not possible at the corresponding site at the N-terminus. Reverse transcriptase molecules isolated from HIV virions have an N-terminal proline. Translation always initiates with a methionine so an N-terminal methionine is required. For convenience, we routinely embed the codon for the initiator methionine of our DNA constructions within the recognition site for the restriction enzyme *Nco*I (CCATGG) (see Fig. 1). This sequence can function as an efficient site for translational initiation in *Escherichia coli*, yeast, and

Use of X-Ray Crystallography in the Design of Antiviral Agents
297

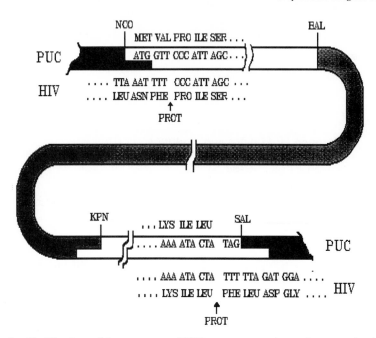

Figure 1. Modifications of the sequences of HIV reverse transcriptase for expression in *E. coli*. The synthetic DNA segments are shown as open boxes, pUC12N segments in black and the portion deriving directly from the HIV genome hatched. The DNA sequence of the original HIV clone and the amino acids encoded near the sites of proteolytic processing (marked PROT) are given below the drawing of the expression plasmid. The sequences of the synthetic DNA segments at the joints between the synthetic DNA and pUC12N (marked pUC) are also given within the boxes representing the synthetic DNA. The amino acids encoded by these synthetic DNA segments are given above the boxes.

higher eukaryotes and allows us to express precisely the same unfused protein in each of these hosts (Hughes *et al.,* 1987; Hizi *et al.,* 1988; Hizi and Hughes, 1988b). The *Nco*I–ATG expression vectors have been prepared for all three systems. Once an appropriate genetically engineered segment with an *Nco*I–ATG has been created, it is a simple matter to move the segment from one expression vector to another. If the *Nco*I site is retained, the first base of the second codon is specified as a G. This precludes the use of certain codons for the second amino acid (including those for proline). The modified HIV reverse transcriptase sgment we have created encodes methionine and valine as the first two amino acids, followed by the proline found at the N terminus in the virion-derived enzyme (Hizi *et al.,* 1988) (see Fig. 2).

If the amino acid at the second position is crucial for function, synthetic DNA segments that have bases other than G at the first position of the

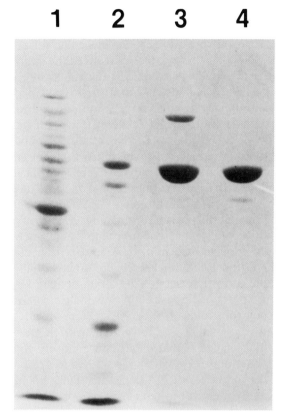

Figure 2. Purification of HIV reverse transcriptase produced in *E. coli*. Twenty-five μg of total protein was loaded onto each lane of an SDS polyacrylamide gel. Proteins were visualized by staining by Coomassie brilliant blue. Lane 1, crude lysate; lane 2, pooled fractions from the first column; lane 3, pooled fractions from the second column; lane 4, pooled fractions from the third column. The minor band is a breakdown product of the predominant 66-kDa protein. Yield from 100 gm of *E. coli* paste (wet weight) was approximately 60 mg of pure HIV-1 reverse transcriptase (fraction 4).

second codon can be ligated to an *Nco*I–ATG vector. However, in such constructions the *Nco*I site is not retained and it is more difficult to move the modified segment from one vector to another. *Nco*I–ATG vectors can be used to express a variety of proteins. We have used *E. coli Nco*I–ATG vectors to express the murine leukemia virus (MuLV) reverse transcriptase (Hizi and Hughes, 1988b) and the integration proteins from HIV-1 and MuLV (Hizi and Hughes, 1988a). We have also expressed several nonviral proteins using both *E. coli* and eukaryotic *Nco*I–ATG vectors (Hughes *et al.*, 1987, and unpublished observations).

We chose to express the HIV reverse transcriptase in *E. coli* because there is no evidence that this enzyme is modified posttranscriptionally. If it had been necessary to express the protein in a eukaryotic host to produce an appropriately modified protein, the *Nco*I site at the initiator ATG would have allowed us to move the modified segment to any one of several eukaryotic expression plasmids. A yeast expression plasmid has been prepared and strains carrying this expression plasmid make significant amounts of enzymatically active HIV reverse transcriptase (C. McGill, A. Hizi, D. Garfinkel, S. Hughes, and J. Strathern, unpublished observations). For expression in *E. coli,* the modified segment encoding the reverse transcriptase was introduced into the expression plasmid pUC12N, a derivative of pUC12 that has two bases near the *lacZ* ATG mutated to create an *Nco*I site (Vieira and Messing, 1982; Norrander *et al.,* 1985). Introduction of the pUC12N plasmid containing HIV reverse transcriptase into *E. coli* results in the synthesis of large amounts of a new 66-kiloDalton (kDa) protein, the expected size of the HIV reverse transcriptase. Lysis of this strain with Triton X-100 releases approximately half of the 66-kDa protein in soluble form. Extracts of the strain contain large amounts of RNA-dependent DNA polymerase activity, which is not found in extracts prepared from control strains of *E. coli* (Hizi *et al.,* 1988).

The observation that the 66-kDa protein has RNA-dependent DNA polymerase activity suggests that its structure closely approximates the viral reverse transcriptase. However, we wished to demonstrate directly that the active site for the HIV RNA-dependent DNA polymerase made in *E. coli* is essentially identical to the active site of the viral enzyme. We probed the structure of the active site with the competitive inhibitors dideoxy-GTP and dideoxy-TTP. The effects of these inhibitors were tested not only with the two forms of HIV reverse transcriptase (prepared from virions and from *E. coli*), but also with Moloney murine leukemia virus (M-MuLV) reverse transcriptase. In this assay the two HIV reverse transcriptases were indistinguishable but were clearly distinguished from the MuLV enzyme, demonstrating that the assay can discern small differences in the active sites of RNA-dependent DNA polymerases (Hizi *et al.,* 1988).

A major reason for expressing the HIV-1 reverse transcriptase in *E. coli* is to provide starting material for purifying large amounts of the enzyme. Under optimal growth conditions the bacteria contain, judged by gel electrophoresis and staining with Coomassie blue, several percent of their total protein as the HIV reverse transcriptase. We have worked closely with Program Resources, Inc. (PRI) and developed large scale growth and purification procedures for this enzyme (manuscript in preparation). Pat Clark (of PRI) has recently prepared approximately 1 gm of essentially pure 66-kDa HIV reverse transcriptase from *E. coli* (see Fig. 2). This material has been used to produce antibodies (see Fig. 3) and is being used for structural and biochemical studies.

More extensive biochemical analyses of the RNA-dependent DNA polymerase activity have been done with HIV-I reverse transcriptase from *E. coli* and from virions. In assays with several specific inhibitors, including AZT triphosphate, the two enzymes behaved identically, suggesting that it is reasonable to use the enzyme from *E. coli* in the evaluation of potential inhibitors of HIV reverse transcriptase *in vitro* (Schinazi *et al.*, 1989). As expected, the HIV-1 reverse transcriptase purified from *E. coli* also has an inherent RNase H activity (data not shown). We have begun a biochemical characterization of the RNase H activity, since it is a potential target for directed drug therapy.

A clear understanding of the structure of the HIV reverse transcriptase should emerge from X-ray crystallographic analyses. Crystallographic analyses are being carried out in the laboratories of Edward Arnold (Rutgers University) and Richard Dickerson (UCLA). We hope that a complete understanding of the structure of the enzyme will make it possible to design drugs to fit one (or both) of the active sites exactly. Selective inhibition of either of the enzymatic functions of the HIV reverse transcriptase should specifically block replication of the virus. However, it is reasonable to expect that HIV-1 reverse transcriptase is capable of structural variation. Treatment of AIDS patients with inhibitors of reverse transcriptase could select for variants of the enzyme partially or completely resistant to the inhibitor.

Given this potential for variation, it will be important not only to determine the strcutures of the active sites of a particular variant of the HIV reverse transcriptase, but also to define the possible limits that the active sites can assume and retain their proper functions. With this knowledge it may be possible to synthesize inhibitors that the enzyme cannot evade. There are two pathways that can be used to explore the acceptable limits of variation of the structure of the active sites. The first is to analyze evolutionary variants of reverse transcriptase. Although there are substantial differences in the amino acid sequences of the reverse transcriptases of different retroviruses, all these enzymes have similar enzymatic activities, implying some conservation of the structure of the active sites. Closer examination of the amino acid sequences of the various known reverse transcriptases reveals substantial regions of significant amino acid homology, reinforcing the idea that the three-dimensional structures of the catalytic sites are similar (Johnson *et al.*, 1986; Barber *et al.*, 1989).

A clear understanding of the structural features of at least some of the reverse transcriptases from other retroviruses will probably be useful in understanding the acceptable limits of structural variation of active sites of

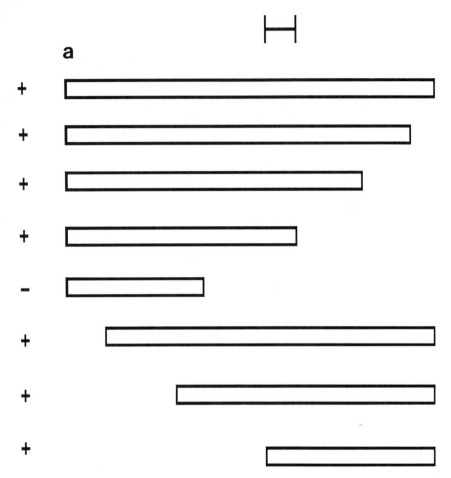

Figure 3. Mapping the recognition sites for monoclonal antibodies. A set of N- and C-terminal deletions of the HIV reverse transcriptase were used to map the epitopes recognized by monoclonals directed against HIV reverse transcriptase. (a) The principle used in this assay is illustrated. The full-length protein is schematically represented by the open box at the top of the figure. The next four smaller boxes represent C-terminal deletions generated by excising portions of the clone encoding various amounts of the C-terminal portion of the enzyme. The bottom three boxes represent N-terminal deletions. To obtain an in-frame N-terminal deletion, synthetic DNA segments were inserted between the *Nco*I–ATG and convenient internal restriction sites. In this example, a particular monoclonal is seen to react with most, but not all, of the deleted forms of HIV reverse transcriptase as indicated by (+) or (−) at the left in the drawing. The region containing the epitope recognized by this hypothetical monoclonal is shown by a small bar at the top of the drawing. (b) The approximate position of the particular epitope recognized by a monoclonal antibody is determined by fractionating the various mutant reverse transcriptase proteins by acrylamide gel electrophoresis, transferring them to nitrocellulose filters, and probing with the antibody. The

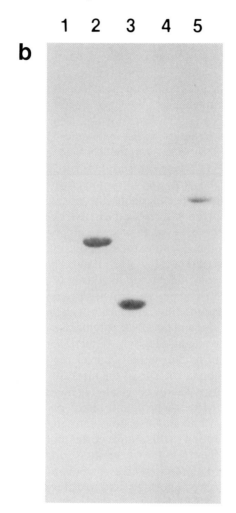

antibody used in this example reacts with carboxy-terminal deletions but fails to react with amino-terminal deletions. This locates the epitope recognized by this monoclonal antibody to the amino-terminal end of the reverse transcriptase. The monoclonal antibody was prepared by Steve Showalter of PRI. Lane 1, N-terminal deletion of 23 amino acids; lane 2, C-terminal deletion of 133 amino acids; lane 3, C-terminal deletion of 250 amino acids; lane 4, Negative control; lane 5, Full-length HIV reverse transcriptase (66 kDa).

the HIV-1 reverse transcriptase. Several laboratories, including ours, have expressed the M-MuLV reverse transcriptase in *E. coli* (Kotewicz *et al.*, 1985; Tanese *et al.*, 1985; Hu *et al.*, 1986; Hizi and Hughes, 1988b) and X-ray crystallographic analyses are planned.

It is also possible to analyze permissible variation in the structure of the HIV reverse transcriptase by making and analyzing mutant forms of the enzyme. The *E. coli* system makes the generation and analyses of these mutant enzymes a relatively simple task. Three types of mutagenic procedures are being used: deletion mutagenesis, site-directed mutagenesis, and random mutagenesis. We have begun with an analysis of the RNA-dependent DNA polymerase function. Deletions have been useful in defining the domain structure of the reverse transcriptase from M-MuLV (Tanese and Goff, 1988). However, the RNA-dependent DNA polymerase function of the HIV-1 reverse transcriptase is relatively sensitive to small deletions in either the N or the C terminus of the protein (Hizi *et al.*, 1988). In an attempt to better define the portion of HIV reverse transcriptase that is involved in forming the active site of the RNA-dependent DNA polymerase, we have made a series of small in-frame insertions at various positions in the *E. coli* expression plasmid in the region encoding HIV-1 reverse transcriptase (Hizi *et al.*, 1989). The mutant plasmids, when introduced into *E. coli,* induce the synthesis of mutant forms of HIV-1 reverse transcriptase. The mutant proteins accumulate in *E. coli* to levels similar to that of the unmodified enzyme, but all the mutations so far tested have reduced RNA-dependent DNA polymerase activity. With one interesting exception, the magnitude of the reduction in RNA-dependent DNA polymerizing activity in the mutants correlates well with the degree of sequence conservation at the site of the insertion. Insertions into regions that are evolutionarily well conserved among the various retroviral reverse transcriptases have a more profound effect on RNA-dependent DNA polymerase activity than do insertions into regions that are less well conserved. The exception to this simple correlation is that a small insertion into the region of the HIV-1 reverse transcriptase sequence encoding RNase H gives rise to a protein with esesntially no RNA-dependent DNA polymerase activity. We have suggested that this mutation may affect the ability of the protein to fold properly (Hizi *et al.*, 1989). This could also explain our previous observation that small C-terminal deletions markedly reduce RNA-dependent DNA polymerase activity (Hizi *et al.*, 1988). The effects of these small insertions on the RNase H function are now being tested, and we have begun to make and to analyze an extensive set of amino acid substitutions, concentrating initially on regions of sequence conservation.

The purified HIV-1 reverse transcriptase was used to generate polyclonal and monoclonal antibodies. The *E. coli* expression system provides a relatively simple means to map the approximate sites of the epitopes recognized by monoclonal antibodies that react well in Western transfer assays (Tisdale *et al.*, 1988; Ferris *et al.*, 1989). We have prepared a nested set of amino- and carboxyl-terminal deletions by making deletions in the plasmid used to express the full-length HIV-1 reverse transcriptase and determined which of the deleted forms of the HIV-1 reverse transcriptase react with individual monoclonal antibodies. Interestingly, the epitopes recognized by the monoclonal antibodies isolated so far are not distributed uniformly along the length of the HIV-1 reverse transcriptase. Most of the antibodies prepared by our group (Ferris *et al.*, 1989) and those reported by Tisdale *et al.* (1988) recognize a relatively small central region of HIV-1 reverse transcriptase. However, monoclonal antibodies have also been obtained that react with the N and the C termini of the molecule. The monoclonal antibodies we have used were prepared with, and originally identified as reacting with, HIV-1 reverse transcriptase in native form. We have suggested that the monoclonal antibodies are recognizing segments on the surface of the properly folded enzyme (Ferris *et al.*, 1989; Barber *et al.*, 1989).

This definition of the epitopes by deletion mapping is operational. If some or all of the antibodies recognize native protein, it is possible that some or all of the binding sites for these monoclonal antibodies are composed of sequences that are not contiguous in the HIV-1 RT primary sequence. Mapping the antibody binding sites using deletion mutants allows us to define segments of the HIV-1 reverse transcriptase that react with the individual monoclonal antibodies, and the antibodies can be used in proteolytic mapping experiments even if these segments are not the complete antibody recognition sites.

Monoclonal antibodies have been used to study proteolytic processing of the 66-kDa protein (Lowe *et al.*, 1988; Ferris *et al.*, 1989). In contrast to the protein purified from *E. coli,* the reverse transcriptase purified from virions is a 1 : 1 mixture of the 66-kDa component and a protein of approximately 51 kDa that has the same amino terminus as the 66-kDa form but is missing a carboxyl-terminal segment that comprises the RNase H domain. The 51-kDa form is, as far as is now known, unique to HIV. Other well-studied retroviruses do not contain an equivalent truncated form of reverse transcriptase. It has been generally assumed that the 51-kDa form is processed directly from the polyprotein, or from the 66-kDa form, by the viral protease, but the involvement of the viral protease in generating the 51-kDa form has not been demonstrated. Individual proteolytic digestion products of the HIV-1 reverse transcriptase can be unambiguously identified in Western transfer assays using monoclonal antibodies that recognize specific segments of the HIV-1 reverse transcriptase. We have used this technique to demonstrate that in an *in vitro* reaction containing only the two purified proteins the viral protease cleaves the 66-kDa protein at a site very near or identical to the site where the cleavage that generates the

51-kDa form in virions occurs (Ferris *et al.*, 1989). We have also tested the specificity of other, unrelated, proteases. Although none of the other proteases we have tested have as much specificity as the HIV-1 viral protease, at least two other proteases, papain and trypsin, have preferential cleavage sites near the site used by the viral protese (Ferris *et al.*, 1989). Similar results with nonviral proteases have been reported by others (Lowe *et al.*, 1988). We interpret these data to suggest that although the cleavage of the 66-kDa form to the 51-kDa form in virions is a result of the action of the viral protease, at least some of the specificity of the cleavage reaction is due to the structure of the 66-kDa form of the reverse transcriptase.

Acknowledgments

We are grateful to Hilda Marusiodis for preparing this manuscript. Our research was sponsored by the National Cancer Institute, DHHS, under contract No. NO1-CO-74101 with Bionetics Research, Inc., and the National Institute of General Medical Sciences. The contents of this publication do not necessarily reflect the views or policies of the Department of Health and Human Services, nor does mention of trade names, commercial products, or organizations imply endorsement by the United States Government.

References

Barber, A., Hizi, A., and Hughes, S. H. (1989). HIV-1 reverse transcriptase: structure predictions for the polymerase domain. Submitted for publication.

Barr, P. J., Power, M. D., Lee-Ng, C. T., Gibson, H. L., and Luciw, P. A. (1987). Expression of active human immunodeficiency virus reverse transcriptase in *Saccharomyces Cerevisiae*. *Bio/Technology* **5**, 486–489.

Farmerie, W. G., Loeb, D. D., Casavant, N. C., Hutchinson, C. A., III, Edgell, M. H., and Swanstrom, R. (1987). Expression and processing of the AIDS virus reverse transcriptase in *Escherichia coli*. *Science* **236**, 305–308.

Ferris, A., Hizi, A., Schowalter, S. D., Pichuantes, S., Babe, L., Craik, C., and Hughes, S. H. (1989). Immunologic and proteolytic analysis of HIV reverse transcriptase structure. Submitted for publication.

Hizi, A., and Hughes, S. H. (1988a). Expression of the Moloney murine leukemia virus and human immunodeficiency virus integration proteins in *Escherichia coli*. *Virology* **167**, 634–638.

Hizi, A., and Hughes, S. H. (1988b). Expression of soluble enzymatically active Moloney murine leukemia virus reverse transcriptase in *Escherichia coli*. *Gene* **66**, 319–323.

Hizi, A., McGill, C., and Hughes, S. H. (1988). Expression of soluble, enzymatically active, human immunodeficiency virus reverse transcriptase in *Escherichia coli* and analysis of mutants. *Proc. Natl. Acad. Sci. USA* **85**, 1218–1222.

Hizi, A., Barber, A., and Hughes, S. H. (1989). Effects of small insertions on the RNA-dependent DNA polymerase activity of the HIV-1 reverse transcriptase. *Virology* **170**, 326–329.

Hu, S., Court, D., Zweig, M., and Levin, J. (1986). Murine leukemia virus *pol* gene products:

Analysis with antisera generated against reverse transcriptase and endonuclease fusion proteins expressed in *Escherichia coli. J. Virol.* **60**, 267–274.

Hughes, S. H., Greenhouse, J. J., Petropoulos, C. J., and Sutrave, P. (1987). Adaptor plasmids simplify the insertion of foreign DNA into helper-independent retroviral vectors. *J. Virol.* **61**, 3004–3112.

Johnson, M. S., McClure, M. A., Feng, D. F., Gray, J., and Doolittle, R. (1986). Computer analysis of retroviral *pol* genes: Assignment of enzymatic function to specific sequences and homologies with nonviral enzymes. *Proc. Natl. Acad. Sci. USA* **83**, 7648–7652.

Kotewicz, M., D'Alessio, M., Driftmier, M., Blodgett, K., and Gerard, G. (1985). Cloning and overexpression of Moloney murine leukemia virus reverse transcriptase in *Escherichia coli. Gene* **35**, 249–258.

Larder, B., Purifoy, D., Powell, K., and Darby, G. (1987). AIDS virus reverse transcriptase defined by high level expression in *Escherichia coli. EMBO J.* **6**, 3133–3137.

LeGrice, S., Benck, V., and Mous, J. (1987). Expression of biologically active T-cell lymphotropic virus type III reverse transcriptase in *Bacillus subtilis. Gene* **55**, 95–103.

LeGrice, S. F. S., Zehnk, R., and Mouse, J. (1988). A single 66-kilodalton polypeptide processed from the human immunodeficiency virus type 2 *pol* polyprotein in *Escherichia coli* displays reverse transcriptase activity. *J. Virol.* **62**, 2525–2529.

Lowe, D. M., Aitken, A., Bradley, C., Darby, G. K., Larder, B. A., Powell, K. L., Purifoy, D. S. M., Tisdale, M., and Stammer, D. K. (1988). HIV-1 reverse transcriptase: Crystallization and analysis of domain structure by limited proteolysis. *Biochemistry* **27**, 8884–8889.

Mous, J., Heimer, E. P., and LeGrice, S. (1988). Processing protease and reverse transcriptase from human immunodeficiency virus type 1 polyprotein in *Escherichia coli. J. Virol.* **62**, 1433–1436.

Norrander, J., Vieira, J., Rubenstein, I., and Messing, J. (1985). Manipulation and expression of the maize zein storage protein in *Escherichia coli. J. Biotechnol.* **2**, 157–175.

Schinazi, R. F., Eriksson, B. F. H., and Hughes, S. H. (1989). Comparison of inhibitory activities of various antiretroviral agents against particle-derived and recombinant human immunodeficiency virus type 1 reverse transcriptase. *Antimicrob. Agents Chemother.* **33**, 115–117.

Tanese, N., and Goff, S. P. (1988). Domain structure of the Moloney murine leukemia virus reverse transcriptase: Mutational analyses and separate expression of the DNA polymerase and RNase H activities. *Proc. Natl. Acad. Sci. USA* **85**, 1777–1781.

Tanese, N., Roth, M., and Goff, S. (1985). Expression of enzymatically active reverse transcriptase in *Escherichia coli. Proc. Natl. Acad. Sci. USA* **82**, 4944–4948.

Tanese, N., Sodroski, J., Haseltine, W., and Goff, S. P. (1986). Expression of reverse transcriptase activity of human T-lymphotropic virus type III (HTLVIII/LAV) in *Escherichia coli. J. Virol.* **59**, 743–745.

Tisdale, M., Ertl, P., Larder, B. A., Purifoy, D. J. M., Darby, G., and Powell, K. L. (1988). Characterization of human immunodeficiency virus type 1 reverse transcriptase by using monoclonal antibodies: Role of the C terminus in antibody reactivity and enzyme function. *J. Virol.* **62**, 3662–3667.

Vieira, J., and Messing, J. (1982). The pUC plasmids, an M13 mp7-derived system for insertion mutagenesis with synthetic universal primers. *Gene* **19**, 259–268.

Weiss, R., Teich, N., Varmus, H. E., and Coffin, J. (1982, 1985). "RNA Tumor Viruses." Cold Spring Harbor Lab., Cold Spring Harbor, New York.

23 Structural Studies on Human Immunodeficiency Virus Reverse Transcriptase

D. K. Stammers, K. L. Powell, B. A. Larder, G. Darby,
D. J. M. Purifoy, M. Tisdale, and D. M. Lowe
Department of Molecular Sciences, The Wellcome Research Laboratories,
Beckenham, Kent BR3 3BS, United Kingdom

D. I. Stuart, E. Y. Jones, G. L. Taylor, E. F. Garman,
R. Griest, and D. C. Phillips
Laboratory of Molecular Biophysics, University of Oxford, Oxford OX1
3QU, United Kingdom

I. Introduction

The threat posed by the AIDS epidemic has led to the search for drugs to combat this disease. The identification of the human immunodeficiency virus (HIV) as the cause of AIDS (Barre-Sinoussi *et al.*, 1983) has focused much effort on the gene products of this virus, as they represent potential targets for antiviral drugs. HIV is a retrovirus, and hence an important feature of its replication is the conversion of its RNA genome into double-stranded DNA. This is carried out by a virus-encoded reverse transcriptase (RT) present in the virion. An equivalent enzyme is not thought to have a role in normal cell function, hence it represents a virus-specific drug target. Indeed 3′ azidothymidine (AZT, Zidovudine) is a selective inhibitor of HIV in its metabolically activated triphosphate (TP) form (Furman *et al.*, 1986). AZTTP inhibits HIV RT with a K_i of 0.04 μM; by contrast the cellular DNA polymerase is inhibited at much higher levels (IC_{50} = 260 μM). The drug competes with thymidine triphosphate for incorporation into DNA by RT and causes chain termination. To date AZT is the only drug that has been approved for treatment of AIDS and its clinical success establishes RT as an excellent anti-HIV target.

There is clearly a need to develop follow-up compounds to AZT. As part of a wide-ranging strategy to develop second generation RT inhibitors we are attempting to determine the three-dimensional structure of HIV RT using X-ray crystallography. This should form the basis for the rational design of novel inhibitors of RT using the method of receptor fit.

Use of X-Ray Crystallography in the Design of Antiviral Agents

We have cloned and expressed the RT gene at high level in *Escherichia coli* and purified large quantities of native and modified enzymes for extensive crystallization experiments.

II. Production and Purification of Recombinant HIV RT

The cloning and expression of the HIV RT gene and the purification of RT have been described previously (Larder *et al.*, 1987a; Tisdale *et al.*, 1988). Briefly, a *Bgl*II–*Eco*RI fragment of a subgenomic clone of HIV containing part of the *pol* gene was cloned into the M13 phage system. Oligonucleotide-directed mutagenesis was used to remove the endonuclease sequence at the 3' end and introduce a *Hin*dIII site in this position, and at the 5' end to remove the protease sequence leaving an *Eco*RI site. This construct coded for a 66-kDa RT polypeptide. To obtain sufficient protein for crystallization studies the RT DNA cassette was inserted in the high level expression plasmid pKK 233-2.

RT was initially purified by DNA-cellulose and anion exchange chromatography (Larder *et al.*, 1987a). This material was then used to raise monoclonal antibodies (MAbs) in mice (Tisdale *et al.*, 1988). Two monoclonal antibodies, with specificity for different epitopes on the RT molecule, were grown on a large scale and used to produce immunoaffinity columns. These were used to purify RT from bacterial extracts.

III. Crystallization of HIV RT

Initial crystallization experiments involved varying pH, temperature, the concentration of various salts, PEGs, and organic solvents and the addition of different metal ions and nonionic detergents.

Various methods of crystallization were used, primarily hanging drop vapor diffusion, although microdialysis and sitting drop vapor diffusion were also tried. Seeding experiments were carried out on occasions when small crystals were obtained.

Attempts to produce better-ordered crystals have focused on three main areas. First, we have tried to optimize the isolation procedure to produce more homogeneous preparations of RT as judged by isoelectric focusing. Second, we have attempted cocrystallization of RT with ligands including DNA oligomers, inhibitors, and Fabs. Third, we have modified RT by protein engineering methods to introduce single-point amino acid changes, truncations of the C-terminus, and deletion of internal regions. In addition we have cloned and expressed the RT gene from different virus isolates and prepared RT from an eukaryotic expression system. We have also carried out RT crystallization experiments under microgravity conditions in collaboration with C. Bugg of the University of Alabama, Birmingham. The different crystallization conditions that have been tried are summarized in Table I.

IV. Results

A. Native RT

Rodlike crystals of HIV RT were obtained from ammonium sulphate at pH7.0 and 4°C (Fig. 1). When these were redissolved and run on SDS gels it was found that they were composed of apparently equimolar amounts of 66-kDa and 51-kDa subunits (Fig. 2) (Lowe *et al.*, 1988). Similar molecular mass forms are found in the whole virus (di Marzo Veronese *et al.*, 1986). The site of cleavage is toward the C-terminal region of the RT subunit as

Table I. Screening Crystallization Conditions for HIV RT

System Variations	Enzyme Variants
Salts	Protease digestion
Peg	C-terminal truncations
Organic solvents	Deletion mutants
pH	Single point mutations
Temperature	Different virus isolates
Crystallization method	Different expression
Vapor diffusion	system
Hanging drop	
Sitting drop	
Microdialysis	
Seeding	
Metal ions	
Nonionic detergents	
Cocrystallization partners	
Fabs	
Substrates	
Inhibitors	
Microgravity	

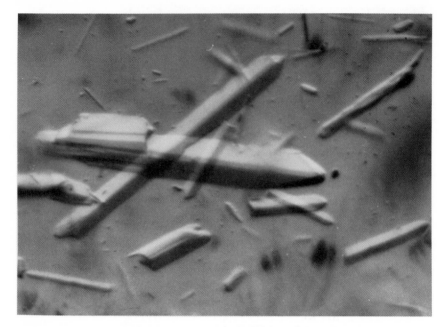

Figure 1.   Crystals of HIV RT containing an equimolar ratio of 66-kDa and 51-kDa subunits.

the N terminus is the same in both 66-kDa and 51-kDa forms. Gel filtration experiments indicate that 66-kDa and 66–51-kDa behave as dimers (Lowe *et al.*, 1988). It thus seems that either cleavage of one subunit of the 66-kDa homodimer causes a conformational change such that the second site is protected or the subunits of the homodimer are not symmetrically arranged, such that only one subunit can be cleaved.

The largest crystals obtained of this form are 0.7 × 0.3 × 0.3 mm and grew over a period of around 9 months. Data have been collected on both the SERC Daresbury synchrotron (Fig. 3) and the Oxford Xentronics area detector to a maximum resolution of 6 Å. The indexing of these data indicated a primitive orthorhombic cell of $a = 148$, $b = 192$, and $c = 180$ Å ($\alpha = \beta = \gamma = 90°$). If a packing density in the normal range is assumed, the asymmetric unit consists of four RT dimers.

Under these same crystallization conditions we have obtained prismlike crystals (Fig. 4) that when redissolved and run on SDS gels show a 3 : 1 ratio of 66-kDa to 51-kDa subunits. They show disorder when examined in the X-ray beam.

Using different conditions we have obtained crystals of 66-kDa, which grow as bundles of needles (Fig. 5).

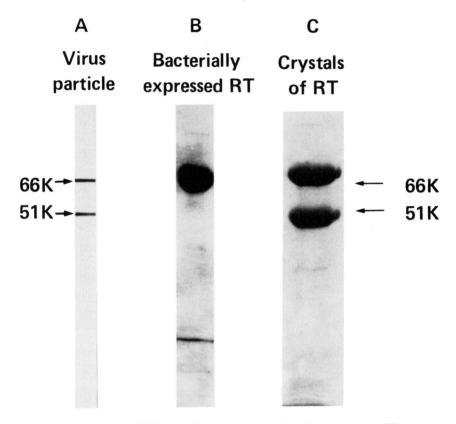

Figure 2. SDS gels of HIV RT. Note that the virus particles (A) were run on a different gel system that (B) and (C). (A) Western blot probed with an anti-RT monoclonal; (B) and (C) Coomassie blue-stained gels. Virus particles run on Du Pont Kit: different gel system.

B. RT–Ligand Complexes

In some instances enzyme–substrate or enzyme–inhibitor complexes can be crystallized more easily than unliganded enzyme. Indeed, using *E. coli* dihydrofolate reductase only enzyme–ligand complexes have been crystallized (Baker *et al.,* 1981).

We have investigated short DNA oligomer template/primers as RT substrates. Weak activity is observed for 15/9-mer, 17/9-mer, 19/9-mer, etc. oligomers with a variety of base compositions. "Hairpin" oligomers in which the two DNA chains are joined by four additional thymidine bases have been prepared and are also active as substrates.

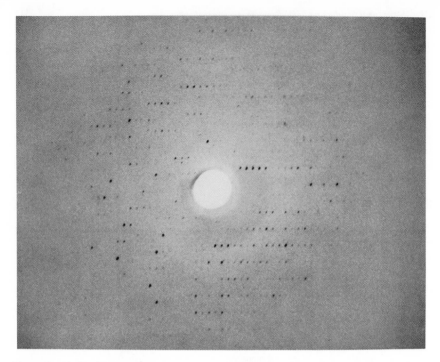

Figure 3. An oscillation photograph of an HIV RT crystal taken at the SERC synchrotron at Daresbury.

Figure 4. Crystals of HIV RT containing a 3 : 1 ratio of 66-kDa and 51-kDa subunits.

Figure 5. Crystals of HIV RT composed of 66-kDa subunits.

Cocrystallization of RT with DNA oligomers has produced platelike crystals which are morphologically distinct from the apoenzyme (Fig. 6). However, these are only ordered to 12 Å resolution. Addition of a chain terminator such as AZTTP, ddTTP, or acyclo-GTP to the RT DNA complex does not give better-ordered crystals. We have also attempted to crystallize a ternary complex of RT–oligonucleotide and α-β-methylene TTP (a nonhydrolyzable analog of TTP synthesised by R. Nichol of The Wellcome Research Laboratories). Platelike crystals similar to the binary RT–oligonucleotide complex have been obtained. A ternary complex of RT–oligomer and the pyrophosphate analog, phosphonoformate, has been crystallized, but these crystals are also disordered.

C. RT–Fab Complexes

A number of protein–Fab complexes have been crystallized. In the case of an influenza neuraminidase, large crystals of a complex with a Fab were grown whereas only microcrystals of the enzyme alone were obtained (Air *et al.,* 1987). It is to be expected that addition of a Fab molecule will alter contacts in the crystal.

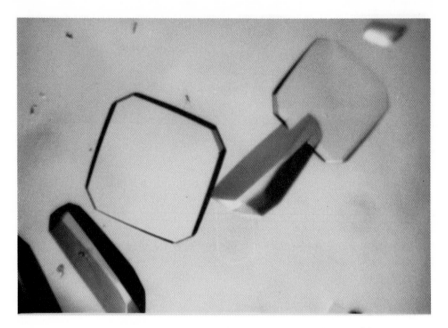

Figure 6.   Crystals of HIV RT complexed with a DNA oligonucleotide.

From a panel of eight anti-RT monoclonals (Tisdale *et al.*, 1988) four have been selected for the preparation of Fab fragments. These vary in the epitope that they recognize. Fabs were prepared by digestion of Mabs by papain followed by purification by gel filtration and MonoQ. Fabs were complexed with both RT homodimers and heterodimers in different ratios depending on the specificity of the Fab for different molecular mass forms of the enzyme. The binding of different Fabs is in some cases not mutually exclusive; hence, RT complexes with more than one Fab have been made for crystallization experiments. To date no crystals of RT–Fab complexes have been obtained.

D. C-Terminal Truncation

It has been observed that some proteins crystallize more readily or give better-ordered crystals when the C-terminal region is truncated. For example, the H-ras p21 oncogene protein gave more ordered crystals when 18 amino acids were removed from the C-terminus (De Vos *et al.*, 1988).

A number of C-terminally truncated RTs have been prepared by insertion of a premature stop codon in the RT gene (Tisdale *et al.*, 1988). Truncations of 6, 11, 17, and 21 amino acids retained essentially full activity, while deletion of 27 amino acids resulted in the loss of 70% of the

RT activity. We have grown crystals of mutants lacking 6, 11, and 17 amino acids from the C-terminus. They have the same morphology as the native protein and are no better ordered.

Deletion of 130 amino acids produces a 51-kDa polypeptide that has only 5% of the specific activity of the 66-kDa material. No crystals from this mutant RT have been obtained.

E. Single Amino Acid Changes

Site-directed mutagenesis has previously been used to probe possible functionally important regions of the RT molecule (Larder *et al.*, 1987b). Further mutants have been prepared in putative surface regions. They are Thr 360→Val, Asp 491→Ala, Gln 515→Glu, and Leu 563→Pro, which retain activity comparable with the native enzyme. Crystals of all four mutants have been obtained which have similar morphology to those of the native protein. X-ray examination of the 360 and 515 mutants show they are, like the native protein, disordered.

F. Deletion Mutants

As noted previously the region between the N-terminal 51-kDa domain and the 15-kDa C-terminal domain appears susceptible to proteolysis (Lowe *et al.*, 1988). We have attempted to eliminate cleavage here by deletion mutagenesis. Mutants in which residues 436–441 or 427–439 have been deleted were prepared. The resulting RTs were much less active, expressing 13% and 21% of the activity of full length protein. Surprisingly, they also appear to break down more rapidly than wild-type RT. This might indicate that the protein is unable to fold correctly.

G. RT from Different Isolates

The RT gene from a different HIV isolate (HTLVIIIB) has been cloned and expressed. RT purified from this clone does not form better-ordered crystals than that from the LAV isolate. Five new clinical isolates of HIV have recently been obtained and for each the RT gene is now cloned. Studies to characterize the RT from these are underway to assess their suitability for crystallization studies.

H. RT from a Eukaryotic Expression System

The RT gene has been cloned and expressed in the baculovirus system. Polypeptides of 66 kDa and 60 kDa are produced. Crystallization trials are underway with this protein.

V. Conclusions

A wide-ranging survey of crystallization conditions for recombinant HIV RT has been undertaken. To date approximately 3000 crystallization conditions have been tried.

This work shows that a number of apparently different crystal forms of HIV RT can be grown that are of distinct morphology and have different subunit compositions. We have observed that 66-kDa RT during the course of crystallization undergoes cleavage at the C-terminus to give a 51-kDa subunit that forms a stable heterodimer in combination with a 66-kDa subunit. Crystals of this form of RT are orthorhombic and grow as long rods. We have observed prismlike crystals that contain a 3 : 1 ratio of 66-kDa to 51-kDa subunits. This probably represents cocrystallization of equimolar amounts of 66–51-kDa heterodimer and 66–66-kDa homodimer. Under different conditions needlelike crystals containing only 66-kDa subunits have been obtained. All these crystals show disorder when examined in the X-ray beam. The maximum resolution of the observed diffraction pattern is 6 Å.

Cocrystallization of RT with DNA oligomers produces well-formed platelike crystals that show disorder. A number of mutant forms of RT produced by site-directed mutagenesis have been crystallized. These involve changes in single amino acids or truncations of the C-terminal region.

Acknowledgments

We would like to thank C. K. Ross, C. Bradley, A. Emmerson, and P. Ertl for their expert technical assistance in this work.

References

Air, M. G., Webster, R. G., Colman, P. M., and Laver, W. G. (1987). Distribution of sequence differences in influenza N9 neuraminidase of tern and whale viruses and crystallisation of the whale neuraminidase complexed with antibodies. *Virology* **160,** 346–354.
Baker, D. J., Beddell, C. R., Champness, J. N., Goodford, P. J., Norrington, F. E. A., Smith, D. R., and Stammers, D. K. (1981). The binding of trimethoprim to bacterial dihydrofolate reductase. *FEBS Lett.* **126,** 49–52.
Barre-Sinoussi, F., Cherman, J. C., Rey, F., Nugeybe, M. T., Charmaret, S., Gruest, J., Dauguet, C., Axler-Blin, C., Vezinet-Brun, F., Rouzioux, C., Rozenbaum, W., and Montagnier, L. (1983). Isolation of a T-lymphotrophic retrovirus from a patient at risk of acquired immune deficiency syndrome (AIDS). *Science* **220,** 868–870.
De Vos, A. M., Tong, L., Milburn, M. V., Matias, P. M., Jancarik, J., Noguchi, S., Nishimura, S., Miura, K., Ohtsuka, E., and Kim, S.-H. (1988). Three dimensional

structure of an oncogene protein: Catalytic domain of human c-H-ras p21. *Science* **239**, 888–893.

di Marzo Veronese, F., Copeland, T. D., De Vico, A. L., Rahman, R., Orozlan, S., Gallo, R. C., and Sarngaharan, M. G. (1986). Characterization of highly immunogenic p66/p51 as the reverse transcriptase of HTLV-III/LAV. *Science* **231**, 1289–1291.

Furman, P. A., Fyfe, J. A., St. Clair, M. H., Weinhold, K., Rideout, J. L., Freeman, G. A., Nusinoff Lehrman, S., Bolognesi, D. P., Broder, S., Mitsuya, H., and Barry, D. W. (1986). Phosphorylation of 3'-azido-3'-deoxy thymidine and selective interaction of the 5'-triphosphate with human immunodeficiency virus reverse transcriptase. *Proc. Natl. Acad. Sci. USA* **83**, 8333–8337.

Larder, B. A., Purifoy, D. J. M., Powell, K. L., and Darby, G. (1987a). AIDS virus reverse transcriptase defined by high-level expression in *E. coli*. *EMBO J.* **6**, 3133–3137.

Larder, B. A., Purifoy, D. J. M., Powell, K. L., and Darby, G. (1987b). Site-directed mutagenesis of AIDS virus reverse transcriptase. *Nature (London)* **327**, 716–717.

Lowe, D. M., Aitken, A., Bradley, C., Darby, G. K., Larder, B. A., Powell, K. L., Purifoy, D. J. M., Tisdale, M., and Stammers, D. K. (1988). HIV-1 reverse transcriptase: Crystallization and analysis of domain structure by limited proteolysis. *Biochemistry* **27**, 8884–8889.

Tisdale, M., Ertl, P., Larder, B. A., Purifoy, D. J. M., Darby G., and Powell, K. L. (1988). Characterisation of human immunodeficiency virus Type 1 reverse transcriptase by using monoclonal antibodies: Role of the C-terminus in antibody reactivity and enzyme function. *J. Virol.* **62**, 3662–3667.

24 Human Immunodeficiency Virus (Type 1) Protease: Enzymology and Three-Dimensional Structure of a New AIDS Drug Target

Paul L. Darke, Chih-Tai Leu, Jill C. Heimbach,
and Irving S. Sigal*
Department of Molecular Biology, Merck Sharp and Dohme Research
Laboratories, West Point, Pennsylvania 19486

James P. Springer, Manuel A. Navia, Paula M. D. Fitzgerald,
and Brian M. McKeever
Department of Biophysical Chemistry, Merck Sharp and Dohme Research
Laboratories, Rahway, New Jersey 07065

I. Introduction

A. Retroviral Proteases

The spread of the retrovirus human immunodeficiency virus (HIV) and the resulting AIDS pandemic has increased interest in the structural, catalytic, and regulatory proteins of retroviruses. A constant feature of the genetic structure of retroviruses such as HIV-1 is the presence of a protease coding region between the regions coding for the structural proteins known as gag and the enzymes reverse transcriptase and integrase. The integrity of the protease region was shown to be essential for replication in the case of murine leukemia virus (Crawford and Goff, 1985; Katoh *et al.*, 1985). More recently, inactivation of the HIV-1 protease through a single base substitution in the 9-kilobase HIV-1 genome was shown to eliminate the infectivity of HIV-1 (Kohl *et al.*, 1988). Thus it appears that the HIV-1 protease is a potential target for the design of agents that suppress viral replication in infected individuals.

Although retroviral proteases have been studied for many years (Vogt *et al.*, 1979; Moelling *et al.*, 1980; Yoshinaka *et al.*, 1986), the mechanisms of catalysis and the structures of these enzymes have remained obscure until recently.

---

* This presentation is dedicated to the memory of Irving S. Sigal, whose insight and intensity guided the work described herein.

Use of X-Ray Crystallography in the Design of Antiviral Agents

B. Identification of the HIV-1 Protease Sequence

The HIV protease was predicted to belong to the aspartic class of proteases, based on sequence homologies (Toh *et al.*, 1985) to such well-characterized enzymes as pepsin and renin (Fig. 1). The conserved sequence Asp-Thr-Gly occurs twice in the pepsins, and in each instance the Asp residue contributes a carboxylate to the active site, as seen from high-resolution crystallographic studies (James and Sielecki, 1983). In addition, a structural similarity between each of the two domains of the pepsins that contain the active site Asp residues was noted (Tang *et al.*, 1978), and it was suggested to be the result of gene duplication. The occurrence of the Asp-Thr-Gly sequence once within the amino acid sequence predicted for the smaller retroviral proteases suggested that the aspartic protease two-domain structure might be reproduced in the viral

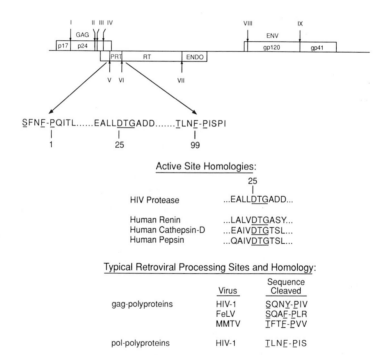

Figure 1.   The open reading frames of HIV which code for polyproteins. The proteolytic processing sites are indicated by vertical bars and Roman numerals. The protease coding region is expanded to show the conserved DTG region characteristic of aspartic proteases, as well as the amino and carboxyl termini produced by proteolysis. In the lower part of the figure are homologies useful in defining the 99-amino acid protease.

enzymes through dimerization of two identical subunits, each contributing one-half of the residues critical for catalysis.

Retroviral enzymes are initially synthesized as a continuous part of much larger polyproteins, which are subsequently proteolytically processed to release the mature proteins. In the case of HIV, translation of the *pol* reading frame encoding the enzymes is initiated through a ribosomal frameshift out of the *gag* open reading frame, resulting in a 160-kDa gag–pol polyprotein (Jacks *et al.*, 1988). Since the processed protease from HIV viral particles was not available for sequencing, it was not clear what the exact amino and carboxyl termini of the mature, active form of the protein would be. However, retroviral cleavage sites in the gag and pol polyproteins often contain a Phe (or Tyr) in the P_1 site, a proline in the P_1' site and a Ser (or Thr) in the P_4 site (Pearl and Taylor, 1987a). The occurrence of similar sequences on either side of the conserved Asp-Thr-Gly, including the known amino terminus of reverse transcriptase (di Marzo Veronese *et al.*, 1986), predicted a 99-amino acid protein as the mature form (Fig. 1). Dimerization of the 99-mer could produce a 198-amino acid enzyme containing two Asp-Thr-Gly sequences, as found in the pepsinlike aspartic proteases.

Potent inhibitors of the aspartic proteases have been known for years. Given the potential therapeutic value of an HIV protease inhibitor, the complete catalytic and structural characterization of this key enzyme in its fully processed form and comparison of those properties to the well-characterized aspartic proteases was initiated.

II. Biochemistry of the HIV-1 Protease

A. Preparation

Since obtaining milligram quantities of an active HIV-1 protease from HIV was judged to be difficult and dangerous, both chemical synthesis and microbial expression were used to obtain catalytically active forms of the predicted 99-amino acid protease. An advantage of chemical synthesis for the early stages of characterization was that the proteolytic activity found was assuredly intrinsic, as opposed to a contaminating protease that might be present in biologically produced samples (Nutt *et al.*, 1988; Schneider and Kent, 1988). Thus it was possible to develop convenient assays facilitating work with the cloned material (Darke *et al.*, 1988). Despite the success of the chemical synthesis, there was the distinct possibility that there would be present small amounts of failure sequences or side products extremely similar to, and not removable from, the desired sequence. Since crystallization attempts and other biophysical measurements could then be frustrated, the cloning of the protease was considered essential. Many laboratories have now expressed active HIV-1 protease as either the 99-mer form (Graves *et al.*, 1988; Darke *et al.*, 1989) or as part of larger proteins from which the protease appears to cleave itself out (Kramer *et al.*, 1986; DeBouck *et al.*, 1987; Giam and Boros, 1988; Hansen *et al.*, 1988; Krausslich *et al.*, 1988; Mous *et al.*, 1988; Seelmeier *et al.*, 1988).

In our attempts to obtain large quantities of the 99-amino acid form of the protease, the coding region of the protease was altered by site-directed mutagenesis to contain an initiating methionine codon prior to the N-terminal proline, and a stop codon after the C-terminal phenylalanine. Due to the apparent toxicity of the active enzyme for host *Escherichia coli*, the resulting coding sequence was expressed behind the tightly regulated *Trp* promoter. Use of only the soluble portion of the expressed enzyme (about one-third of the total) was thought to be the most likely way to obtain protein amenable to crystallization. Purification of that fraction involved common ion exchange and hydrophobic interaction chromatographic procedures (McKeever *et al.*, 1989). Fortunately, bacterial enzymes removed the N-terminal methionine, yielding only the 99-amino acid protein.

B. Catalytic Properties

1. Cleavage Sites Represented as Peptides

Peptides representing known cleavage sites in the gag and pol polyproteins of HIV-1 were demonstrated to be substrates for the protease, and were cleaved at the anticipated positions shown in Fig. 2. The amino acids immediately flanking the cleaved bonds appear to be quite diverse, although in general are hydrophobic. Only a few of the sequences (sites I, V, and VI) actually conform to the cleavage site motif described in the introduction. In addition, it has been shown that a peptide with the sequence Acetyl-Thr-Phe-Gln-Ala-Tyr-Pro-Leu-Arg-Glu-Ala-NH$_2$ is not a substrate (Krausslich *et al.*, 1989). Thus, the structural determinants necessary for susceptibility to HIV-1 protease cleavage are likely to involve several of the residues flanking the cleavage site and are still ill-defined.

2. Minimal Peptide Substrate Sequence

The minimal length of peptide necessary for efficient cleavage by the HIV-1 protease was estimated to be seven amino acids when the sequence representing the gag p17–p24 cleavage site was used (Darke *et al.*, 1989). Other cleavage sites can be represented as heptamer peptides, with the bond cleaved being four amino acids from the N-terminus of the peptide (Billich *et al.*, 1988). Similar results have been obtained with another

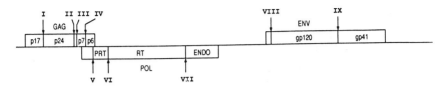

HYDROLYSIS OF PEPTIDES
CORRESPONDING TO PROTEIN CLEAVAGE SITES

Site	Sequence	K_m (mM)	V_{max} (nmol/min/mg)
HIV-1 Sites	Peptides Hydrolyzed		
I GAG 124-138	HSSQVSQNY-PIVQNI	N.D.[a]	> 275
II GAG 357-370	GHKARVL-AEAMSQV	2.3	100
III GAG 370-383	VTNTATIM-MQRGNF	0.16	682
VI GAG 440-453	SYKGRPGNF-LQSRP	13.9	382
V POL 59-62	DRQGTVSFNF-PQit	0.70	954
VI POL 162-174	GCTLNF-PISPIET	N.D.[a]	> 120
VII POL 721-734	AGIRKIL-FLDGIDK	6.1	145
HIV-2 Site			
GAG 129-142	SEKGGNY-PVQHVGG	2.3	295
	Peptides Not Hydrolyzed		
HIV-1 Site			
IX ENV 511-524	RVVQREKR-AVGIGA	–	> 2
AMV-Site			
p12/p15[c]	PAVS-LAMTMEHK	–	> 2

[a] K_m not determined due to low solubility.

Figure 2. Peptides which represent proteolytic cleavage sites. Kinetic parameters for hydrolysis of the sequences shown were determined with the chemically synthesized HIV protease. The relative susceptibility of the peptides to cleavage were confirmed with the purified, microbially expressed protein. The anticipated point of cleavage for each peptide is indicated by a dash. Cleavage sites for the peptides hydrolyzed were confirmed by sequencing.

retroviral protease, avian sarcoma leukosis virus protease (Kotler *et al.*, 1988), as well as for another aspartic protease, renin (Skeggs *et al.*, 1968). Pepsin, however, can catalyze reactions with much smaller substrates, so it must be stated that the minimal peptide length for cleavage by the retroviral enzymes could well be sequence-specific.

3. Other Catalytic Properties

Aspartic proteases were at one time referred to as acid proteases because of the generally acidic optima observed for catalysis. The pH optimum for HIV-1 protease cleavage of the gag p17–p24 peptide was determined to be slightly acidic, pH 5.5, with a broad peak (Darke *et al.*, 1989). It should be noted that an important exception in this group of enzymes is renin, with a pH optimum of 7.0.

Another common feature of the aspartic proteases is inhibition by pepstatin. The K_i for pepstatin inhibition of peptide cleavage by the HIV-1 enzyme is 1.4 μM (Darke *et al.*, 1989), and others have observed the inhibition in various assays (Seelmeier *et al.*, 1988; Hansen *et al.*, 1988).

III. Crystal Structure of the HIV-1 Protease

A. Data Acquisition

The crystallization of HIV-1 protease has been described elsewhere (McKeever *et al.*, 1989). Suitable crystals (tetragonal bipyramids) for structural analysis formed in hanging drop experiments at pH 7.0 in 100 mM imidazole buffer with 250 mM NaCl, 10 mM DTT, and 3 mM of NaN$_3$. The space group symmetry was P4$_1$2$_1$2 with $a = b = 50.29$ Å and $c = 106.80$ Å with one molecule of 99 amino acids in the asymmetric unit. Diffraction data were collected on a Rigaku RU-200 rotating anode X-ray generator using CuK$_\alpha$ radiation and a Nicolet X-ray area detector. The R$_{sym}$ of the native data set of 2099 unique reflections having an average sixfold redundancy to 3.0 Å resolution was 0.063. The two heavy atom derivatives used to solve the structure, phenyl mercury acetate and lead nitrate, were processed to preserve the anomalous scattering signal and therefore had an average threefold redundancy and R$_{sym}$ of 0.062 and 0.076, respectively. Initial phases were generated using standard multiple isomorphous replacement techniques. A contoured minimap of electron density was calculated and the chain traced through this map. The C$_\alpha$ guide points were used to build a complete model, which was initially adjusted using FRODO (Jones, 1985). This was followed by refinement of the coordinates using CORELS (Sussman, 1985) to an R-factor of 0.37, which is the basis for the present structural discussion. The programs XPLOR (Brunger *et al.*, 1987) and PROLSQ (Hendrickson, 1985) have been subsequently used to refine the model to its present R-factor of 0.25 with good geometry. No bound solvent was included in this refinement. Further rebuilding of the model and refinement are in progress. A more detailed description of the structure solution has been given elsewhere (Navia *et al.*, 1989).

B. Structural Description

1. Folding Pattern

Strong, continuous electron density was observed for all 99 amino acids of HIV-1 protease except for the 5 N-terminal residues. Since amino acid sequencing of washed crystals of the protease demonstrated the presence of these residues, we concluded that these five residues were not well ordered in the crystal lattice. The possible greater flexibility of this region may have significance in the mechanism of action of the enzyme (see below). A schematic description of the protease structure is given in Fig. 3. The active site of the protease molecule is composed of two monomer components related to each other by a 180° rotation about a crystallographic twofold axis. Two stereo views of the dimer are shown in Figs. 4a and 4b. There are a number of interactions holding the two monomers together: these involve residues Asp 25, Thr 26, and Gly 27 from one monomer with Gly 27, Thr 26, and Asp 25 of the other; Thr 91, Gln 92, and Ile 93 of one monomer with Ile 93, Gln 92, and Thr 91 of the other; Trp 6 of one monomer with Phe 99 of the other monomer; and Ile 50 of one monomer with Ile 50 of the other monomer.

The proposed binding cleft between the two monomers is long enough to accomodate the seven amino acids shown to be necessary for efficient hydrolysis of the gag p17–p24 cleavage site peptides. The cleft also ap-

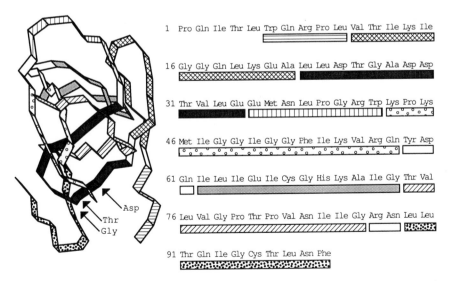

1 Pro Gln Ile Thr Leu Trp Gln Arg Pro Leu Val Thr Ile Lys Ile

16 Gly Gly Gln Leu Lys Glu Ala Leu Leu Asp Thr Gly Ala Asp Asp

31 Thr Val Leu Glu Glu Met Asn Leu Pro Gly Arg Trp Lys Pro Lys

46 Met Ile Gly Gly Ile Gly Gly Phe Ile Lys Val Arg Gln Tyr Asp

61 Gln Ile Leu Ile Glu Ile Cys Gly His Lys Ala Ile Gly Thr Val

76 Leu Val Gly Pro Thr Pro Val Asn Ile Ile Gly Arg Asn Leu Leu

91 Thr Gln Ile Gly Cys Thr Leu Asn Phe

Figure 3. A C$_\alpha$ tracing of HIV-1 protease monomer broken down into eight extended and bent structural elements differentiated by unique shading schemes.

328

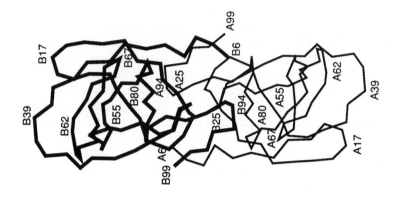

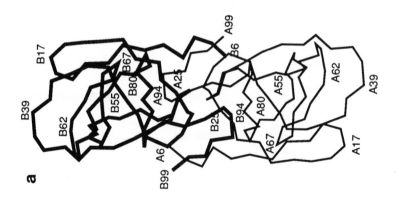

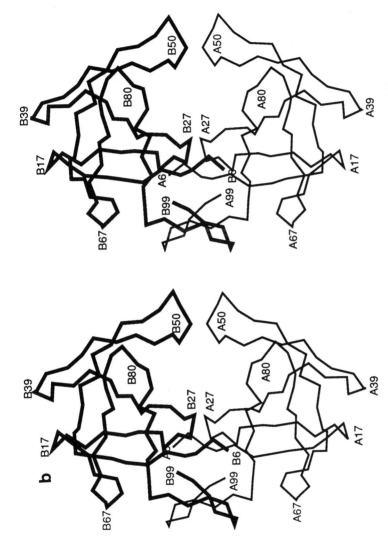

Figure 4. A C_α representation of the structure of residues 6–99 of the HIV-1 protease dimer, with one monomer (arbitrarily labeled A) drawn in thin lines and the second monomer (arbitrarily labeled B) drawn in thick lines. (a) The view is along the twofold symmetry axis of the protease dimer. (b) The view represents a rotation of 90° about the vertical axis of the view of (a). Reprinted by permission from *Nature* **337**, 615–620 (1989), copyright © 1989, Macmillan Magazines Ltd.

329

pears to have no charged amino acids other than Asp 25, consistent with the generally hydrophobic nature of cleavage sites.

2. Comparison to Other Aspartic Proteases

Toh and co-workers first reported the sequence homology between active sites of the pepsinlike aspartic proteases and portions of the retroviral proteases (Toh *et al.*, 1985). The pepsinlike aspartic proteases are composed of two pseudosymmetric domains, each of which contributes residues to the active site. Subsequently, a model proposing a completely symmetric structure for the retroviral proteases was presented (Pearl and Taylor, 1987b). The structure of HIV-1 protease was compared to the structure of the three aspartic proteases whose coordinates are available in the Protein Data Bank (Bernstein *et al.*, 1977): rhizopuspepsin (Suguna *et al.*, 1987a), penicillopepsin (James and Sielecki, 1983), and endothiapepsin (Pearl and Blundell, 1984). A total of 47 C_α of the HIV-1 protease monomer can be aligned with the N-terminal domains of the three microbial proteases with an rms tolerance of 1.29 Å or less while only 29 residues of the monomer can be aligned to the C-terminal domains with a tolerance of 1.45 Å or less. When the dimer is compared with the fungal proteases, the tolerance is somewhat larger at 1.71 Å or less, reflecting the fact that the HIV-1 protease is completely symmetric while the microbial proteases are pseudosymmetric; that is, the two domains in penicillopepsin have been described as being rotated by 174.6 Å and translated by 1.0 Å. While we conclude that residues in the active site of HIV-1 protease have structural analogs to the pepsinlike aspartic proteases, a significant difference at the dimer interface does exist. HIV-1 protease has a four-stranded β-sheet which differs in orientation and topology from the six-stranded β-sheet found in the other aspartic proteases.

C. Implications for the Viral Maturation Process

The HIV-1 protease is synthesized as part of a polyprotein gene product of the *gag* and *pol* genes (Jacks *et al.*, 1988), and the protease appears to release itself from this polyprotein autocatalytically, as judged from results of expression in a variety of systems (Farmerie *et al.*, 1987; DeBouck *et al.*, 1987; Giam and Boros, 1988). This implies that two of the polyproteins must dimerize to form an active protease, since the proposed mechanisms for the aspartic proteases require the participation of two active site aspartic acids (Suguna *et al.*, 1987b; James and Sielecki, 1985; Pearl, 1987). The structure presented above suggests a mechanism by which these autocatalytic events might take place. The position of Trp 6 is such that the five disordered N-terminal residues could easily fit into the active site. This and the implied flexibility of the five residues suggest that the first proteolytic event may be a peptide bond cleavage at the N terminus. Because of the geometry at the C termini, Phe 99 cannot be moved into the active site of the same protease dimer without unravelling a significant portion of the structure and destroying important contacts that hold the dimer together. Therefore, the hydrolysis at the C-termini must be intermolecular. During the process of viral assembly, the gag–pol polyprotein is believed to be anchored to the cytoplasmic side of the host cell membrane by a covalently linked N-terminal myristic acid (Mervis *et al.,* 1988). Making the N-terminal cuts first would effectively solubilize the protease and allow it to be released to make the remainder of the proteolytic cuts necessary for the formation of the other structural proteins and enzymes encoded by the *gag* and *gag–pol* genes. The necessity of first forming an active dimeric protease from two of the gag–pol polyproteins may serve as a regulatory mechanism in virus assembly by controlling the timing of the proteolytic release of the structural proteins and enzymes from the polyproteins.

D. Future Directions

The immediate active site of HIV-1 protease appears to be quite similar to the previously characterized microbial aspartic proteases. Therefore, insights derived from previous studies of these proteases as well as the comparison of the present structure to another retroviral protease, that of Rous sarcoma virus (Miller *et al.,* 1989), may help in understanding the details of the functioning of HIV-1 protease. However, understanding the ability of HIV-1 protease to cleave the peptide bond between Phe and Pro (see above), an unusual cleavage sequence, must wait further refinement of the structure. Also, details of how substrates and inhibitors bind to the enzyme and the conformational changes that take place upon binding must await structural studies of protease–ligand complexes. These studies are actively being pursued. It is hoped that insights derived from these studies may help to design HIV-1 protease inhibitors which could be used as effective agents in the control of AIDS.

Note added in proof: Higher resolution data have allowed the reinterpretation of the electron density. Positions for the 5 N-terminal residues have been determined and an α-helix has been found at residues 86 to 94. A similar conclusion has also been reached by Wlodawer, A. *et al.* (1989). *Science* **245,** 616–621.

Acknowledgments

We would like to thank Victor M. Garsky for the synthesis of substrate peptides and Ruth F. Nutt and Daniel F. Veber for supplying us with the chemically synthesized enzyme.

References

Bernstein, F. C., Koetzle, T. F., Williams, G. J. B., Meyer, E. F., Brice, M. D. Rodgers, J. R., Kennard, O., and Shimanouchi, T. (1977). The Protein Data Bank: A computer-based archival file for macrmolecular structures. *J. Mol. Biol.* **112,** 535–542.

Billich, S., Knoop, M.-T., Hansen, J., Strop, P., Sedlacek, J., Mertz, R., and Moelling, K. (1988). Synthetic peptides as substrates and inhibitors of human immune deficiency virus-1 protease. *J. Biol. Chem.* **263,** 17905–17908.

Brunger, A. T., Kuriyan, J., and Karplus, M. (1987). Crystallographic R factor refinement by molecular dynamics. *Science* **230,** 458–460.

Crawford, S., and Goff, S. P. (1985). A deletion mutation in the 5′ part of the *pol* gene of Moloney murine leukemia virus blocks proteolytic processing of the gag and pol polyproteins. *J. Virol.* **53,** 899–907.

Darke, P. L., Nutt, R. F., Brady, S. F., Garsky, V. M., Ciccarone, T. M., Leu, C.-T., Lumma, P. K., Freidinger, R. M., and Sigal, I. S. (1988). HIV-1 protease specificity of peptide cleavage is sufficient for processing of gag and pol polyproteins. *Biochem. Biophys. Res. Commun.* **156,** 297–303.

Darke, P. L., Leu, C.-T., Davies, L. J., Heimbach, J. C., Diehl, R. E., Hill, W. S., Dixon, R. A. F., and Sigal, I. S. (1989). Human immunodeficiency virus protease. Bacterial expression and characterization of the purified aspartic protease. *J. Biol. Chem.* **264,** 2307–2312.

DeBouck, C., Gorniak, J. G., Strickler, J. E., Meek, T. D., Metcalf, B. W., and Rosenberg, M. (1987). Human immunodeficiency virus protease expressed in *Escherichia coli* exhibits autoprocessing and specific maturation of the gag precursor. *Proc. Natl. Acad. Sci. USA* **84,** 8903–8906.

di Marzo Veronese, F., Copeland, T. D., DeVico, A. L., Rahman, R., Oroszlan, S., Gallo, R. C., and Sarngadharan, M. G. (1986). Characterization of highly immunogenic p66/p51 as the reverse transcriptase of HTLVIII/LAV. *Science* **231,** 1289–1291.

Farmerie, W. G., Loeb, D. D., Casavant, N. C., Hutchison, C. A., III, Edgell, M. H., and Swanstrom, R. (1987). Expression and processing of the AIDS virus reverse transcriptase in *Escherichia coli. Science* **236,** 305–308.

Giam, C.-Z. and Boros, I. (1988). *In vivo* and *in vitro* autoprocessing of human immunodeficiency virus protease expressed in *Escherichia coli. J. Biol. Chem.* **263,** 14617–14620.

Graves, M. C., Lim, J. J., Heimer, E. P., and Kramer, R. A. (1988). An 11-kDa form of human immunodeficiency virus protease expressed in *Escherichia coli* is sufficient for enzymatic activity. *Proc. Natl. Acad. Sci. USA* **85,** 2449–2453.

Hansen, J., Billich, S., Schulze, T., Sukrow, S., and Moelling, K. (1988). Partial purification and substrate analysis of bacterially expressed HIV protease by means of monoclonal antibody. *EMBO J.* **7,** 1785–1791.

Hendrickson, W. A. (1985). Stereochemically restrained refinement of macromolecular structures. *Methods Enzymol.* **115,** 252–270.

Jacks, T., Power, M. D., Masaiarz, F. R., Luciw, P. A., Barr, P. J., and Varmus, H. E. (1988). Characterization of ribosomal frameshifting in HIV-1 gag–pol expression. *Nature (London)* **331,** 280–283.

James, M. N. G., and Sielecki, A. R. (1983). Structure and refinement of penicillopepsin at 1.8 angstrom resolution. *J. Mol. Biol.* **163,** 299–361.

James, M. N. G., and Sielecki, A. R. (1985). Stereochemical analysis of peptide bond hydrolysis catalyzed by the aspartic proteinase penicillopepsin. *Biochemistry* **24,** 3701–3713.

Jones, T. A. (1985). Interactive computer graphics: FRODO. *Methods Enzymol.* **115,** 157–171.

Katoh, I., Yoshinaka, Y., Rein, A., Shibuya, M., Odaka, and Oroszlan, S. (1985). Murine leukemia virus maturation. Protease region required for conversion from immature to mature core form and for virus infectivity. *Virology* **145,** 280–292.

Katoh, I., Yasunaga, T., Ikawa, Y., and Yoshinaka, Y. (1987). Inhibition of retroviral protease activity by an aspartic proteinase inhibitor. *Nature (London)* **329,** 654–656.

Kohl, N. E., Emini, E. A., Schleif, W. A., Davis, L. J., Heimbach, J. C., Dixon, R. A. F., Scolnick, E. M., and Sigal, I. S. (1988). Active human immunodeficiency virus protease is required for viral infectivity. *Proc. Natl. Acad. Sci. USA* **85,** 4686–4690.

Kotler, M., Katz, R. A., Danho, W., Leis, J., and Skalka, A. M. (1988). Synthetic peptides as substrates and inhibitors of a retroviral protease. *Proc. Natl. Acad. Sci. USA* **85,** 4185–4189.

Kramer, R. A., Schaber, M. D., Skalka, A. M., Ganguly, K., Wong-Staal, F., and Reddy, E. P. (1986). HTLV-III gag protein is processed in yeast cells by the virus pol-protease. *Science* **231,** 1580–1584.

Krausslich, H. G., Schneider, H., Zybarth, G., Carter, C. A., and Wimmer, E. (1988). Processing of *in vitro*-synthesized gag precursor proteins of human immunodeficiency virus (HIV) type 1 by HIV proteinase generated in *Escherichia coli. J. Virol.* **62,** 4393–4397.

Krausslich, H. G., Ingraham, R. H., Skoog, M. T., Wimmer, E., Pallai, P. V., and Carter, C. A. (1989). Activity of purified biosynthetic proteinase of human immunodeficiency virus on natural substrates and synthetic peptides. *Proc. Natl. Acad. Sci. USA* **86,** 807–811.

McKeever, B. M., Navia, M. A., Fitzgerald, P. M. D., Springer, J. P., Leu, C.-T., Heimbach, J. C., Herber, W. K., Sigal, I. S., and Darke, P. L. (1989). Crystallization of the asprtyl protease from the human immunodeficiency virus, HIV-1. *J. Biol. Chem.* **264,** 1919–1921.

Mervis, R. J., Ahmad, N., Lillehoj, E. P., Raum, M. G., Salazar, F. H. R., Chan, H. W., and Venkatesan, S. (1988). The *gag* gene products of human immunodeficiency virus type 1: Alignment within the *gag* open reading frame, identification of posttranslational modifications, and evidence for alternative *gag* precursors. *J. Virol.* **62,** 3993–4002.

Miller, M., Jaskolski, M., Mohana Rao, J. K., Leis, J., and Wlodawer, A. (1989). Crystal structure of a retroviral protease proves relationship to aspartic protease family. *Nature (London)* **337,** 576–579.

Moelling, K., Scott, A., Dittmar, K. E. J., and Owada, M. (1980). Effect of p15 associated protease from an avian RNA tumor virus on avian virus-specific polyprotein precursors. *J. Virol.* **33,** 680–688.

Mous, J., Heimer, E. P., and Le Grice, S. F. J. (1988). Processing protease and reverse transcriptase from human immunodeficiency virus type I polyprotein in *Escherichia coli. J. Virol.* **62,** 1433–1436.

Navia, M. A., Fitzgerald, P. M. D., McKeever, B. M., Leu, C.-T., Heimbach, J. C., Herber, W. K., Sigal, I. S., Darke, P. L., and Springer, J. P. (1989). Three-dimensional structure of aspartyl protease from human immunodeficiency virus, HIV-1. *Nature (London)* **337,** 615–620.

Nutt, R. F., Brady, S. F., Darke, P. L., Ciccarone, T. M., Colton, C. D., Nutt, E. M., Rodkey, J. A., Bennett, C. D., Waxman, L. H., Sigal, I. S., Anderson, P. S., and Veber, D. F. (1988). Chemical synthesis and enzymatic activity of a 99-residue peptide with a sequence proposed for the human immunodeficiency virus protease. *Proc. Natl. Acad. Sci. USA* **85,** 7129–7133.

Pearl, L. H. (1987). The catalytic mechanism of aspartic proteinases. *FEBS Lett.* **214,** 8–12.

Pearl, L., and Blundell, T. (1984). The active site of aspartic proteases. *FEBS Lett.* **174,** 96–101.

Pearl, L. H., and Taylor, W. R. (1987a). Sequence specificity of retroviral proteases. *Nature (London)* **328,** 482.

Pearl, L. H., and Taylor, W. R. (1987b). A structural mdoel for the retroviral proteases. *Nature (London)* **329,** 351–354.

Schneider, J., and Kent, S. B. H. (1988). Enzymatic activity of a synthetic 99-residue protein corresponding to the putative HIV-1 protease. *Cell* **54,** 363–368.

Seelmeier, A., Schmidt, H., Turk, V., and von der Helm, K. (1988). Human immunodeficiency virus has a protease that can be inhibited by pepstatin A. *Proc. Natl. Acad. Sci. USA* **85,** 6612–6616.

Skeggs, L. T., Lentz, K. E., Kahn, J. R., and Hochstrasser, H. (1968). Kinetics of the reaction of renin with nine synthetic peptide substrates. *J. Exp. Med.* **128,** 13–34.

Suguna, K., Bott, R. R., Padlan, E. A., Subramanian, E., Sheriff, S., Cohen, G. H., and Davies, D. R. (1987a). Structure and refinement and 1.8 Å resolution of the aspartic proteinase from *Rhizopus chinensis*. *J. Mol. Biol.* **196,** 877–900.

Suguna, K., Padlan, E. A., Smith, C. W., Carlson, W. D., and Davies, D. R. (1987b). Binding of a reduced peptide inhibitor to the aspartic protease from *Rhizopus chinensis*. Implications for a mechanism of action. *Proc. Natl. Acad. Sci. USA* **84,** 7009–7013.

Sussman, J. L. (1985). Constrained-restrained least-squares (CORELS) refinement of proteines and nucleic acids. *Methods Enzymol.* **115,** 271–303.

Tang, J., James, M. N. G., Hsu, I. N., Jenkins, J. A., and Blundell, T. L. (1978). Structural evidence for gene duplication in the evolution of the acid proteases. *Nature (London)* **271,** 618–621.

Toh, H., Ono, M., Saigo, K., and Miyata, T. (1985). Retroviral protease-like sequence in the yeast transposon Ty 1. *Nature (London)* **315,** 691.

Vogt, V. M., Wight, A., and Eisenman, R. (1979). *In vitro* cleavage of avian retrovirus gag proteins by viral protease p15. *Virology* **98,** 154–167.

Yoshinaka, Y., Katoh, I., Copeland, T. D., Smythers, G. W., and Oroszlan, S. (1986). Bovine leukemia virus protease purification chemical analysis and *in-vitro* processing of gag precursor polyproteins. *J. Virol.* **57,** 826–832.

25 Oligomeric Structure of Retroviral Envelope Glycoproteins

David A. Einfeld and Eric Hunter
Department of Microbiology, University of Alabama at Birmingham,
Birmingham, Alabama 35294

I. Introduction

Retroviral envelope glycoproteins play an essential role in virus infectivity, binding specifically to receptors on target cells and initiating fusion of viral and cellular membranes. In the case of influenza virus the hemagglutinin molecule (HA) carries out both of these functions (reviewed in Wiley and Skehel, 1987). This glycoprotein assembles into a trimer which undergoes a conformational change at reduced pH, a condition mimicking that encountered by the virus within endosomes during the natural process of infection. This structural change exposes a hydrophobic sequence previously sequestered within the trimer and membrane fusion occurs. The degree to which this model applies to retroviral envelope glycoproteins is not clear, and to date information on their three-dimensional structure is not available. We have therefore initiated studies aimed at delineating the ternary organization of these retroviral surface components. Like HA, the Rous sarcoma virus (RSV) envelope glycoprotein in its mature form consists of two polypeptides cleaved from a single precursor. For both viruses the amino-terminal product which bears the receptor binding site is anchored to the transmembrane carboxyl-terminal fragment by disulfide bonds, and a stretch of amino acids near the cleavage-generated, extracellular amino terminus of the transmembrane component is thought to initiate fusion. Previous studies in this laboratory employing sedimentation analyses have shown that the RSV envelope glycoprotein is also an oligomer, most likely a trimer (Einfeld and Hunter, 1988). Whether this oligomeric structure undergoes a modification analogous to that of HA as a means of activating its fusogenic properties is not known.

In the present study chemical cross-linking has been applied to the RSV glycoprotein in order to confirm the nature of the oligomer by SDS-polyacrylamide gel electrophoresis. In addition the envelope glycoprotein of the Mason–Pfizer monkey virus (M-PMV), a type D retrovirus, has been analyzed using sucrose gradient analysis. While the M-PMV glyco-

Use of X-Ray Crystallography in the Design of Antiviral Agents

protein also forms an oligomer, its stability, in contrast to that of RSV, is markedly pH-sensitive.

II. Materials and Methods

Cross-linked phosphorylase-b molecular weight markers were obtained from Sigma. The cross-linking agents *bis*[2-(succinimidooxycarbonyloxy)ethyl]sulfone (BSCOES) and ethylene glycol*bis*(succinimidylsuccinate) (EGS) were purchased from Pierce, Rockford, Illinois.

The RSV envelope protein was expressed from an SV40 vector as described by Wills *et al.* (1984). Cells on 35-mm plates were pulse-labeled for 15 min with 200 μCi [^{3}H]leucine (60 Ci/mmol) (Amersham, Arlington Heights, Illinois). At 72 hr postinfection cells were lysed in 0.5 ml volume of 50 mM HEPES, pH 7.8, 150 mM NaCl, and 40 mM octyl glucoside. Solutions of the cross-linking agents EGS and BSCOES were prepared in DMSO and 1 μl added to 100-μl aliquots of lysate to yield the final concentrations indicated below. After 30 min at room temperature 25 μl of 0.2 M glycine, pH 8.0, was added. Immune complexes were prepared as described (Einfeld and Hunter, 1988) and analyzed on 5% acrylamide gels with 4% SDS included in both the gel and the electrophoresis buffer.

An M-PMV-expressing HeLa cell line produced by transfection with an infectious clone of the virus (Rhee and Hunter, 1987) provided a source of the M-PMV envelope glycoprotein for sucrose gradient analysis. Cells on 35-mm plates were pulse-labeled for 15 min with 200 μCi [^{3}H]leucine and lysed immediately or following a 3-hr chase in complete medium. Lysis buffer consisted of 150 mM NaCl, 40 mM octyl glucoside, and either 50 mM Tris, pH 7.5, or 50 mM citrate, pH 6.0. Sucrose gradients were prepared, centrifuged, and fractionated as described (Einfeld and Hunter, 1988) except that 50 mM citrate, pH 6.0, was substituted for Tris to analyze lysates prepared at pH 6.0. Immunoprecipitates were isolated using a rabbit anti-gp70-specific antiserum or an anti-gp20 monoclonal antibody together with staph A and analyzed by SDS-PAGE.

III. Results

A. Cross-linking of RSV Envelope Glycoprotein

We have previously characterized the size of the RSV envelope glycoprotein oligomer by comparing its sedimentation with those of oligomeric proteins of known size on parallel sucrose gradients. The results of that comparison suggested that the RSV envelope protein forms a trimer. We wished to confirm the multiplicity of the RSV oligomer using chemical cross-linking as an independent approach to this question. Pulse-labeled lysates from cells expressing the RSV glycoprotein were cross-linked with increasing concentrations of the cross-linkers EGS and BSCOES (Fig. 1). At the higher concentrations both agents gave rise to a specific band of ~290 kDa. A less intense band of ~195 kDa is also detected at lower concentrations of both cross-linking agents. The appearance of these higher molecular weight bands coincides with disappearance of the monomeric form of the 95-kDa precursor (pr95) and their size correlates with that expected for dimers and trimers of pr95.

There are two notable features of these results. First, the trimer appears to be the fully cross-linked species. The dimer apparently is an intermediate product, disappearing under conditions of more extensive cross-linking. By contrast the trimer accumulates rather than giving way to a

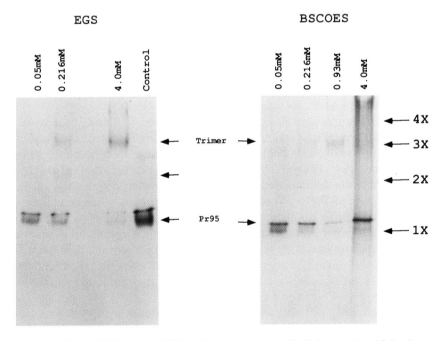

Figure 1. Cross-linking of the RSV envelope precursor, pr95. Cells were lysed following a 15-min pulse labeling and exposed to varying concentrations of cross-linkers. The control sample received DMSO without cross-linking agent. The center arrows indicate immunoprecipitated material migrating at the expected size of monomers, dimers, and trimers of pr95. The arrows on the right show the migration in a parallel lane of cross-linked phosphorylase b complexes having the indicated number of subunits.

higher molecular species. The ~470-kDa band arising after treatment with 4 mM EGS is not observed in BSCOES-treated lysates. Second, it appears that the bulk of pr95 can be cross-linked in these lysates.

B. Oligomeric Structure of M-PMV Envelope Glycoprotein

Since retroviral envelope glycoproteins exhibit common structural features such as cleavage from a precursor polyprotein and presence of an extracellular fusogenic sequence, it might be expected that the oligomeric structure seen in RSV is a general feature of these proteins. The RSV envelope glycoprotein assembles intracellularly into an oligomeric structure that can be detected on sucrose gradients buffered at pH 7.5. Processing of the M-PMV envelope glycoprotein was analyzed under the same conditions (Fig. 2). The 86-kDa precursor protein (pr86) detected in lysates of pulse-labeled cells exhibits a heterogeneous sedimentation on sucrose gradients. After a 3-hr chase the mature 70-kDa SU protein can also be detected and is primarily restricted to fractions 14 and 15. In this same gradient pr86, although also detected lower in the gradient, is present primarily in fractions 13 and 14. This difference in sedimentation behavior between gp70 and the majority of pr86 is consistent with the difference in their molecular masses. The heterogeneity in the sedimentation of pr86 could result from formation of an oligomer that is unstable during lysis or subsequent gradient centrifugation.

Previous studies on vesicular stomatitis virus (VSV) G protein had shown that the pH to which the protein complex is exposed during the fractionation procedure could affect the stability of its quaternary structure (Doms *et al.*, 1987). We have therefore investigated whether lowering the pH of both the lysis buffer and gradient buffer affected the results obtained with M-PMV. When lysis and centrifugation were carried out at pH 6.0 there was a dramatic change in the sedimentation of gp70 (Fig. 3). Rather than being present in fractions 14 and 15, virtually all of gp70 is detected in fractions 8–10. In addition, the transmembrane (TM) protein gp20 is also observed in fractions 8–10. Thus the reduction in pH appears to stabilize an oligomeric complex containing multiple copies of gp70–gp20.

Since an anti-gp70 antiserum that does not react with gp20 was employed in these immunoprecipitations, the presence of both gp70 and gp20 in fractions 8–10 indicates retention of the noncovalent association that is known to exist between gp20 and gp70. In contrast gp20 did not coprecipitate with gp70 from the pH 7.5-buffered gradients of chase products. Immunoprecipitations of fractions from gradients buffered at pH 6.0 with an anti-gp20 monoclonal antibody confirmed the identification of the low molecular weight band in fractions 8–10 as gp20 (data not shown). This

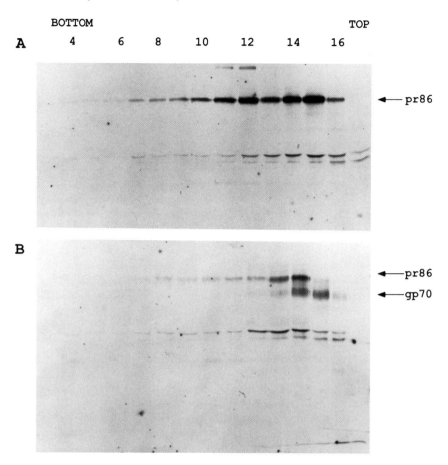

Figure 2. Sedimentation of M-PMV glycoprotein on pH 7.5-buffered gradients. Cells expressing M-PMV were lysed at pH 7.5 following a 15-min pulse labeling (A) or a 15-min pulse and 3-hr chase (B). Gradient fraction numbers are indicated at the top. An anti-gp70 rabbit antiserum was used for immunoprecipitations.

antibody also detected gp20 in fractions 14 and 15 of pH 7.5 gradients, while the anti-gp70 antiserum precipitated no gp20 from these fractions. This indicates that under neutral pH conditions dissociation of gp70 from gp20 occurs either during lysis or during centrifugation. The fact that gp70 and gp20 are present in the same fractions on the pH 7.5 gradients but cannot be coprecipitated with either anti-gp70 or anti-gp20 antibodies raises the possibility that the gp20 detected in these fractions is itself an oligomer while gp70 is monomeric. This would explain why gp70 and gp20 sediment more slowly (i.e., act as smaller molecules) than pr86, a result

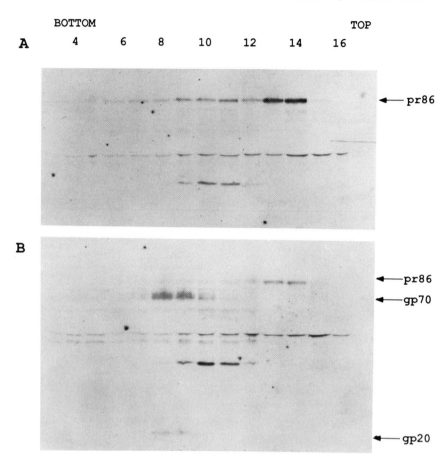

Figure 3. Sedimentation of M-PMV glycoprotein on pH 6.0-buffered gradients. After a 15-min pulse (A) or 15-min pulse and 3-hr chase (B) cells were lysed at pH 6.0. Gradient fractions were immunoprecipitated as for Fig. 2.

which would not be expected if gp70 were associated with gp20 in the neutral pH gradients.

IV. Discussion

Cross-linking of the RSV envelope glycoprotein indicates that the oligomer formed by this protein is a trimer. This is consistent with our estimate of the size of this complex based on comparison of its sedimentation with those of oligomeric proteins of known size. Interestingly, the majority of precursor molecules (pr95) from 15-min pulse-labelings could be cross-linked. The kinetics of oligomerization detected by sucrose gradient analysis suggested that this process is relatively slow ($t_{1/2}$ is approximately 80 min). The latter finding probably reflects the time required for acquisition of a ternary structure resistant to dissociation by detergent and gradient centrifugation. Cross-linking, however, may reveal an association of pr95 molecules at a stage that is not resistant to gradient analysis. Indeed, we have shown previously that virtually no oligomeric pr95 can be detected in sucrose gradients of lysates from cells pulse-labeled for 5 min, but if these cells are incubated on ice for 60 min prior to lysis a significant amount of the oligomer is found, suggesting oligomerization prior to stable trimer formation (Einfeld and Hunter, 1988). The cross-linking results are consistent with this interpretation.

The results reported here reveal that the M-PMV envelope glycoprotein forms a pH-sensitive oligomeric complex. When analyzed at pH 6.0 the gp70 and gp20 from lysates of M-PMV expressing cells cosediment to fractions lower in the gradient than those containing monomeric gp70. The magnitude of this difference suggests that the oligomer of gp70–gp20 is a trimer. The exact nature of the influence of pH on the structure of the oligomer is not clear, but it appears that lowering the pH to 6.0 stabilizes the association between gp70 and gp20 such that both cosediment in sucrose gradients and can be immunoprecipitated by anti-gp70 antiserum alone. Dissociation of gp70 from gp20, which involves a noncovalent association, in lysates analyzed at pH 7.5 presumably accounts for the sedimentation of gp70 as a monomer. Since under these conditions gp20 sediments with the same apparent mass as gp70, it may retain its oligomeric structure in an analogous fashion to that which we have observed with the RSV envelope glycoprotein. In that case reduction of the disulfide linkage between gp85 and gp37 caused gp85 to sediment as a monomer while gp37 sedimented as an oligomer. These observations are summarized in Fig. 4.

Earlier observations of oligomeric complexes of retroviral envelope glycoproteins following chemical cross-linking pointed to the presence of trimeric or higher order complexes in virions (Takemoto *et al.*, 1978; Pinter and Fleissner, 1979; Racevskis and Sarkar, 1980). Sucrose gradient analysis of the RSV glycoprotein indicated that oligomerization occurs intracellularly at the stage of the precursor protein and may be required for transport out of the endoplasmic reticulum (Einfeld and Hunter, 1988). For most retroviruses the site of membrane fusion during virus entry is not known. In the case of HIV this process appears to be pH-independent and does not require endocytosis of the receptor, CD4 (Stein *et al.*, 1987; McClure *et al.*, 1988; Maddon *et al.*, 1988). Thus fusion probably occurs at

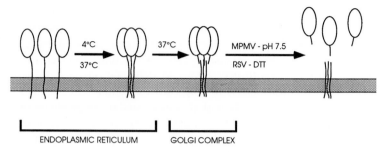

Figure 4. Schematic representation of stages in the maturation of the envelope glyco-proteins of RSV and M-PMV. The precursor forms an oligomer, apparently within the ER since this association can also occur at 4°C and thus does not require transport. At 37°C the oligomer reaches the Golgi, where the protein backbone is cleaved and the oligosaccharide chains modified to produce the mature glycoprotein. Disruption of the disulfide bond linking the extracellular (SU) and transmembrane (TM) proteins of the RSV with dithiothrietol (DTT) results in the appearance of monomeric SU and oligomeric TM on sucrose gradients. An analogous dissociation occurs between the M-PMV SU and TM proteins at pH 7.5.

the plasma membrane rather than the endosomal site used by influenza virus. The extent to which this is true of other retroviruses is unclear, since HIV has the unusual ability to cause syncytia formation, the cell-to-cell fusion mediated by interaction of gp120 and CD4 which is thought to mimic the virus-to-cell fusion that occurs during virus entry. In the case of influenza virus HA, it is the lower pH of the endosomal compartment which triggers an opening up of the trimeric structure to expose a fusogenic domain. If retroviral glycoproteins have also adopted an oligomeric con-figuration to sequester a fusion sequence, the trigger which elicits expres-sion of the fusion activity has yet to be defined.

References

Doms, R. W., Keller, D. S., Helenius, A., and Balch, W. E. (1987). Role for adenosine triphosphate in regulating the assembly and transport of vesicular stomatitis virus G protein trimers. *J. Cell Biol.* **105,** 1957–1969.
Einfeld, D., and Hunter, E. (1988). Oligomeric structure of a prototype retrovirus glyco-protein. *Proc. Natl. Acad. Sci. USA* **85,** 8688–8692.
McClure, M. O., Marsh, M., and Weiss, R. A. (1988). Human immunodeficiency virus infection of CD4-bearing cells occurs by a pH-independent mechanism. *EMBO J.* **7,** 513–518.
Maddon, P. J., McDougal, J. S., Clapham, P. R., Dalgleish, A. G., Jamal, S., Weiss, R. A., and Axel, R. (1988). HIV infection does not require endocytosis of its receptor, CD4. *Cell* **54,** 865–874.
Pinter, A., and Fleissner, E. (1979). Characterization of oligomeric complexes of murine and feline leukemia virus envelope and core components formed upon crosslinking. *J. Virol.* **30,** 157–165.

Racevskis, J., and Sarkar, N. H. (1980). Murine mammary tumor virus structural protein interactions: Formation of oligomeric complexes with cleavable cross-linking agents. *J. Virol.* **35,** 937–948.

Rhee, S. S., and Hunter, E. (1987). Myristylation is required for intracellular transport but not for assembly of D-type retrovirus capsids. *J. Virol.* **61,** 1045–1053.

Stein, B. S., Gowda, S. D., Lifson, J. D., Penhallow, R. C., Bensch, K. G., and Engleman, E. G. (1987). pH-independent HIV entry into CD4-positive cells via virus envelope fusion to the plasma membrane. *Cell* **49,** 659–668.

Takemoto, L. J., Fox, C. F., Jensen, F. C., Elder, J. H., and Lerner, R. A. (1978). Nearest-neighbor interactions of the major RNA tumor virus glycoprotein on murine cell surfaces. *Proc. Natl. Acad. Sci USA* **75,** 3644–3648.

Wiley, D. C., and Skehel, J. J. (1987). The structure and function of the hemagglutinin membrane glycoprotein of influenza virus. *Annu. Rev. Biochem.* **56,** 365–394.

Wills, J. W., Srinivas, R. V., and Hunter, E. (1984). Mutations of the Rous sarcoma virus *env* gene that affect the transport and subcellular location of the glycoprotein products. *J. Cell Biol.* **99,** 2011–2023.

26 Tumor Necrosis Factor: A Nonviral Jelly Roll

E. Y. Jones and D. I. Stuart
Laboratory of Molecular Biophysics, Oxford OX1 3QU United Kingdom

N. P. C. Walker
Hauptlaboratorium, BASF Aktiengesellschaft, 6700 Ludwigshafen, Federal Republic of Germany

I. Introduction

Tumor necrosis factor (TNF) is one of the cytokine family of polypeptide mediators, a group which includes the interferons and interleukins. It was first characterized and named on the basis of its *in vivo* anti-tumor properties and *in vitro* cytotoxicity to certain transformed cell lines (Carswell *et al.*, 1975). However, the availability of pure recombinant human TNF has enabled numerous additional activities to be identified; these indicate that TNF plays a central role in inflammation and the cellular immune response (Old, 1987). To date the broad spectrum of physiological effects associated with TNF includes fever, septic shock, cachexia (wasting), rheumatoid arthritis, inflammatory tissue destruction, and antiviral and antimalarial activity (Beutler and Cerami, 1988). Thus TNF is of great pharmacological interest. Currently it is under investigation both as an anticancer agent (phase II clinical trials) and conversely as a target in the control of severe septic shock.

The gene for TNF lies within the major histocompatibility complex (Muller *et al.*, 1987). Human TNF is synthesized as a 233-amino acid precursor and secreted by macrophages as an unglycosylated polypeptide of 157 amino acids (17 kDa). Primary structure is highly conserved between species, and there is over 30% sequence identity with the related cytokine lymphotoxin (TNF β) (Pennica *et al.*, 1984). The active form of the protein is thought to be trimeric (Smith and Baglioni, 1987); it binds to a high-affinity cell surface receptor, a glycoprotein of about 75kDa (Aggarwal *et al.*, 1985; Kull *et al.*, 1985), and is then probably internalized via the endosome pathway (Fiers *et al.*, 1986). After signal transduction two independent mechanisms may be activated; one is involved in cytotoxicity, the other can profoundly affect gene regulation.

We have determined the structure of the TNF trimer at 2.9 Å resolution

Use of X-Ray Crystallography in the Design of Antiviral Agents

by X-ray crystallography (Jones *et al.*, 1989). The main chain fold for a single TNF subunit conforms to the "jelly roll" structural motif of viral capsid proteins, making it unique among the currently known nonviral proteins. We discuss here common and contrasting features of this motif as exhibited in TNF and viruses, in particular rhinovirus (HRV14) (Rossmann *et al.*, 1985); foot-and-mouth disease virus (FMDV) (Acharya *et al.*, 1989); influenza hemagglutinin (HA) (Wilson *et al.*, 1981); and adenovirus hexon (Roberts *et al.*, 1986).

II. Structure Determination

Recombinant human TNF (17 kDa) crystals of space group $P3_121$ (unit cell dimensions $a = b = 166$ Å, $c = 93$ Å) with two trimers per asymmetric unit diffract to 2.9 Å resolution. Data was collected on film at the SERC Daresbury synchrotron and at Oxford on a Xentronics multiwire area detector. A variety of self-rotation function calculations failed to locate the noncrystallographic threefold axes of the two trimers. Thus, with six molecules of unknown orientation within the asymmetric unit, the analysis of difference Patterson maps for putative heavy atom derivatives was very difficult. The heavy atom derivatives were solved by the use of GROPAT, a suite of programs specifically written for this project (Jones and Stuart, 1989), but no useful heavy atom phase information was available beyond 4 Å. Very gradual phase extension to 2.9 Å was achieved by solvent flattening (Wang, 1985) and threefold averaging (Bricogne, 1976) performed separately about the two trimers. The resultant electron density map was of high quality; apart from two regions of high mobility there were no main chain breaks, side chain density was well defined, and a high percentage of main chain carbonyl bulges were visible. An unambiguous chain trace was rapidly established and confirmed by fitting the amino acid sequence. The model comprised 153 residues, complete except for the first 4 N-terminal amino acids, which lacked any clear electron density in the map. The structure has now been refined using X-PLOR (Brunger *et al.*, 1987) on a CONVEX C210. The current R-factor is 22.2% (on all data from 6 to 2.9 Å) after restrained B-factor refinement with excellent stereochemistry (rms deviation from ideal bond lengths of 0.013 Å) and no model for the solvent structure.

III. Structure

A. The Main Chain Fold

The TNF monomer forms an antiparallel β-sandwich with the topology of the classic jelly roll motif (Fig. 1). The general shape is typical of the viral

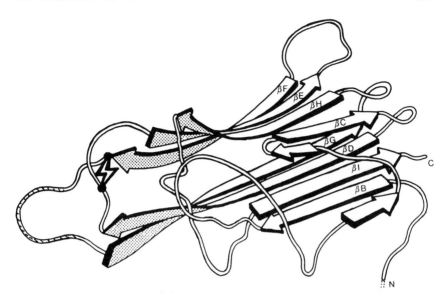

Figure 1. Diagrammatic sketch of the subunit fold: β Strands are shown as thick arrows in the amino→carboxy direction and labeled according to the viral convention. The disulfide bridge is denoted by a lightning flash and a region of high flexibility is cross-hatched. The TNF trimer threefold axis would be horizontal in this orientation.

proteins: wedgelike with length about 50 Å and maximum width of 30 Å. The standard eight β-strands are all present but with an insertion between βB and βC which adds a short β-strand onto the edge of both β-sheets and truncates the N-terminal half of βC. A similar addition of a pair of β-strands to the β-barrel occurs in the comovirus bean pod mottle virus (Chen *et al.*, 1989), although in this case the insertion is between βC and βD. For the viral proteins the β-strands of the [F E H C] sheet tend to be somewhat shorter than those of the [G D I B] sheet (Roberts and Burnett, 1987); this is also a feature of the TNF β-barrel. The N-terminal region prior to βB is highly flexible in TNF, again a property often observed of its longer counterparts in the viral capsid proteins.

With a polypeptide chain of 157 amino acids, TNF is at the short end of the range in sequence length covered by proteins incorporating the jelly roll motif. Plant viruses such as satellite tobacco necrosis virus (STNV) Liljas *et al.* (1982), with 195 residues, provide the other examples of relatively short polypeptides; the lengths increase for the picornaviruses from 220 to 238 for VP3 to between 213 and 306 for VP1. It has been suggested that the elaborate loops which embellish the standard jelly roll motif in the picornaviruses have arisen out of their need to evade the surveillance of the host immune system, a pressure which is absent in plant viruses (Luo *et al.*, 1987). This argument may be extended to TNF, which is itself part of the immune system and thus has no need to evade its surveillance.

The structural similarity between TNF and the viral capsid proteins may be quantified using SHP (Stuart *et al.,* 1979), a program which allows automatic definition of the "equivalent" residues based on a superposition of C_α positions. Typically, over 60% of TNF residues may be matched with rms deviations of less than 3.5 Å (Jones *et al.,* 1989), a level of similarity comparable to that obtained between animal and plant viruses or FMDV and HA (Acharya *et al.,* 1989). The closest match occurs between TNF and STNV; 71% of TNF residues are structurally equivalent to residues in STNV with an rms deviation of 3.3 Å.

B. Amino Acid Distribution

The core of the TNF β-sandwich has the expected filling of tightly intercalating, large apolar residues. Superposition of TNF onto VP1 of HRV14 reveals no obvious equivalent of the WIN binding pocket (Smith *et al.,* 1986), although the occurrence of structural rearrangement sufficient to accommodate a short WIN compound cannot be ruled out in the absence of experimental data. In general the level of sequence identity between TNF and the viral proteins is very low (less than 10% based on structural alignment), again a situation analogous to that observed between animal and plant viruses. Some examples of possible sequence conservation will be discussed in Section III,D. The disulfide bridge (residues 69 and 101) is not a feature of viral jelly rolls and is not necessary for biological activity in TNF (Narachi *et al.,* 1987).

C. Oligomeric Packing

The jelly roll motif provides a structural building block which may be used in a variety of ways. In the small plant virus and picornavirus capsids the jelly roll subunits pack side by side with their β-strands lying tangential to the capsid surface. The stabilizing interactions are mainly provided by the extended N-termini. For all these capsids β-sheet [F E H C] forms the solvent accessible exterior surface while β-sheet [G D I B] faces the RNA interior. In contrast, TNF monomers associate tightly to form a trimer with the strands of the β-sandwich running essentially parallel to the threefold axis. The resulting interactions of the β-sandwiches are edge to face; the edge of one monomer, consisting of strands F and G, lies across sheet [G D I B] of a threefold related subunit (Fig. 2). The stabilizing forces are predominantly hydrophobic. However, although the subunit interactions used for capsid and trimer formation are very different it is again the [F E H C] sheet which is solvent-accessible.

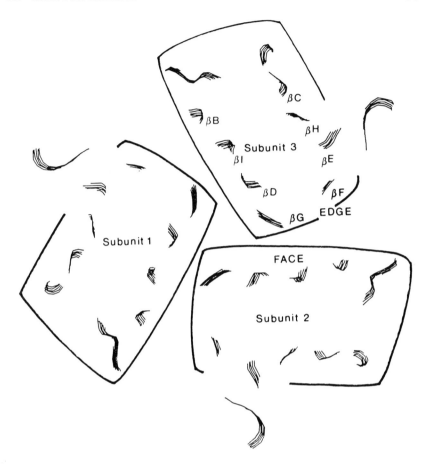

Figure 2. Edge-to-face packing of β-sandwiches in the TNF trimer. The view, down the threefold axis, shows a narrow slab of the trimer with β-strands represented by ribbons running into and out of the page.

HA is trimeric; however, a major contribution to oligomeric stability is provided by the coiled-coil core of the fibrous domain. The jelly rolls in the globular heads are oriented with their β-strands perpendicular to the threefold axis, the exact opposite of the arrangement in TNF. Finally, in the adenovirus hexon the β-strands of the jelly rolls (P1 and P2) are oriented normal to the capsid surface. Thus the threefold clusters of P1 or P2 which occur about each corner of the hexon on formation of the capsid would appear to have a general mode of packing common to that of the TNF trimer.

D. Receptor Binding

In Color Plate 29 the positions of regions implicated in the receptor binding of the various viral proteins are mapped onto the TNF jelly roll. The very different characteristics of each of these sites on the original protein are created by the myriad possible variations of the inter-β-strand loops on the jelly roll. In the picornavirus HRV14 the hypothesis of a receptor binding site located in the canyon on VP1 (Rossmann *et al.*, 1985) and hence inaccessible to antibody is supported by the site-directed mutagenesis results of Colonno *et al.* (1988). Strands H and E plus the so-called FMDV loop between strands G and H contribute to the base of the canyon. In FMDV a tripeptide Arg-Gly-Asp (residues 145–147), a recognition sequence for integrin binding, on the highly flexible FMDV loop in VP1 is implicated in receptor binding (Acharya *et al.*, 1989). Outside the picornavirus family the area of interest shifts to the sialic acid binding site of HA (Weis *et al.*, 1988).

The amino acids which comprise the HA sialic acid binding site are virtually all nonequivalenced in structural comparisons with the TNF jelly roll. The site is built from large interstrand loops and is effectively a structural feature unique to HA.

In general, TNF represents a rather austere form of the jelly roll, as it lacks the canyon walls erected by the more elaborate loops of HRV14. However, residues 155, 220, and 223 of HRV14 VP1 at the base of the canyon, which were highlighted by Colonno *et al.* (1988), are equivalent to residues on the TNF β-barrel. For one of these, a Ser on βH (residue 223 in HRV14 and 133 in TNF), both sequence and structural identity are conserved. However, it is on the FMDV loop that the most surprising juxtapositions of equivalenced residues occur. Here in TNF and TNF β (which are thought to share a common receptor) the dipeptide Gly-Asp is strictly conserved at residues 129–130. Residue 128 varies between Lys, Gln, and, in bovine TNF β, Arg. It is noteworthy that these residues should occur on the same (though in TNF rigid) loop as the Arg-Gly-Asp tripeptide implicated in FMDV receptor binding. These same Gly-Asp residues are structurally matched against Gly-240 and Asp-241 in HA although for HA they are not concerned in receptor binding. Finally, in HRV14 VP1 His 220 on the canyon floor is structurally equivalent to Asp 130 in TNF while Pro 155 is matched to residue Ile 83 in TNF. This region, comprising the FMDV loop, the end of strand E, and the loop to strand F, may be considered a putative site for receptor binding in TNF on the basis of preliminary evidence from neutralizing antibodies (Yone *et al.*, 1989). It is therefore very surprising to find that the same subset of residues may have been incorporated into apparently very different receptor binding sites.

IV. Conclusions

TNF, a polypeptide mediator of the cellular immune system, has the structural motif of a viral jelly roll. In TNF three such subunits tightly associate to form the biologically active trimer in a form of packing echoed in adenovirus. There is obviously much of interest to be learnt about receptor binding both of viruses and of host mediators such as TNF. It is certainly a strange twist in the tale to note that the jelly roll motif in rhinovirus binds to ICAM-1 (S. Marlin, presented at this meeting), expression of which is controled by TNF (Pober, 1987), a jelly roll mediator of the host immune system. Further studies of these systems may highlight common structural and/or functional principles of use in the pharmacological applications of TNF, itself possibly a natural antiviral agent (Mestan *et al.,* 1986; Wong and Goeddel, 1986), and in the search for anti-viral agents directed against the viral jelly roll structures.

References

Acharya, R., Fry, E., Stuart, D. I., Fox, G., Rowlands, D., and Brown, F. (1989). The three-dimensional structure of foot-and-mouth disease virus at 2.9 Å resolution. *Nature (London)* **337,** 709–716.

Aggarwal, B. B., Eessalu, T. E., and Hass, P. E. (1985). Characterization of receptors for human tumour necrosis factor and their regulation by gamma-interferon. *Nature (London)* **318,** 665–667.

Beutler, B., and Cerami, A. (1988). Tumor necrosis, cachexia, shock, and inflammation: A common mediator. *Annu. Rev. Biochem.* **57,** 505–518.

Bricogne, G. (1976). Methods and programs for direct-space exploitation of geometric redundancies. *Acta Crystallogra., Sect. A* **A32,** 832–847.

Brunger, A. T., Kuriyan, J., and Karplus, M. (1987). Crystallographic R-factor refinement by molecular dynamics. *Science* **235,** 458–460.

Carswell, E., Old, L. J., Kassel, R. L., Green, S., Fiore, N., and Williamson, B. (1975). An endotoxin induced serum factor that causes necrosis of tumors. *Proc. Natl. Acad. Sci. USA* **72,** 3666–3670.

Chen, Z., Stuaffacher, C., Li, Y., Schmidt, T., Bomu, W., Kamer, G., Shanks, M., Lomonossoff, G., and Johnson, J. (1989). Protein–RNA interactions in an icosahedral virus at 3.0 Å resolution. *Science* **245,** 154–159.

Colonno, R. J., Condra, J. H., Mizutani, S., Callahan, P. L., Davies, M.-E., and Murcko, M. (1988). Evidence for the direct involvement of the rhinovirus canyon in receptor binding. *Proc. Natl. Acad. Sci. USA* **85,** 5449–5453.

Fiers, W., Brouckaert, P., Devos, R., Fransen, L, Leroux-Roels, G., Remaut, E., Suffys, P., Tavernier, J., Van der Heyden, J., and Van Roy, F. (1986). Lymphokines and monokines in anti-cancer therapy. *Cold Spring Harbor Symp. Quant. Biol.* **51,** 587–595.

Jones, E. Y., and Stuart, D. I. (1989). In preparation.

Jones, E. Y., Stuart, D. I., and Walker, N. P. C. (1989). Structure of tumour necrosis factor. *Nature (London)* **338,** 225–228.

Kull, F. C., Jr., Jacobs, S., and Cuatrecasas, P. (1985). Cellular receptor for ^{125}I-labeled tumor necrosis factor; specific binding, affinity labeling and relationship to sensitivity. *Proc. Natl. Acad. Sci. USA* **82,** 5756–5760.

Liljas, L., Unge, T., Jones, T. A., Fridborg, K., Lovgren, S., Skoglund, U., and Strandberg, B. (1982). Structure of satellite tobacco necrosis virus at 3.0 Å resolution. *J. Mol. Biol.* **159**, 93–108.

Luo, M., Vriend, G., Kamer, G., Minor, I., Arnold, E., Rossmann, M. G., Boege, U., Scraba, D. G., Duke, G. M., and Palmenberg, A. C. (1987). The atomic structure of Mengovirus at 3.0 Å resolution. *Science* **235**, 182–191.

Mestan, J., Digel, W., Mittnacht, S., Hillen, H., Blohm, D., Moller, A., Jacobsen, H., and Kirchner, H. (1986). Antiviral effects of recombinant tumour necrosis factor *in vitro*. *Nature (London)* **323**, 816–819.

Muller, U., Jongeneel, C. V., Nedospasov, S. A., Lindahl, K. F., and Steinmetz, M. (1987). Tumour necrosis factor and lymphotoxin genes map close to H-2D in the mouse major histocompatibility complex. *Nature (London)* **325**, 265–267.

Narachi, M. A., Davis, J. M., Hsu, Y.-R., and Arakawa, T. (1987). Role of single disulfide in recombinant human tumor necrosis factor-α. *J. Biol. Chem.* **262**, 13107–13110.

Old, L. J. (1987). Polypeptide mediator network. *Nature (London)* **326**, 330–331.

Pennica, D., Nedwin, G. E., Hayflick, J. S., Seeburg, P. H., Derynck, R., Palladino, M. A., Kohr, W. J., Aggarwal, B. B., and Goeddel, D. V. (1984). Human tumour necrosis factor: Precursor structure, expression and homology to lymphotoxin. *Nature (London)* **312**, 724–729.

Pober, J. S. (1987). *Tumour Necrosis Factor Relat. Cytotoxins, Ciba Found. Symp.* No. 131, pp. 170–184.

Roberts, M. M., and Burnett, R. M. (1987). Adenovirus hexon: a novel use of the viral beta-barrel. *In* "Biological Organization: Macrmolecular Interactions at High Resolution," pp. 113–124. Academic Press, Orlando, Florida.

Roberts, M. M., White, J. L., Grutter, M. G., and Burnett, R. M. (1986). Three-dimensional structure of the adenovirus major coat protein hexon. *Science* **232**, 1148–1151.

Rossmann, M. G., Arnold, E., Erickson, J. W., Frankenberger, E. A., Griffith, J. P., Hecht, H.-J., Johnson, J. E., Kamer, G., Luo, M., Mosser, A. G., Rueckert, R. R., Sherry, B., and Vriend, G. (1985). Structure of a human common cold virus and functional relationship to other picornaviruses. *Nature (London)* **317**, 145–153.

Smith, R. A., and Baglioni, C. (1987). The active form of tumor necrosis factor is a trimer. *J. Biol. Chem.* **262**, 6951–6954.

Smith, T. J., Kremer, M. J., Luo, M., Vriend, G., Arnold, E., Kamer, G., Rossmann, M. G., McKinlay, M. A., Diana, G. D., and Otto, M. J. (1986). The site of attachment in human rhinovirus 14 for antiviral agents that inhibit uncoating. *Science* **233**, 1286–1293.

Stuart, D. I., Levine, M., Muirhead, H., and Stammers, D. K. (1979). Crystal structure of cat muscle pyruvate kinase at a resolution of 2.6 Å *J. Mol. Biol.* **134**, 109–142.

Wang, B. C. (1985). Resolution of phase ambiguity in macromolecular crystallography. *Methods Enzymol.* **115**, 90–111.

Weis, W., Brown, J. H., Cusack, S., Paulson, J. C., Skehel, J. J., and Wiley, D. C. (1988). Structure of the influenza virus hemagglutinin complexed with its receptor, sialic acid. *Nature (London)* **333**, 426–431.

Wilson, I. A., Skehel, J. J., and Wiley, D. C. (1981). Structure of the haemagglutinin membrane glycoprotein of influenza virus at 3 Å resolution. *Nature (London)* **289**, 366–373.

Wong, G. H. W., and Goeddel, D. V. (1986). Tumour necrosis factors α and β inhibit virus replication and synergize with interferons. *Nature (London)* **323**, 819–822.

Yone, K., Suzuki, J., Tsunekawa, N., and Ichikawa, Y. (1989). Neutralizing monoclonal antibodies recognize the specific region of TNF-α. *Poster Abstr. Int. Conf. Tumor Necrosis Factor Relat. Cytokines, 2nd* p. 22.

Index